The Advanced TEXbook

Springer
New York
Berlin
Heidelberg
Barcelona
Budapest
Hong Kong
London
Milan
Paris
Tokyo

David Salomon

The *Advanced* *T_EXbook*

Springer

David Salomon
Department of Computer Science
California State University, Northridge
18111 Nordhoff Street
Northridge, CA 91330-8281

Cover photo © Uniphoto New York, NY.

Library of Congress-in-Publication Data
Salomon, D. (David), 1938–
        The advanced TeXbook / David Salomon.
            p.   cm.
        Includes bibliographical references (p.   ) and index.
        ISBN 0-387-94556-3 (softcover : alk. paper)
        1. TeX (Computer file)   2. Computerized typesetting.   I. Title.
Z253.4.T47S25   1995
686.2′2544–dc20   95-19423

Printed on acid-free paper.

Production managed by Frank Ganz; manufacturing supervised by Jacqui Ashri.
Photocomposed pages prepared from the author's TEX files.
Printed and bound by Hamilton Printing Co., Rensselaer, NY.
Printed in the United States of America.

9 8 7 6 5 4 3 2 1

ISBN 0-387-94556-3 Springer-Verlag New York Berlin Heidelberg

Preface

"Why is TeX so hard to use?" is the most frequent comment/complaint made by (frustrated) TeX users. The answer: Because it is programmable (has many features commonly found in programming languages), because it pays attention to detail, and because its creator has developed it for his own use (perhaps also his administrative assistant's) and not for general use.

The material presented here is a direct result of this complaint and is an attempt to make it easier for inexperienced users to get the kind of high-quality typesetting that is possible with TeX. The material is based on classes taught since 1985, and on the author's personal experience with TeX, which includes writing three books and numerous articles, handouts, and letters. Both introductory and advanced material is included here. There are many examples as well as a detailed discussion of topics, such as \valign and \emergencystretch, that are only briefly touched upon in *The TeXbook*. Chapter 20 describes the macros used to typeset this book; it also lists the METAFONT programs for the special characters used.

Among the advanced topics, two should be specifically mentioned, the line- and page-break algorithms, and output routines (OTRs). The former discusses the parameters affecting line and page breaks; it should be of use to anyone who feels defeated by \tolerance, \hbadness, and their relatives. The latter contains many OTR examples as well as a detailed discussion of insertions. These two topics reflect the special interest of the author. They are treated in detail and may be of immediate benefit to readers looking for answers to specific problems.

The intended audience of this book are readers with a little TeX experience who want to become advanced users. They may find the examples useful and may benefit from the detailed discussion of practical topics such as leaders and the space factor. For the hard-working reader, many exercises are sprinkled throughout the text, almost all with solutions provided.

As is always the case with such a book, many topics are mentioned before they are properly introduced. Some topics are discussed in unexpected places (e.g., badness is discussed in connection with boxes). Also, several topics are interleaved. Therefore, when searching for a topic, the reader should consult the extensive index.

There is a chapter on advanced math typesetting techniques. Notably absent is a chapter on basic features of math typesetting, because this material is easy to master just by reading *The TEXbook* and doing the exercises.

This book does not pretend to replace *The TEXbook* (Ref. 1), just to complement it with many examples, exercises and diagrams, and with detailed discussion of advanced topics. As a result, the TEXbook is referred to many times, and a special notation is used: [123], [ex. 15.1], [Ch. 7] refer to page 123, exercise 15.1, and Chapter 7 in *The TEXbook*, respectively. The notation §123 stands for section 123 in the WEB code of TEX (Ref. 2).

▶ **Exercise 1:** What if the notation [123] is part of a topic being discussed, and not a reference?

A minor but interesting point is that this book was prepared with Textures on the Macintosh computer. Textures has several bells and whistles not commonly found on other TEX implementations (notably the use of windows, the use of formats, extensive use of \special, and the use of POSTSCRIPT fonts and pictures). Also, it does not distinguish between terminal output and log file output. As a result, this book is sometimes not clear on what gets output on both the terminal and the log file, and what's written on the log file only.

▶ **Exercise 2:** What do you do if you don't understand something in this book, or if you think you have discovered an error?

■ Certain methods described in this book are not completely general and may fail in certain situations. Others are more general and seem robust. However, none of them has been thoroughly tested, and bugs may be discovered at any time.

■ To make it easy to read and understand the macros described here, they were deliberately kept as simple as possible. As a result, they may not be general and may not handle every possible situation. The point is that the macros should not be copied and used verbatim. Rather, they should be studied and modified for specific problems.

The keyboard symbol indicates a point that can best be understood by performing an experiment on the computer. At such a point, the serious reader should stop reading and should perform the experiment.

This arrow is used to indicate points in the text where especially important definitions or concepts are introduced.

Following is a short overview of the book:

■ Chapter 1 is an introduction to basic concepts of TEX, such as its input and output files, the common fonts used, and terms such as magnification, boxes, glue, and penalties.

- A more advanced introduction is the subject of Chapter 2. It presents many advanced topics, such as modes, registers, the anatomy of TeX, category codes, and the handling of end-of-lines.

- The very long Chapter 3 introduces the most important concepts of boxes and glue. They are included in the same chapter since they are normally used together. The emphasis is on examples, of which this chapter has about one hundred. The most important example shows how boxes can be used to place accents above and below text.

- Chapter 4 deals with paragraphs. The main concepts discussed are paragraph structure, paragraph shapes, and the space factor.

- Chapter 5 discusses the all important concept of macros. Macro definition and expansion are covered, as are related concepts and commands such as `\futurelet`, `\let`, and recursion.

- Conditionals are the topic of Chapter 6. They are important since they are the way to make decisions in a TeX program. This way, each time the program is run, it may behave differently.

- Since macros are confusing, and examples are important, an entire chapter, Chapter 7, is devoted to examples of macros. Examples of macros using different techniques and commands are presented and explained in detail.

- Two separate topics, tokens and file I/O, are discussed in Chapter 8. Tokens can be used as strings of characters. File I/O has proved to be an important tool in advanced applications.

- Chapter 9 illustrates the important technique of a multipass TeX job, by means of typical examples, such as cross-referencing and table-of-contents preparation.

- The `\special` command is discussed in Chapter 10, together with examples that show how TeX can be used for arithmetic calculations.

- Leaders are a little known and greatly underused tool. The job of Chapter 11 is to change this, mostly by means of fun examples.

- Many users have been originally attracted to TeX because of two features: tables and math typesetting. Here it is relatively easy to get results that far surpass most of today's commercial word processors and page layout software. Tables are discussed in Chapter 12; advanced concepts of math typesetting, in Chapter 13.

- Chapter 14 is a detailed presentation, with examples, of the line-break and page-break algorithms. Related concepts, such as demerits and hyphenation, are also included.

- Chapter 15 is very practical. It teaches how to understand and respond to error messages.

- The difficult concept of OTRs is the subject of the following four chapters (16–19). The first is an introduction; the second and third, a presentation of techniques for solving common OTR problems; and the fourth, a discussion of insertions (perhaps the least understood feature of TeX).

- Chapter 20 presents the macros used to typeset this book. The most important ones are the cross-reference macros, and the macros used to create the index and the table of contents. Also listed are the METAFONT programs used to create the few special characters used in this book. These programs are not written in TeXand are included here for completeness (read: to avoid having readers pester me with requests for the METAFONT source codes).

■ The TeX bibliography shows how popular this topic has become. The answers to the exercises are most useful, since many of the exercises touch on important or subtle points. The index is very extensive.

I would like to acknowledge the many students who helped improve the book by finding errors, suggesting improvements, and pointing out weak sections. In addition, two more persons should be acknowledged. C. G. van der Laan has read most of the material and made numerous suggestions and corrections. Johnny Tolliver has implemented the popular MakeIndex program on the Macintosh computer, making it possible for me to prepare the extensive index in a reasonable amount of time.

The reader should note the quotations at the end of each chapter. There are standard sources for such quotations, such as *Bartlett's* and the *Oxford Dictionary of Quotations*. The traditional sources, however, provide quotations for terms such as *marriage*, *family*, and *leadership*. How does one go about finding quotations for terms such as *horizontal*, *output routine*, and *boxes*? I used the following:

1. Extensive personal reading. This is how I found the quotations from Coward and Maugham.

2. The Bible and Shakespeare's works. These are now available from many archives (try world.std.com and ftp.spies.com) and can easily be searched for quotations.

3. Project Gutenberg has placed many books (among them *The New Hacker's Dictionary*) in its archive (mrcnext.cso.uiuc.edu), where they are available by anonymous FTP.

4. The various books on TeX are a good source, as well as Knuth's other books (see quotation below).

5. If all of the above fail, the dictionary definition of a term can be used as a quotation (see Appendix A).

6. Similarly, an anagram can be used instead of a quotation. How about "go blue sex" or "eels bug ox" as a quotation for "boxes & glue"?

7. (Tongue in cheek) If even this fails, the required quote can almost certainly be found in *Finnegans Wake*. However, the entire book has to be read, since it is not commonly available electronically.

"So you can create these tables?" she asked him. "In a form suitable for a TeX file?" TeX, pronounced like "Tech," is a computer program that's used for typesetting technical papers and books. "No," said Turner. "I don't do TeX. I do Troff (a comparable program). I guess I'll have to learn it, though. It's an invention of the devil."

— Lemonick, M. D., The Light at the Edge of the Universe (1993)

Disclaimer

The macros and programs listed in this book have been tested by the author but are not guaranteed. They are meant to be read, understood, and modified by the reader for specific applications. They are not meant to be copied and used verbatim. The publisher does not offer any warranties or representations, nor does it accept any liabilities with respect to them.

Some of the material presented here has originally been published in *TUGboat*, the publication of the TEX Users Group.

Trademark Notice

TEX is a trademark of the American Mathematical Society
METAFONT is a trademark of the Addison-Wesley Publishing Company
PostScript is a trademark of Adobe Inc.

This is that book.
— *Knuth, D. E., 3:16, Bible Texts Illuminated (1990)*

This page illustrates the capabilities of TeX; its purpose is to whet the reader's appetite. The next page shows the commands used to create this page. The rest of the book is intended for readers who want to learn TeX well enough to be able to create pages like this one themselves.

The most important feature of TeX is its line-breaking capabilities, illustrated by the paragraphs on this page. A careful observer will note how the spaces between words are controlled. They vary in size from 2.2 points to 5 points, and so are not too small and not too large. This feature is the key to creating beautiful documents. The size of the interword space is determined by the font designer and is different for different fonts. The user, however, can override the default size and specify any size for the interword space. Note how spaces following punctuation marks are somewhat wider.

The second most important feature of TeX is its programmability. The user can define new commands (called macros) based on existing ones. This is of great help when a complex document has to be designed. Here are two examples:

1. When a book is written, it is organized in chapters. Each chapter has a different name and number, but they are all typeset in the same format. The book designer may decide to start each chapter on a new page, to skip two inches from the top of the page, to typeset the name and number of the chapter in a certain font, and to skip 0.2 inches below that, before starting the text itself. All the necessary commands can be grouped and defined as a new command whose name is \chapter. When this command is used, the user has to provide the name and number of the specific chapter. Typical uses are '\chapter 3{The Voyage}', '\chapter {12}{Coming Home}'.

2. TeX can typeset horizontal and vertical rules. These can be combined to surround a given piece of text $\boxed{\text{in a box.}}$ Again, the necessary commands can be grouped and defined as a macro (a new command) named \boxit. The new command can then be used, as in '\boxit{\hbox{in a box.}}3'.

Other important features include tables and math expressions. Here is an example of a simple table:

First	Last	English	First	Last	English	Pronounced[1]
Gÿorgi	Gábor[2]	George	Zoltán	Kodály		Ko-Die
Miklós	Károlyi	Mike	Laci	Virag	Bloom	La-Tsee
Janos	Hatvany	John	Egón	Mikes	Eugene	Mee-Cash
Mátyás	Rákosi	Mathias	Ferenc	Puskás	Francis	Push-Cash
István	Farkas	Steven	Sándor	Petőfi	Sender	

[1] 's' is normally pronounced 'sh', while 'sz' is pronounced 's'.   [2] Also a first name.

All this illustrates the design philosophy of TeX. It involves attention to detail, and doing things automatically as much as possible, while giving the user the capability to override any decisions and to change any default values.

> A single bad habit will mar an otherwise faultless character,
> as an ink-drop soileth the pure white page.
>
> — *Hosea Ballou*

Here is how the previous page was generated. The first two paragraphs contain the text and just a few commands.

```
This page illustrates the capabilities of
\TeX;^[\TeX!its capabilities] its purpose
...
pages like this one themselves.

The most important feature of \TeX\ is its line-breaking
capabilities,
...
punctuation marks are somewhat wider.
```

(Note how the original line breaks are ignored. Also note how a blank line denotes a new paragraph.)

The command '`\TeX`' typesets the logo TEX. If we want this logo to be followed by a space, we type the command '`\ `' (a control space). The circumflex '`^`' indicates an index item; the command '`^[\TeX!its capabilities]`' will create, in the final index, the item 'TEX', followed by the subitem 'its capabilities'.

The paragraphs on programmability include two interesting features:

1. The use of the vertical bar '`|`' as a verbatim command (section 7.10). For example, the text: '`\chapter 3{The Voyage}`' was typeset by saying '`|\chapter 3{The Voyage}|`'.

2. The use of macro `\boxit` (page 75). The text $\boxed{\text{in a box}}$ was typeset by '`\boxit{\hbox{in a box}}3`'.

The table was created by the `\halign` command (Chapter 12). Here is part of it:

```
\halign{\tabskip=1em &# \cr
\underbar{First} &\underbar{Last} &\underbar{English}
&\underbar{First} &\underbar{Last}
&\underbar{English} &\underbar{Pronounced$^1$}\cr
\noalign{\smallskip}
G\H yorgi & G\'abor$^2$ & George & Zolt\'an & Kod\'aly & & Ko-Die \cr
Mikl\'os & K\'arolyi & Mike & Laci & Virag & Bloom & La-Tsee\cr
...
Istv\'an & Farkas & Steven &&S\'andor & Pet\H ofi & Sender \cr
\noalign{\smallskip\hrule width1in\smallskip}
\multispan5{$^1$ \sevenrm`s' is normally pronounced `sh',...\hfil}&
\multispan2{$^2$ \sevenrm Also a first name.\hfil}\cr}
```

The quote at the end of the page was typeset by the special macro `\quote` (section 20.2).

```
\quote
{A single bad habit will mar an otherwise faultless character,

as an ink-drop soileth the pure white page.}
Hosea Ballou.^[Ballou, H.]
```

Contents

Character—A graphic symbol (as a hieroglyph
or alphabet letter) used in writing or printing.

— *Webster's Dictionary*

1. Introduction

TeX is not written TEX, it is not spelled 'T' 'E' 'X', and it is not pronounced "tex". It is written TeX (a trademark of the AMS), it is spelled with the Greek letters Tau Epsilon Chi, and it is pronounced "tech" or rather "teck".

TeX is not a word processor. It is a program that sets text in lines and paragraphs—but its design philosophy, its methods, and its approach are all different from those of a word processor.

A modern, state-of-the-art word processor is a WYSIWYG-type program; TeX is not: It uses one-dimensional input to generate its two-dimensional output. With a word processor it is easy to generate special printing effects and easy to underline text, to italicize and emphasize; TeX can do those things—and with a lot more attention to detail—but the user has to work harder.

The same is true for applications that mix text with graphics. Good, modern word processors can also create diagrams, whereas TeX does not support graphics and some work is necessary to insert diagrams into its output. Other features that are standard and easy to use in a word processor are either nonexistent or hard to use with TeX.

If TeX is not a word processor, then what is it?

1.1 Basics

From the point of view of the user, TEX is a typesetting program that can be extended to a complete computerized typesetting system by adding a printer driver, the METAFONT program, and some utilities.

From the point of view of the computer, however, TEX is a *compiler* whose main task is to *compile* a source program. The source language has typed variables (called registers), block structure, two executable statements (assignment and if), and a powerful macro facility that makes it possible to extend the language. Looping is done by recursion. The main output of the compiler is a file containing not machine instructions, but detailed instructions on what characters to place on a page, and exactly where to place them.

When a compiler sees something in the source file which does not belong in the source language, it issues an error message. TEX, on the other hand, simply typesets any material in the source file that's not part of the language. The source file for TEX contains the text to be typeset, embedded commands, and comments. It's a text file and can thus be prepared with any editor or word processor. TEX generates three outputs: a log file—with run-time information and error messages; terminal output (output considered important enough to be seen by the user immediately); and a .dvi file (for Device Independent), containing the page coordinates of each character to be typeset. A separate printer driver program later reads the .dvi file and generates printer commands (which are different for different printers) to print the text. Most TEX implementations also include a screen preview program that makes it possible to view the typeset pages before they go on paper.

Two font files must exist for each font used in the text, a .tfm file and a bitmap file (usually called .gf or .pk file). The .tfm file (for TEX Font Metric) contains the dimensions of each character in the font (width, height, and depth) and some other information. The bitmap file contains the actual shape of the characters. TEX uses only the .tfm file. When the position of a character on the page is determined, TEX uses the character's width to move the reference point to where the next character should appear. As a result, TEX is ignorant of the actual shape of the characters, and its final output is a list of commands that specify what should be placed on the page and where. A typical command is a triplet of the form:

<char. # in current font, x-coord., y-coord.>

The bitmap file is only used by the printer driver, which sends the individual pixels of each character to the printer.

Figure 1.1 summarizes the relationships between TEX, METAFONT, the printer driver, and the files involved.

To come back to the point of view of the user, the main aim of TEX is to produce *beautifully typeset documents*, especially documents containing mathematics. As a result, TEX is ideal for book publishing, for documents that should look beautiful—such as concert programs and invitations—and for technical publishing. Since typesetting is not a simple process (it is in fact very complex, especially the typesetting of tables and mathematics), TEX is *not easy to use*. It has several features that are complex and hard to master. As a result, new users tend to use TEX only if they need high-quality output. When such users need a quick letter, a simple memo, or a short note, they may go back to their familiar word processor. However,

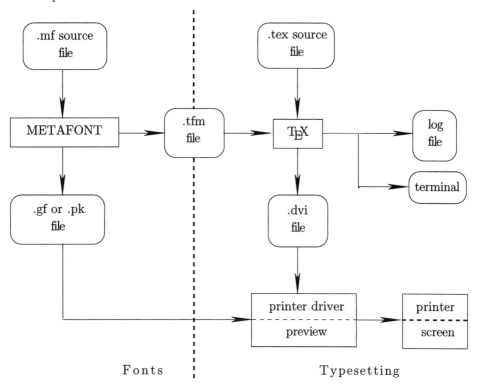

Figure 1.1

advanced users usually have macros that help produce all kinds of documents easily, thus using TeX's power all the time. Either way, if you like beautiful text, or if you write mathematics, TeX is by far the best typesetting tool available today.

To understand the power of TeX, let's consider the differences between using a typewriter and using a computer keyboard. On a typewriter, the keys for '1' and 'l' are identical. On a computer keyboard, they are different. Most typewriters only have one, nonoriented, single quote ('), but computer keyboards normally have two oriented single quotes, a left (') and a right ('). Similarly, when switching from word processing to TeX, a few more adjustments become necessary in order to get book quality. The most important ones are the following:

■ Traditionally, computer keyboards have just one kind of doublequote ("), but finely printed books have two kinds, a left doublequote (") and a right one ("). To produce them in TeX, you simply type two single quotes (both left or both right) in a row. Thus to typeset "my thanks." you need to type `''my thanks.''`.

■ Another important difference is the dash (or hyphen). Computer keyboards have only one dash, namely (-); in a carefully typeset book, however, there are four different symbols:

a hyphen (-);   an en-dash (–);   an em-dash (—);   a minus sign (−);

Hyphens are used for compound words like "right-quote" or "X-Register". En-dashes are used for number ranges, like "exercises 12–34". They are obtained by typing two consecutive dashes '--'. Em-dashes are used for punctuation in sentences—like the ones around this section—they are what we usually call dashes. To get one, just type three consecutive dashes. Minus signs are, of course, used in typesetting mathematics.

■ The main task of TEX can be described as converting a one-dimensional stream of text into a two-dimensional page where all lines have the same width. This is done by stretching or shrinking the spaces between words on each line. In fine typesetting, however, those interword spaces have limited flexibility, so sometimes words have to be hyphenated. TEX uses a sophisticated hyphenation algorithm, so hyphenation is automatic. In cases where the algorithm does not work, the user can specify the correct hyphenation of words.

■ Well-printed books use ligatures and kerning. Certain combinations of letters, like 'ff', 'fl', 'fi', 'ffl', look better in the traditional roman type when the letters are combined. Such a combination is called a **ligature**. Foreign languages may have other ligatures, such as 'ij' in Dutch. Compare, for instance, the word "fluffier" set by TEX to the same word "fluffier" generated by a word processor. When a font is designed for TEX, the designer should specify all the special combinations that should be replaced by ligatures, and **design** the ligatures. That information goes into the font's `.tfm` file and is used by TEX to substitute the ligatures automatically. TEX can even remove a ligature later, for example, if it decides to hyphenate the word at that point.

Kerning refers to certain letter combinations that should be moved closer (negative kerning) or apart (positive kerning) for better appearance. An 'A' adjacent to a 'V' is a good example (compare 'AVAV' to 'AVAV'). Other examples are 'away', 'by', 'ox', 'ov', 'xe', and 'OO' (the last one has positive kerning). Again, the font designer decides what the kerning should be, and that information also goes into the font's `.tfm` file. TEX, of course, allows the user easily to override ligatures and to change the kerning to any desired value.

Ligatures and kerning improve the relationship between adjacent letters. Together with hyphens and flexible interword spaces, they are the main participants in the delicate balancing act required for the line-break decisions.

Considering the power and flexibility of TEX, it is surprising that its basic algorithms are based mostly on three concepts— **boxes, glue**, and **penalties**.

■ A box in TEX is an indivisible unit of material to be typeset. TEX will not break the contents of a box across lines or between pages.

To begin with, each character in a font is enclosed in a box whose dimensions (width, height, and depth) become the dimensions of the character. When TEX typesets a word, it pastes the individual character boxes side by side, with no spaces in between (except when kerning demands shifting boxes horizontally, or when the user wants boxes moved—which is how the TEX logo is produced). The result is a box containing the typeset word. The width of such a box is the sum of the widths of the boxes inside, plus the kernings, which, of course, may be negative. The height of the wordbox is the maximum height of the component boxes, and the same is true for the depth.

Similarly, when TEX decides to break a line, it pastes the individual word boxes side by side, with appropriate spaces between them, and generates a new horizontal box, a linebox. The width of the linebox is the sum of widths of the word boxes inside it plus the sum of the spaces (glue) between the individual word boxes. The height and depth of the line box are the maximum heights and depths of the component boxes.

To set an entire page, TEX accumulates enough horizontal lineboxes; it then pastes them vertically, one below the other, with appropriate interline spaces, to generate a vertical pagebox. The next step is to call an *output routine*, which is either written by the user or supplied by TEX. The output routine adds finishing touches such as heading, footing, marginal notes, and page numbers. It may even decide to trim the page, to return part of the bottom of the page to TEX (to become the top of the next page), or to do anything else to the page. The output routine should normally execute `\shipout`, which translates the contents of the page box to `.dvi` file specifications. TEX then continues to read the source to build the next page.

■ The term "glue" refers to the spaces between boxes. In order to justify the text, TEX adjusts the spaces between words (but not the spaces between characters in a word). Those spaces must, therefore, be flexible. The same applies to interline spaces on the page. A glob of glue in TEX is a triplet (w, y, z), where w specifies the natural size of the glue, y is the amount of stretch in the glue, and z is the amount by which it can shrink. For interword glue, these values depend on the font and are specified by the font designer. For the common cmr10 roman font used by TEX, the values are $(3.3333\,\mathrm{pt}, 1.66666\,\mathrm{pt}, 1.1111\,\mathrm{pt})$. The line-break algorithm considers various alternatives of combining words into lines, and for each alternative it calculates the amount by which the individual globs of glue have to be stretched or shrunk. After considering all the possible line breaks for an entire paragraph, TEX chooses the best line breaks (best in a sense explained in Chapter 14), sets the entire paragraph, and reads the next item from the source file (usually the start of the next paragraph).

■ The third mathematical construct used by TEX is the *penalty*. A penalty can be inserted into the text at any point either explicitly, by the user, or automatically, by TEX. The penalty specifies the desirability of breaking a line or a page at that point, and it is used to discourage bad breaks, to encourage good ones, to force certain breaks, and to avoid others. Penalty values are in the range $[-10000, 10000]$. Any value ≤ -10000 is treated as $-\infty$, and any value ≥ 10000 is considered $+\infty$. The user may, for example, insert '`\penalty100`' at a certain point in a paragraph, which has the effect that, if TEX tries to break the line at that point, a penalty of 100 will be added to the line, making that breakpoint less desirable. This tends to discourage line breaking at that point. A negative penalty is actually a merit, and it encourages a break. Infinite penalty prohibits a line break where it is specified, and $-\infty$ forces one. The commands (control sequences) `\break` and `\nobreak` can be used to insert those infinite penalties at any point in the text. There are two common examples of penalties. The first occurs at any hyphenation. When TEX decides to hyphenate a word, it inserts a penalty of 50 at any potential hyphenation point. The second has to do with psychological bad breaks. In a text containing "...`Appendix G`" breaking the line between the two words is bad because it interrupts the smooth flow of reading. To prevent such a break, a TEX user should type '`Appendix~G`'.

The tilde (~) acts as a "tie". It ties the two words such that in the final document there will be a space between them but no line break.

Similar examples are as follows:

```
Table~G-5   Figure~18    dimension~d   Louis~XVI
Register~Y  modulo~eˣ    Rev.~Henry    HRH~Prince Abdul
1,~2, or~3  from 0 to~1
```

Each tie is converted by TeX into a penalty of $+\infty$ to prevent a break.

This mechanism of boxes, glue, and penalties to arrange text in lines and pages has proved extremely flexible and powerful. It makes it possible to set text in non-standard ways to achieve special effects. Features such as narrow paragraphs, newspaper formats, paragraphs with variable line widths, hanging punctuations (see [394]), ragged right margins, centered text, and complex indentations can all be achieved with TeX. This mechanism is one of the main innovations introduced by TeX.

1.2 Line Breaking and Page Layout

To achieve a straight right margin, TeX adjusts the spaces between words, but not the spaces between characters in a word. The boxes defining each character are juxtaposed with no intervening spaces, for the following reasons:

▪ It makes for a more uniform appearance of the page. Because of the nature of human vision, spaces between words disturb the eye less than spaces between characters in a word.

▪ In an underlined font, spacing the characters would break the underline into separate segments.

This is perhaps a good point to discuss underlining in TeX. Underlining is one of the features that distinguish TeX from word processors, and it can give an insight into the design philosophy of TeX. In a word processor it is usually easy to underline text; in TeX, underlining is discouraged. The reason is that TeX is not a word processor but instead a typesetter designed to produce beautiful books. In a book, underlining is rare, and emphasizing is done using either **boldface**, *italics*, or *slanted* fonts. If a certain text requires a lot of underlining, the best way to do it in TeX is to design a special font in which all the letters are underlined. Such a font requires no spaces between the characters in a word.

To achieve a uniform page layout, TeX uses two principles: 1. The vertical distance between lines is kept as constant as possible, using the rules below. 2. When a good page break requires squeezing another line on a page, or removing a line from a page, TeX changes the vertical distance between **paragraphs** (or between math formulas), not between lines. This again has to do with the way our eyes see, and it guarantees that all the pages would appear to have the same proportion of black to white areas.

To determine the vertical distance between consecutive lines, three parameters a, b, c are used. Their values depend on the font, and for the common cmr10 font they are $a = 12\,\mathrm{pt}$, $b = 0$, $c = 1\,\mathrm{pt}$. Typically, consecutive lines are not juxtaposed vertically but are spaced such as to make the distance between consecutive baselines equal to a. However, if the distance [top of lower line]−[bottom of upper line] is less than b, then the lines are spaced such that that distance is set equal to c. This may happen if the line contains a large (say, 18pt) character. Parameters a, c are of type `<glue>` so they may have flexibility. Typically this flexibility is zero, but the user may want to define, for example, $a = <12, 2, 1>$. Such a value makes sense for a short, one-page document. The flexibility of the glue would make it easier for TEX to fit the text on one page.

Any discussion of TEX's line-break algorithm should start with an outline of a typical line-break algorithm used by a modern, commercial word processor. The method uses three values—for the natural, minimum, and maximum spacings between words—and proceeds by appending words to the current line, assuming natural spacing. If, at the end of a certain word, the line is too long, the algorithm tries to shrink the line. If that is successful, the next word will start the next line, and the current line is printed. Otherwise, the word processor discards the last word and tries to stretch the line. If that is successful, the discarded word becomes the first one of the next line. If neither shrinking nor stretching works (both exceed the preset parameters), a good word processor tries to hyphenate the offending word, placing as much of it on the current line as would fit. The user may be asked to confirm the hyphenation, and the rest of the hyphenated word is placed at the start of the next line. A word processor that does not hyphenate has to resort to overstretching and may generate very loose lines.

The important feature of all such methods is that once a breakpoint is determined, the algorithm does not memorize it but starts afresh with the next line. We can call such an algorithm "first fit" and its main feature is that it does not look at the paragraph as a whole. Such an algorithm produces reasonably good results. For a high-quality typesetting job, however, it is not fully satisfactory. For such a job, an algorithm is needed that considers the paragraph as a whole. Such an algorithm makes only tentative decisions for line breaks and may, if something goes bad toward the end of the paragraph, go back to the first lines and change their original, tentative, breakpoints. TEX's algorithm determines several feasible breakpoints for each line and calculates quantities called the *badness* and *demerits* of the line for each such breakpoint. After doing this for the entire paragraph, the final breakpoints are determined in a way that minimizes the demerits of the entire paragraph. Mathematically, the problem is to find the shortest path in an acyclic graph.

1.3 Fonts

Traditionally, the word *font* refers to a set of characters of type that are all of the same size and style, such as Times Roman 12 point. A *typeface* is a set of fonts of different sizes but in the same style, like Times Roman. A *typeface family* is a set of typefaces in the same style, like Times.

The size of a font is normally measured in points (more accurately, *printer's points*), where 72.27 points equal 1 inch. (See section 1.6 for a description of all the valid dimensions in TEX.) The *style* of a font describes its appearance. Traditional styles are roman, **boldface**, *italic*, *slanted*, `typewriter`, and sans serif. In TEX, a font can have up to 256 characters, although most fonts have only 128 characters.

▶ **Exercise 1.1:** Why use numbers such as 128 or 256, instead of nice round numbers such as 120 or 250?

When typesetting text there is, at any time, a single *current font* that's used to typeset the characters being input (the situation is different when typesetting mathematics; see Chapter 13). To switch fonts, they must first be assigned names, which is done by the \font command whose general form is

\font⟨internal name⟩=⟨external name⟩. The ⟨internal name⟩ is a control sequence, such as \abc, \it, or \beeg. The ⟨external name⟩ is the name of the font file. It obeys the rules of the operating system and is not part of TeX. Examples include \font\tenrm=cmr10 and \font\sstwe=cmss12.

The \font command can also take one of the optional parameters at and scaled. Thus '\font\twerm=cmr10 scaled 1200' scales the font on file cmr10 by a factor of 1.2 and assigns it the internal name \twerm. Similarly, the command '\font\fivrm=cmr10 at5pt' creates a 5 point version of the font on file cmr10. Such transformations are not as good as the real thing, and should be used only as a last resort (see also section 1.7 for a discussion of magnification and magsteps).

If \font\abc=.. is executed inside a group, \abc will not be known outside the group. It is possible to say, however, \global\font\abc... .

⟶ Note that the \font command does not change the current font. To switch fonts, it is necessary to say the name of the new font. It is a common error to say '\font\abc=..' and expect the font to change to '\abc'. To switch to this font, the user should say '\abc' or, if the switch is temporary, '{\abc...}'.

1.4 The CM Fonts

Computer Modern (CM) is a metafont, developed in METAFONT, from which many different fonts have been derived, by different settings of parameters. The fonts all look different, but they blend together. They are called the CM fonts, and their names start with 'cm'. The standard CM fonts are

- cmr or Roman. These are used for plain text. The standard sizes are (the '*' indicates fonts that are automatically loaded by the **plain** format) cmr17 cmr12 cmr10* cmr9 cmr8 cmr7* cmr6 cmr5*.
- cmsl or Slanted. They are slanted versions of the cmr characters. The standard sizes are cmsl12 cmsl10* cmsl9 cmsl8.
- cmdunh or Dunhill. Same as cmr but with higher ascenders. Only cmdunh10 is standard.
- cmbx or Bold Extended. These are used for boldface characters. cmbx12 cmbx10* cmbx9 cmbx8 cmbx7* cmbx6 cmbx5*.
- cmb or Bold. This is bold but too narrow for normal use. cmb10.
- cmbxsl or Bold Extended Slanted. cmbxsl10.
- cmtt or Typewriter. Fixed-space, resembling typewriter style. cmtt12 cmtt10* cmtt9 cmtt8.
- cmvtt or Variable Typewriter. Same as cmtt but with proportional spacing. cmvtt10.
- cmsltt or Slanted Typewriter. cmsltt10.
- cmss or Sans Serif. Used for titles and for formal texts. cmss17 cmss12 cmss10 cmss9 cmss8.

- cmssi or Sans Serif Italics. cmssi17 cmssi12 cmssi10 cmssi9 cmssi8.
- cmssbx or SS Bold Extended. cmssbx10.
- cmssdc or SS Demibold Condensed. Normally used for titles. cmssdc10.
- cmssq or SS Quote. Special SS font for quotations. cmssq8.
- cmssqi or SS Quote Italics. Again used for quotations. cmssqi8.
- cminch or Roman Inch. These are used for titles. cminch.
- cmfib or Fibonacci. In this version the parameters have relative sizes determined by the Fibonacci sequence. cmfib8.
- cmff or Funny Font. Different from the other cm fonts. Rarely used. cmff10.
- cmti or Text Italics. This is the normal italics font. cmti12 cmti10* cmti9 cmti8 cmti7.
- cmmi or Math Italics. Slightly different from text italics, and without spaces (spaces are automatically supplied in math mode). cmmi12 cmmi10* cmmi9 cmmi8 cmmi7* cmmi6 cmmi5*.
- cmbxti or Bold Extended Text Italics. cmbxti10.
- cmmib or Math Italics Bold. For math mode. cmmib10.
- cmitt or (text) Italics typewriter. cmitt10.
- cmu or Unslanted (text) Italics. Unslanted version of cmti. Used for Editor's Notes in *TUGboat*. cmu10.
- cmfi or Funny Italics. An italicized version of cmff. cmfi10.
- cmsy or Math Symbols. Contains the math symbols normally used by TEX. cmsy10* cmsy9 cmsy8 cmsy7* cmsy6 cmsy5*.
- cmbsy. A bold version of the math symbols. cmbsy10.
- cmex or Extension. More math symbols. cmex10*.
- cmtex or TEX Extended. An extended ASCII font. cmtex10 cmtex9 cmtex8.
- cmcsc or Caps and Small Caps. Contains small caps instead of lowercase letters. cmcsc10.
- cmtcsc. A Typewriter version of cmcsc. cmtcsc10.
- cc or Concrete. These were specially developed for the book *Concrete Mathematics* (Ref. 32) to blend with the Euler math fonts.

Some of these fonts are rarely used, but were easy to generate by trying various settings of parameters. Any special sizes not mentioned above should also be easy to derive.

In plain TEX, the default font is cmr10. Also, macros \bf, \it, \sl, and \tt are set [351] to select the different styles in 10 point size. If large parts of the document should be typeset in, say, 12 point, the definitions of \bf and its relatives should be changed accordingly. To do this, the different 12 point fonts should be loaded, at the start of the document, and assigned names by

```
\font\twerm=cmr12
\font\twebf=cmbx12
\font\tweit=cmit12
\font\twesl=cmsl12
\font\twett=cmtt12
```

Macro \twelve should then be defined as

```
\def\twelve{\def\rm{\twerm}\def\bf{\twebf}\def\it{\tweit}%
```

```
\def\sl{\twesl}\def\tt{\twett}%
\rm}
```

Any text areas that should be typeset in 12 points have to be bounded by `\begingroup\twelve` and `\endgroup`.

The CM family (Ref. 4) represents the most ambitious attempt so far to develop a general metafont. It is based on an earlier version called AM (almost modern), which is now obsolete. Two interesting adaptations of CM are outline fonts (Ref. 5) and the Pica fonts. The reader should refer to [Chs. 2, 4, 9 and App. F] for more information on the CM fonts.

Many special fonts have been developed in METAFONT. Examples are exotic languages, music notes, chess figures, astronomical symbols, and logic gates. Reference 6 is a detailed listing. The MetaFoundry (Ref. 7) developed many fonts in the early 1980s. However, very few other metafonts exist, the most well known of which are

- Pandora, developed by N. Billawala (Ref. 8).
- The Euler family, designed by Herman Zapf, and developed at Stanford University (Ref. 9). It is not a true metafont, as the characters were digitized.
- The Gothic family, including Fractur and Schwabacher, developed by Y. Haralambous (Ref. 10).

1.5 Font Examples

The following examples include some non-CM fonts commonly available on personal computers.

This is an example of font cmr10 (roman)

This is an example of font cmbx10 (bold extended)

Some math symbols $\cap [] \sqrt{/} \sqrt{} \amalg \{ \tilde{} \uparrow \amalg | \sqrt{} \tilde{\lceil}$

Notice the absence of spacing in the next two lines. Math italics is automatically used and spaced in the math mode.

$Thisisanexampleoffontcmmi\text{10}{\leftarrow}mathitalics{\leftarrow}$

$Thisisanexampleoffontcmmi\text{5}{\leftarrow}mathit\text{5}pt{\leftarrow}$

This is an example of font cmti10 (text italics)

`This is an example of font cmtt10 (typewriter)`

This is an example of font Helvetica (sans serif)

`This is an example of font Courier (fixed spacing)`

This is an example of font Palatino

This is an example of font Times (roman)

This is an example of font Chancery (cursive)

THIS IS AN EXAMPLE OF FONT CMCSC10 (CAPS AND SMALL CAPS)

This is an example of font cmdunh10 (ascenders)

This is an example of font cmssdc10 (sans serif demibold condensed)

1.6 Dimensions

In high-quality typesetting it is important to be able to specify dimensions in several units. Accordingly, TEX accepts nine different units of dimensions. They are summarized below:

in	inch, common in the USA and the UK
cm	centimeter ($2.54\,$cm $= 1\,$in)
mm	millimeter ($10\,$mm $= 1\,$cm)
pc	pica ($1\,$pc $= 12\,$pt)
bp	big point ($72\,$bp $= 1\,$in)
pt	point ($72.27\,$pt $= 1\,$in)
dd	didot point ($1157\,$dd $= 1238\,$pt). European version of pt
cc	cicero ($1\,$cc $= 12\,$dd). European version of pc
sp	scaled point ($65536\,$sp $= 1\,$pt). The basic TEX dimension

Dimensions don't have to be integers; measurements such as 3.14cm, 2,78in, or 1.1111pt are common (note that a comma can be used instead of a decimal point).

▶ **Exercise 1.2:** Dimensions can also be negative. When does this make sense?

The sp unit is so small that the following quote (from [58]) properly illustrates its size: "…the wavelength of light is approximately 100 sp (in fact: violet = 75 sp, red = 135 sp!)." The sp is used to represent all other dimensions internally.

▶ **Exercise 1.3:** How is the dimension 10.9 sp represented internally?

Dimensions are stored internally as 32-bit integers in sp. Since dimensions can be negative, one bit is used for the sign, leaving 31 magnitude bits. The largest possible dimension is thus $2^{31} - 1\,$sp, but the largest *allowed* dimension is only half that (so that two dimensions could be added without overflow, see §104). It therefore equals $2^{30} - 1\,$sp $\approx 16384\,$pt $\approx 18.892\,$feet $\approx 5.7583\,$meters. This is the value of parameter \maxdimen.

The pt dimension was used by printers for many years until, in 1886, it was officially defined (by the American Typefounders Association) as exactly equal to .013837 in. This made the inch equal to 72.2700007227 pt, but TEX uses the value 72.27 to make it easier to convert other dimensions to sp on many different computers.

▶ **Exercise 1.4:** Estimate the error introduced by changing the official value of the pt to 72.27.

The following illustrates the actual sizes of a few possible dimensions:

⊣   1 in   ⊢
⊣   100 pt   ⊢
⊣   100 dd   ⊢
⊣   2 cm   ⊢
⊣   6500000 sp   ⊢

In addition to the absolute dimensions mentioned above, two relative dimensions, em and ex, can be used. Their size depends on the current font, and they are determined by the font designer. The value of em is the width of 'm', and the value of ex is the height of 'x', in the current font.

It should be noted that TeX uses dimensions to create its final output, the
`.dvi` file. The position of each character on the page is stored in this file in units
of sp, so they are very precise. The printer driver, however, may introduce errors,
as may the printer itself, if it is a low-quality device.

1.7 Scaling and Magnification

The entire document or parts of it can be magnified (or shrunk). There is a
`\magnification` command that must apply to the entire document. It must appear
at the beginning (before the first page is shipped out), and it cannot be changed.
Saying `\magnification=1400` will magnify all dimensions by 1.4, including characters and spaces. Magnifying characters is done simply by using a bigger font, so the
user must make sure that all the necessary fonts are available to the printer driver
at the magnified sizes. Magnifying the spaces is easy. Note that the line and page
breaks are not affected by the magnification (see the exception below). The final
document will be bigger or smaller than the original, but the typesetting will be
the same.

Exception: The user may want certain dimensions to stay the same, regardless
of the magnification used. If the user says `\hskip 1in` somewhere in the document,
and the document is magnified by 1400, the space will grow to 1.4 inches. To keep it
at 1 inch, the user has to divide $1/1.4 = 0.714$ and say `\hskip 0.714in`. A better
way is to say `\hskip 1 true in` (or `\hskip 1truein`). This way, the division
above is automatically done, to keep the space at exactly 1 inch, regardless of the
magnification. When true dimensions are used, the document will look different at
different magnifications.

The `plain` format says [364]:
`\hsize=6.5truein \vsize=8.9truein \dimen\footins=8truein`.
Therefore, a document using `\magnification` may want to redefine these quantities
without the keyword `true`.

The `\magnification` command is a `plain` macro that assigns the magnification
factor to the parameter `\mag`. The value of `\mag` is written on the `.dvi` file at the
beginning and at the end. It is used by the printer driver to perform the actual
magnification. The value of `\mag` is also used by TeX when `true` dimensions are
specified by the user.

Portions of the document may be selectively magnified by magnifying fonts.
`\font\new=cmr10 at 5pt` (or, equivalently `\font\new=cmr10 scaled 500`) creates a font `\new` where everything, including dimensions of characters and spaces
between them, is exactly half that of font cmr10. However, font `\new` is not identical to font cmr5 because, in a high-quality metafont, the different font sizes are
individually designed and dimensions are not simply scaled.

Many installations use outline fonts that are stored at a nominal 1 pt size and
can easily be scaled to any size. In such a case, one may simply say, for example,
`\font\abc=Times at24pt`.

Experience shows that even small magnifications of characters are very noticeable to the eye, so big magnifications are rarely used, and a more practical
magnification scale uses powers of 1.2. In `plain` TeX, the following quantities
are defined [349] to make it easy to select the desired magnification: `\magstep0`

is set to 1000; \magstep1 has the value $1000 \times 1.2 = 1200$; \magstep2 is assigned $1000 \times 1.2^2 = 1440$, and so on, through 1728 and 2074, up to \magstep5, whose value is $1000 \times 1.2^5 = 2488$. There is also \magstephalf, with a value of $1000 \times \sqrt{1.2} = 1095$.

The use of magsteps has two advantages: (1) they provide for small magnification steps; (2) they are multiplicative. If we magnify the entire document by \magstep2, and certain parts by an additional \magstep2, those parts will come out magnified by \magstep4.

▶ **Exercise 1.5:** How can the control sequence \magstep1 be made of letters and digits?

Her name was Poulten-Morse, his own was Lisle-Spruce, but there was obviously going to be an understanding that the hyphens cancelled one another out.

— *Paul Scott, The Bender, 1963*

2. Advanced Introduction

This chapter is an advanced introduction to TeX, stressing the main parts, main operations, and certain advanced concepts. As with most other presentations of this type, some terms have to be mentioned before they are fully introduced, so the best way to benefit from this material is to read it twice.

2.1 Registers

A register is temporary storage, a place where data can be saved for later use. Using a register thus involves two steps. In the first step something is stored in the register; in the second step, the contents of the register is used. With the exception of box registers, the contents of a register can be used and reused indefinitely. There are six classes of registers, summarized in the following table.

Class	Contents	Default value
\count	integer	0
\dimen	dimension	0pt
\skip	glue	0pt plus0pt minus0pt
\muskip	muglue	0mu plus0mu minus0mu
\toks	token string	empty
\box	a box	void

Note that each class of registers can only be used for data of a certain type. The registers are thus similar to strongly typed variables used in many programming languages. Each class contains 256 registers, so there are, for example, the 256 count registers \count0 through \count255.

The use of registers involves five aspects:

1. Allocation. This is done by the \new*xx* and the \def*xx* commands. Each \new*xx* command places a short message (of the form \abc=\skip0) in the log file. If a job allocates hundreds of registers, these messages may slow the computer down considerably, so they can be suppressed by redefining macro \alloc@ [347].

2. Assignment. Each assignment erases the old value.

3. Use of contents. A register can be used many times. Exception: A box register can be used and optionally emptied.

4. Display. The contents of a register can be typeset in the document or shown in the log file. This is particularly useful with box registers, where it is done with the \showbox command.

5. Write. A register can be written on a file. Box registers are again an exception, since a box has a structure and dimensions. (It is interesting to note that the \showbox command does write a box on a file, the log file, but in a form that's not easy to read.)

Reserved Registers. Certain registers are reserved by TeX for special purposes. \box255 is used by the OTR. \count0 through \count9 are used by the plain format for the page number. \count10 through \count20 are used by the \new macros that allocate registers, file numbers, and a few other things. Those registers should not be used unless you know what you are doing. Also, \count10 through \count20 are used by the various \new*xx* macros. For example, if \newdimen decides to allocate \dimen21, then it sets \count11 to 21.

Certain registers are available for temporary use. They are registers 0–9 of every class (except count), and registers 255 of every class except box. Commands such as \toksdef\abc=0 or \dimendef\abc=0 assign the name \abc to the \toks0 or \dimen0 register.

Allocating registers. Since it is not a good idea to refer to registers explicitly, the plain macro \new should be used to allocate registers and assign them names. Thus \newcount\temp is a typical use. It allocates the next available count register and assigns it the name \temp.

Five of the six register classes are similar, but box registers are different. Boxes have dimensions and internal structure, they can be nested by other boxes, and they tend to consume memory space. As a result, there are differences between box registers and other registers.

■ The command \newcount\temp creates a quantity \temp whose value is, for example, \count18. In contrast, \newbox\Temp (note the uppercase 'T') creates \Temp as the number 18. Saying 'A\Temp B' is thus identical to 'A\char18 B'. (This can be verified by \show\Temp.)

■ Assigning a new value to a register is done by an assignment statement of the form \temp=1234. Assigning a new value to a box register is done by means of '\setbox\Temp=...'.

■ Box registers cannot be assigned to any other type of register. It is possible to assign a count register to a dimen register, or a toks register to a box register. It is even possible to assign a skip register to a toks or a box register, but it is impossible

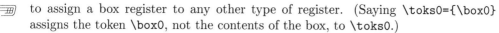

to assign a box register to any other type of register. (Saying \toks0={\box0} assigns the token \box0, not the contents of the box, to \toks0.)

■ A box register is emptied when it is used. The register thus becomes void. In contrast, other registers cannot be void; they must always contain a value.

■ The contents of any register, except a box register, can be written on a file. We can say, for example, \write\abc{\the\skip0}, but we cannot use the similar command \write\abc{\the\box0}. The reason is that boxes are complex structures.

■ The primitive \the can be used to produce the value of any register except a box register.

■ At the start of a job, boxes are normally void, but there is a subtle point, involving \box0 and the plain format, that users should keep in mind. The plain format says (on [361]) \setbox0=\hbox{\tenex B}. This is used in the definition of macro \bordermatrix. As a result, \box0 is not void at the start of a plain job. This point is mentioned again, in connection with \lastbox.

2.2 \the

The primitive \the can be used to produce the values of certain internal quantities. TEX novices are always confused by it. A typical complaint is "Why do I say '\box0' to typeset \box0 but '\the\count0' to typeset \count0?" The answer is—to make it easier for TEX to compile the document and to detect errors.

The command \the must be followed by an internal quantity, such as a register. It produces tokens that represent the value of the quantity. Thus after saying '\newcount\temp \temp=1234', the command \the\temp creates the tokens '1234'. Thus '\hskip\the\temp pt' will skip 1234 points, and 'ABC\the\temp GHK' will typeset '1234' between the C and the G.

The command \temp, in contrast, does not create tokens. TEX considers it the start of an assignment, unless it expects an integer at this point. For example, consider the command '\hskip\temp pt'. After TEX has read the \hskip, it expects a dimension (a number followed by a valid unit of dimension). This command will thus skip 1234 points.

The two examples above suggest that \temp and \the\temp produce the same results, but this is not generally true. Consider the assignment '\skip0=3pt plus 2pt'. Saying '\hskip\the\skip0 minus 1pt' will skip by '3pt plus 2pt minus 1pt', but saying '\hskip\skip0 minus 1pt' will skip by '3pt plus 2pt' and will consider the 'minus1pt' text to be typeset.

In the former case, the tokens produced by \the\skip0 blend with the rest of the document and become part of the \hskip command. In the latter case, TEX expects the \hskip to be followed by a valid glue and, on finding \skip0, is satisfied and executes the command, not looking for any more arguments.

There is also a \showthe command, used to display the values of registers and internal parameters on the log file and the terminal.

As usual, box registers are different. Assignment is done by \setbox, and \the is not used.

2.3 Modes

Of all the advanced concepts, the idea of modes [Ch. 13] is perhaps the most important. At any given time, TeX is in one of six modes, and its behavior depends on the current mode. The mode can frequently change and can also be nested. The six modes are as follows:

- Horizontal, or H, mode. TeX is in this mode when it reads the text of a paragraph.
- Vertical, or V, mode. This is where TeX usually spends its time between paragraphs, executing commands.
- Restricted horizontal (or RH) mode. TeX switches to this mode when it builds an \hbox. This mode is very similar, but not identical, to H mode.
- Internal vertical (or IV) mode. Commands such as \vbox or \vtop force TeX to go into this mode, which is very similar, but not identical, to V mode.
- Inline math mode, where a math inline formula is built.
- Display math mode, where a display formula is constructed.

When TeX starts, it is in V mode (between paragraphs). It reads the input file and, when it sees the first character to be typeset (or anything that's horizontal in nature, such as \noindent, \vrule), it switches to H mode. In this mode, it first executes the tokens in \everypar, then reads the entire paragraph into memory. The paragraph is terminated by \par, by a blank line, or by anything that doesn't belong in H mode (such as \vskip or \hrule). TeX then switches to V mode, where it sets the paragraph and takes care of page breaking.

A paragraph is set by breaking it into lines that are appended—each as an \hbox—to the *main vertical list* (MVL). After appending the lines of a paragraph to the MVL, TeX determines whether there is enough material in the MVL for a full page. If there is, the page-breaking algorithm is invoked to decide where to break the page. It moves the page material from the main vertical list to \box255 and invokes the output routine. Some material is usually left in the MVL, eventually to appear at the top of the next page. The routine can further modify \box255, can add material to it, or can return some material from it to the main vertical list. Eventually, the output routine should invoke \shipout to prepare the .dvi file. TeX stays in V mode and continues reading the input file.

Modes can be nested inside one another. When in V mode, between paragraphs, TeX may be asked to build a \vbox, so it enters IV mode. While in this mode, it may find characters of text, which send it temporarily to H mode. In that mode, it may read a '$', which causes it to switch to inline math mode. The curious example on [88] manages to nest all the modes at once.

2.4 Anatomy of TeX

A quote, from [38], is in order. "It is convenient to learn the concept (of tokens) by thinking of TeX as if it were a living organism." We will develop this concept further, and try to get a better understanding of the overall organization of TeX by considering the functions performed by its main "organs," namely eyes, mouth, gullet, stomach, and intestines (see anatomical diagram on [456]).

- TeX uses its "eyes" to read characters from the current input file. The \input command causes TeX to "shift its gaze" and start reading another input file.

■ In its "mouth," TEX translates the input characters into tokens, which are then passed to the "gullet." Spaces and end-of-line characters are also converted into tokens and sent to the gullet. A token is either a character of text or a control sequence (but see Chapter 8 for a third type). Thus the name of a control sequence, which may be long, becomes a single token. The process of creating tokens involves attaching a category code (section 2.9) to each character token, but not to a control sequence token.

■ The "gullet" is the place where tokens are expanded and certain commands executed. Expandable tokens [373] are macros, \if...\fi tests, and some special operations such as \the and \input.

A token arriving at the gullet is sent to the stomach, unless it is expandable. Expanding a token results in other tokens that are, in turn, either expanded, if they are expandable, or sent to the stomach. This process continues until no more expandable tokens remain in the gullet, at which point the next token is moved from the mouth to the gullet, starting the same process (the word "regurgitation", on [267], properly describes this process). Certain commands, such as \expandafter and \noexpand, affect the expansion of tokens, so they are executed in the gullet.

In general, a token is either expandable or unexpandable. The former category consists of macros, active characters, and a few primitives, such as conditionals, \number and \jobname (see list on [213–214]).

The latter category includes everything else, such as character tokens, anything defined by \chardef, and most primitives.

Expansion takes place in the gullet. For macros with no parameters, the expansion is a simple replacement (also for some primitives, such as \jobname). Normally, however, the token being expanded depends on arguments, which have to be read before the expansion can take place.

Exceptions: (1) The construct \expandafter⟨token₁⟩⟨token₂⟩ is treated by the gullet in a special way. It is replaced by ⟨token₁⟩⟨expansion of token₂⟩, which is then scanned again by the gullet. (2) A \noexpand⟨token⟩ prevents the gullet from expanding the token.

■ As a result, there is a constant stream of tokens arriving at the "stomach," where they get executed. Most tokens are characters of text and are simply typeset (appended to the current list). Tokens that are TEX primitive commands are executed, except that the "stomach" may have to wait for the arguments of the command to arrive from the gullet. Recall that nonprimitive commands are macros and are expanded in the gullet.

Another way of explaining stomach operations is this: The stomach executes tokens coming from the gullet. It classifies all tokens into two groups, tokens used to construct lists, and tokens that are mode independent. The former group includes characters of text, boxes, and glue. They are mode sensitive, can change the mode, and are appended to various lists. The latter can be assignments, such as \def, or other tokens, such as \message or \relax. The \relax control sequence deserves special mention. It is a primitive and is thus unexpandable. The gullet passes it to the stomach, where its execution is trivial. It is an important control sequence, however, because it serves as a delimiter in both the gullet and the stomach.

The result of executing tokens in the stomach is larger and larger units of text (lists). Individual characters are combined to make words, which are combined to form lines, which are combined to make pages of text. When the next page of text is ready, it is sent to the "intestines."

■ The output routine corresponds to the "intestines" of TₑX. It receives a page of text and, after some processing, translates it into .dvi commands that are appended to the .dvi file. The processing may consist of adding something to the page (such as a header, a footer, footnotes or margin notes) or deleting something from it (certain lines or even a big chunk). The deleted material either is discarded or is returned to the stomach.

The .dvi file is the final output of TₑX, and it comes out of the intestines in pages. It consists of commands that tell where each character is to be placed on the page. Reference 11 is a detailed description of the .dvi file format.

The entire anatomy is summarized in the diagram below.

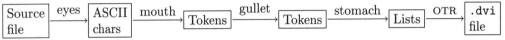

The advantage of this anatomical description is that we can think of the process as a pipeline. Material (mostly tokens) advances from organ to organ and is processed in stages. However, the individual organs sometimes affect each other. The best example is the \catcode primitive. It is executed in the stomach and immediately starts affecting the way the mouth assigns catcodes to tokens. Another example is the \def primitive. When a \def\abc is executed, the definition becomes available for the gullet to use whenever \abc has to be expanded. As soon as another \def\abc is executed in the stomach, the gullet is affected and will use the new definition in future expansions of \abc.

It should be mentioned at this point that any control sequence, even a primitive, can be redefined. If this is done locally, in a group, the old definition is automatically revived when the group ends. If the redefinition is global, however, the old definition cannot be used, even though it is still stored somewhere in memory.

2.5 Characters

TₑX inputs characters from the input file and outputs characters to the .dvi file, so characters are important for an overall understanding of TₑX. Characters usually come from the input file (through the eyes), but may also come from the expansion of macros (in the gullet). These are the only character sources of TₑX. Most characters are simply typeset, but some, such as '\' and '$', have special meanings and trigger TₑX on special tasks.

A character is input as an ASCII code, which becomes the *character code*. The character then gets a *category code* attached to it (in the mouth), which determines how TₑX will process the character. When a macro is defined (in the stomach), catcodes are attached to each character (actually, to each character token) in the definition and are used when the macro is later expanded (in the gullet). Advanced users would like to know that a certain amount of processing goes on even in the mouth. Among other things (see complete description on [46–49]), consecutive spaces are compressed into one space, a copy of \endlinechar is inserted at the end of each line, consecutive spaces at the beginning of a line are ignored, and spaces following a control word are ignored (they never become space tokens).

There is a simple way to find the ASCII code of a character. Just write a left quote followed by the character [44]. Thus `b has the value 98, and \number`b will actually typeset a 98.

▸ **Exercise 2.1:** What will be typeset by \number`1 and what by \number`12?

▸ **Exercise 2.2:** What is the result of \number`{0} and what of \number`%?

The ASCII table provides 128 codes, but present keyboards don't have that many keys. The \char command or the '^^' notation [45–46] can be used to refer to a keyless character. Thus the notation '^^x', where x is any character, refers to the character whose ASCII code is either 64 greater than, or 64 less than, the ASCII code of x. Examples are '^^M' (⟨return⟩), '^^J' (line feed), '^^@' (null), '^^I' (horizontal tab), and the other ASCII control characters. The \char command is easy to use; if the current font is cmr10, then \char98 is the code of 'b'. In general, \char127 is the code of the character in position 127 of the current font.

The difference between the two notations is that \char is easier to use but is executed in the stomach (i.e., late); in contrast, a '^^x' is preprocessed into a single token in the mouth (i.e., as soon as it is input). As a result, the concoction \def\a^^"c{...} defines a macro \abc, but \def\a\char98c{...} defines a macro \a with the string \char98c as a delimiter.

2.6 Upper- and Lowercase Characters

It is very easy to convert between upper- and lowercase characters by means of the two primitives \uppercase and \lowercase [41]. Each takes an argument that is a list of tokens. It scans its argument, ignoring the control sequence tokens and converting the character tokens. Thus the result of '\uppercase{abc}' is 'ABC', but the result of the test '\toks0={abc} \uppercase{\the\toks0}' is 'abc' (\uppercase scans its argument and finds it to consist of two control sequences, so it does nothing). To convert the characters in \toks0, '\the\toks0' has to be expanded before \uppercase is executed. This, naturally, is achieved by saying '\uppercase\expandafter{\the\toks0}'.

How does TEX know the uppercase equivalent of a given character? By means of \uccode and \lccode. Each character has two values associated that are its \uccode and \lccode. \uppercase scans its argument and converts each character token to its \uccode value. If the \uccode is zero, there is no conversion. This is similar for \lccode. INITEX sets the \uccode of all the letters to A through Z; it sets the \lccodes of the letters to a through z, and sets all other codes to zero.

The two commands and two codes above can be used for other purposes. Imagine text stored in a toks register. We want to scan it and change all the occurrences of x to character 255. The first step is to change the \uccode of a to a (instead of A), and repeat for all the letters. The \uccode of x is then changed to 255 (figure 2.1). Because of the \edef, macro \temp is defined as the string '\uppercase{abcx1}', and this becomes the value of \toks1. The actual conversion is done later, by means of '\the\toks1'.

▸ **Exercise 2.3:** Rewrite the above using macros, instead of toks registers, to save the text before and after the conversion.

```
                                   \ifnum\t<'z
\toks0={abcx1}                     \repeat
\begingroup                        \uccode'x=255
\newcount\t \t='a                  \edef\temp{\uppercase{\the\toks0}}
\loop                              \toks1=\expandafter{\temp}
\uccode\t=\t                       \the\toks1
\advance\t by 1                    \endgroup
```

Figure 2.1

▶ **Exercise 2.4:** A token can be either a character or a control sequence, but there is a third type, namely a macro parameter. What is the result of '\def\abc#1{\uppercase{ab#1}} \abc x'?

▶ **Exercise 2.5:** What are the results of '\lowercase{\uppercase{\def\abc{x}}\abc}' and of '\lowercase{\uppercase{\def\abc{X}}\abc}'?

2.7 End-of-Line

We intuitively think of a line of text as ending with a carriage return. TeX, as usual, offers a more general treatment of this feature.

➠   **End-of-line rule:** The source file is made up of lines, each terminated by a special character (or characters) depending on the operating system. Upon reading the line, TeX deletes these characters (and also all trailing blanks) and adds its own end-of-line character, denoted ⟨return⟩, which is normally the token 13_5. Note that if the line has a '%', TeX does not read beyond it and, consequently, does not append the ⟨return⟩.

If the catcode of ⟨return⟩ is indeed 5, it will be turned into a space. (There are two exceptions: (1) if the line is empty, the ⟨return⟩ is turned into a \par; (2) if the last thing on the line is a control word, the ⟨return⟩ is ignored.)

If the catcode of ⟨return⟩ has been changed (by \catcode\^^M=..), it is handled according to its new catcode. The most common case is to make the ⟨return⟩ active, as in \obeylines.

If the source line has '^^M' in it, it will become a ⟨return⟩, and the rest of the line will be ignored.

To make the end-of-line treatment even more general, the token inserted at the end of a line is the value of parameter \endlinechar [48]. This parameter is set by INITEX to a ⟨return⟩ (^^M) but can be changed by the user. If \endlinechar is negative or is greater than 255, no end-of-line character is inserted. In this case, the input is considered one long line. The same effect can be obtained by ending every line with a comment character.

2.8 Numbers

Not everything can be typeset by \the. The definition of a macro (its re-placement text), for example, can only be typset by \meaning. To print a numeric quantity, the primitive \number can always be used, even if the number is the value of a macro. Defining the macro \def\ctst{10}, we cannot say \the\ctst, but we can say \number\ctst. Also, \dimen registers are printed differently by \the and by \number.

The left-quote character usually acts as a normal character of text. Sometimes, however, it signifies the start of a number. Similarly, the right-quote and double-quote characters sometimes have special meanings; they signify the start of an octal or hexadecimal number. When these characters arrive at the stomach, they are tokens with catcode 12 (other). If the stomach is expecting a number, it will assign them their special meanings and expect them to be followed by digits (decimal, octal, or hex). Otherwise, they will simply be typeset.

The following are all valid representations of the decimal number 98: 98 +98 098 ʼ142 "62 ʻb ʻ\b. They may be used with any command requiring an argu-ment of type ⟨number⟩ (see definition of ⟨number⟩ on [269]). Thus \number‘b will typeset 98, and \catcode+98="D will convert the letter ‘b’ into an active character. (Incidentally, once you do that, ‘b’ isn’t a letter any longer, so something like \bye will be interpreted as the control symbol \b [52], followed by the letters ‘ye’.)

▸ **Exercise 2.6:** What is the result of \catcode101=14 \number‘e?

Both notations ‘b and ‘\b produce the character code of ‘b’. The difference between them is that the latter can be used with any character. Thus \number‘\% produces 37 but \number‘% will treat the ‘%’ as a comment and will look for an argument on the next line. Similarly, \number‘\^^M produces 13 (ASCII ⟨return⟩), but \number‘^^M will consider the ^^M an end-of-line, will replace it with a space, and will produce 32 (ASCII ⟨space⟩).

Noninteger numbers can only be used with dimensions and are considered mul-tipliers. Thus, for example, 1pt multiplies the basic unit of a pt by one, and 2.5\baselineskip mutiplies the current natural size of \baselineskip (not its stretch and shrink components) by 2.5.

2.9 Category Codes

One of the main considerations behind the design of TEX was to make it as general and flexible as possible, so it could be adapted to many different tasks. Thus the character ‘\’ is normally used to start the name of a control sequence (it is the *escape character*), but if the ‘\’ is needed for other purposes, any character can be defined as the escape character. The same thing is true for the other special characters, namely { } $ and # ^ _ and %.

Each of those characters is special only because it is assigned a special *category code*. It is easy to change the category codes, thereby changing the meanings of characters. There are 16 category codes [37] numbered 0 to 15 (Figure 2.2).

The category code of the character ‘A’ can be typeset by \the\catcode‘A. It can be displayed in the log file by \showthe\catcode‘A. The category code is assigned to a character in the mouth, and that assignment is permanent. The pair

0 Escape character	4 Alignment tab	8 Subscript	12 Other character
1 Begin. of group	5 End of line	9 Ignored char.	13 Active character
2 End of group	6 Parameter	10 Space	14 Comment char.
3 Math shift	7 Superscript	11 Letter	15 Invalid character

Figure 2.2. The 16 Catcodes

⟨character code, category code⟩ becomes a *token*. The mouth also converts control sequences into tokens, but they do not get a catcode.

When a macro is defined (in the stomach), each token of the replacement code gets a catcode assigned. When the macro is later expanded (in the gullet), the replacement code is copied, with arguments replacing the parameters, and the resulting tokens are sent to the stomach.

The catcode of a character can be changed by the \catcode primitive. The most common example is \catcode'\@=11, which makes the '@' a letter. It can now be used in control words, as any other letter, and the plain format [App. B] makes heavy use of this in order to define "private" macros, inaccessible to the user. Another example is \catcode'\<=1, \catcode'\>=2, which defines the characters '<' and '>' as group delimiters. This does not change the definition of the braces, so now a group may be specified by {...>.

Here are some notes about catcodes:

■ A character created by '^^' is assigned a catcode in the mouth, but a character created by \char is not assigned a catcode at all! This is because \char is a primitive and thus is executed in the stomach. Executing \char always creates a character of text, which is typeset.

■ Code 14 (comment) causes the rest of the input line (including the end-of-line character) to be ignored, and is itself ignored.

■ A space following an active character is not ignored.

■ Category 9 is always ignored, but can be used to delimit a control sequence name. Thus if we change the catcode of 'x' to 9 (by \catcode'\x=9), future occurrences of 'x' will be ignored, and the string 'Nx0' will be typeset as 'NO'. However, a macro \abc can now be expanded by saying \abcx.... The ignored character 'x' serves to delimit the name \abc the way a space normally does. When TEX starts, the only ignored character is the ASCII ⟨null⟩ (^^@).

▶ **Exercise 2.7:** What's the reason for category 9 (in other words, why type a character and then ignore it)?

■ Each blank line becomes a \par token, so there may be several consecutive such tokens, of which only the first is executed by the stomach.

Category 15 (invalid character) is initially assigned to the ASCII ⟨delete⟩ character (^^?). TEX complains when it reads an invalid character.

2.10 Commands

Many different commands are available in TEX, and they can be classified in two ways:

■ A command can be classified as a primitive, a character, or user-defined. The latter category is further divided into a macro or an active character.

■ A command can also be classified as either horizontal, vertical, or neither.

Beginners learn very quickly that a command should start with a '\'. However, an active character is also a command, and even a character of text (catcodes 11, 12) is one (see [267]). When a command is used very often, we want its name to be as short as possible, so we define it as an *active character* (catcode 13). It then becomes a one-character command, without even a '\'.

▶ **Exercise 2.8:** What's another reason for having active characters?

It is useful to consider a character of text (other than a space) a command. Such a command tells TEX to start a new paragraph or, if it is already in H mode, to typeset the character. A character is thus a *horizontal command*. If we want TEX to treat a character as not horizontal, we can place it in an \hbox. Interestingly, an \hbox is not inherently horizontal (but neither is it inherently vertical).

A command starting with a '\' is called a *control sequence*. If it consists of letters only (catcode 11), it is called a *control word*; if it consists of a nonletter (any catcode ≠ 11), it is called a *control symbol*. Any spaces (or end-of-line) following a control word are ignored by TEX since it assumes that they are there only to delimit the word. However, a control symbol can only have one character following the '\', so there is no need for any delimiters. Spaces following a control symbol are not ignored, and a situation such as '\?1' is interpreted as the control symbol \? (normally undefined), followed by a '1' (which is normally typeset).

The ignore-space rule above, however, applies only to characters coming from the input file; if a control sequence comes from a token list [39], a space following it will not be ignored.

▶ **Exercise 2.9:** Devise tests to prove the preceding statements!

The concept of a delimiter is worth a little discussion. If we want to expand a macro \abc, we can say '\abc␣'. However, \abc1... is also okay. The '\' tells TEX that a control sequence starts, the 'a' tells it that this is a control word (just letters), and the '1' delimits the string of letters. After TEX reads the '1' it backspaces over it, expands \abc, and rereads the '1'. The point it that the '1' is read twice and, when it is first read, it is not assigned a catcode.

This is a subtle point that may, sometimes, lead to errors. Consider the following:

```
\def\abc{\catcode`\%=11 }
\abc%
\abc%
```

The first line defines \abc as a macro that makes the '%' a letter. The second line uses '%' to delimit the string of letters abc. The '%' is thus read twice. When it is first read, it is not assigned a catcode, and \abc has not been expanded yet. When the '%' is read again, \abc has already been expanded, so the '%' is assigned a catcode of 11, which causes it to be typeset. When the third line is input, the '%' already has catcode 11, so it is considered a letter. TEX thus ends up with the string 'abc%' and tries to expand macro \abc%, which is normally undefined. The result is the message ! Undefined control sequence \abc%.

▶ **Exercise 2.10:** Why is there a space following the '11' in the definition of \abc above, and what happens without that space?

There is an important difference between a character and a control sequence. A character has a catcode attached to it, which tells the gullet and the stomach what to do with the character. A control sequence has no catcode and may be redefined at any time, so the gullet has to look up the current definition before it can expand the control sequence.

A command may have arguments. Thus \kern must be followed by a dimension, and a '^' in math mode must be followed by the superscript. Sometimes the arguments are optional (the '=' in an assignment is a typical example), and sometimes there is a choice of arguments (\leaders should be followed [281] by either a ⟨box⟩ or a ⟨rule⟩, so it can be followed by one of the following: \box15 \copy16 \vsplit17 \lastbox \hbox \vbox \vtop \hrule or \vrule). The \font command [16] makes for an interesting example. It starts with \font\⟨name⟩=⟨file name⟩, followed by the optional arguments 'at ⟨size⟩' or 'scaled ⟨factor⟩'. The words 'at', 'scaled' are called *keywords* and don't have a '\'. See [61] for a complete list of keywords.

In the command \setbox0=\hbox{...}, the arguments are '0=\hbox{...}'.

2.11 Assignments

An assignment [275] is any command that assigns a new meaning to a control sequence or to an internal quantity. Examples are \def, \hsize=..., \font\abc=..., \setbox0=..., and \advance\x.... Note that the '=' is always optional. Assignments are examples of commands that are executed in the same way regardless of the current mode.

▶ **Exercise 2.11:** What other commands are executed in a mode-independent way?

2.12 Lists

This is another important concept. The contents of a box is a list, as is the contents of a math formula. A list is made up of items such as boxes, glues, and penalties.

TeX is assembling a horizontal list when it is in H mode (building paragraphs) or in RH mode (building an \hbox). The items of such a list are strung horizontally, left to right, and must be H mode material [95]. In H mode, a list is terminated when TeX reads a \par (or a blank line), or anything that's vertical in nature (such as a \vskip). In RH mode, TeX terminates a list when it finds the '}' of the \hbox. If it finds inherently vertical material in this mode, it issues an error. Examples of horizontal commands are \vrule, \valign, \char, and a character of text (see [283] for the complete list).

A character of text is a horizontal command whose meaning is: Add me to the current horizontal list or, if there is no current H list, start one with me as the first item.

TeX is assembling a vertical list when it is in V mode (between paragraphs) or in IV mode (building a \vbox). The items of such a list are stacked vertically, top to bottom, and must be V mode material [110]. In V mode, a list is terminated when TEX sees an inherently horizontal command, such as an \hskip or the first character of the next paragraph. In IV mode, TEX terminates a list when it finds the '}' of the \vbox. If it finds inherently horizontal material in this mode, it issues an error. Examples of vertical commands are \hrule, \halign, and \end (see [286] for the complete list).

Kern is an interesting example of an item that may appear in either horizontal or vertical lists. Even though the same command, \kern, is used, it has different meanings in those lists. Kern is essentially rigid glue with the property that TEX does not break a line or a page at a kern. Thus if two boxes are separated by a kern, they will remain tied. Of course, if the kern is followed by glue, a break is possible at the glue.

2.13 Whatsits

A *whatsit* is an item that may appear in either a horizontal or a vertical list. It has no dimensions and signifies an operation that should be delayed. The paragraph builder and the page builder scan lists submitted to them and execute certain whatsits. There are three types of whatsits:

- \special. This command takes one argument. It is initially stored in the MVL, then ends up being written, at \shipout time, to the .dvi file. The argument is interpreted and executed by the printer driver (and, perhaps, the screen preview program). Individual printer drivers support different \special arguments, which makes \special an example of a *non-compatible command*. A document using it may be printed only by certain printer drivers.

A typical example is \special{postscript ⟨postscript commands⟩}. When the printer driver finds this in the .dvi file, it sends the POSTSCRIPT commands to the printer, so that special printing effects can easily be achieved. (Note: POSTSCRIPT is powerful but is not always easy to use.)

- The three nonimmediate output commands \openout, \closeout, and \write. They are also stored in the MVL and are executed later, at \shipout time. The reason for their delayed execution is that they may have to write the page number on an output file, and that number is only known in the output routine (see chapter 16 about the asynchronous nature of the OTR).

- The \language and \setlanguage commands [455] also produce whatsits each time the language is switched in the midst of a paragraph (e.g., from Icelandic to Serbo-Croatian). These whatsits are stored in memory with the rest of the current paragraph, while the paragraph text is being read in H mode. When the paragraph builder typesets the paragraph (determines the line breaks), the whatsits are used to select the set of hyphenation rules appropriate for each language. There can be up to 256 sets of hyphenation rules.

Even though a whatsit does not have dimensions, placing it in a box may sometimes affect dimensions in an unexpected way. Imagine the (admittedly unusual) construct '\vtop{\hbox{\write16{Ap}}\hbox{a}}'. As far as dimensions are concerned, this construct is identical to '\vtop{\hbox{}\hbox{a}}'. All the dimensions of the top \hbox are zero, but there is \baselineskip glue below it, which pushes the bottom box (with 'a') down. Saying
'Start \vtop{\hbox{\write16{Ap}}\hbox{a}} End' results in 'Start   End'.

a

2.14 Parameters

Many numeric values are used by TEX that a user may want to examine or modify. Those values are, therefore, given names and are considered *parameters* of TEX (different from macro parameters). They are all listed on [272–275]. A typical example of a parameter is \hsize. Its value is a dimension, so it may be used whenever a dimension is necessary, as in \hbox to\hsize or \hskip\hsize. It may also be assigned a new value by \hsize=3.5in (however, the '=' is optional).

The rule is that the value of a parameter is used if its name appears in a context where such a value is needed. The value is changed if the name appears in any other context. A common error is a sentence such as, "The width of a line is normally \hsize but, ..." When TEX sees \hsize, it is in the midst of typesetting our sentence, so it is in a context where it does not need a dimension. It assumes, therefore, that \hsize is an attempt to modify \hsize, and reads ahead, expecting the new value. Finding the word "but" instead, it complains of a missing number. The correct sentence should, of course, be "The width of a line is normally \the\hsize\ but, ..."

2.15 Discardable Items

Considering the amount of work done by TEX, it should come as a surprise to the user to realize that certain quantities are sometimes discarded. However, it is easy to show why this is necessary.

Imagine a command such as '\vskip 2in' placed between two paragraphs. Normally, there will be 2 inches of space (glue) left between the paragraphs, but if TEX decides to break a page between the paragraphs, the glue is discarded. The reason is that it is pointless to leave 2 inches of space at the bottom of the current page or at the top of the next page.

To create a nondiscardable space, it is possible to say '\vbox to2in{}', but note that a box is indivisible. If it happens to be placed close to the bottom of a page, where there is less than 2 inches of space, it will be moved to the next page, and the current page would become underfull.

Sometimes it is desirable to start a page with 2 inches of space at the top, such as at the start of a new chapter. This is also possible to achieve by saying either '\null\vskip 2in' or '\hrule height0pt\vskip 2in'. The latter combination is created by the \vglue macro.

The same ideas apply inside a paragraph. If a line is broken at a glue, the glue is discarded. The best way to start a line with a 1 inch space is to say \hglue 1in, which places a strut before the glue.

Discardable items [95] are glue, kern, penalty and, in an \hbox, also "math-on, math-off" items.

When boxes are stacked vertically, the normal interline glue is automatically added between them. However, when an \hrule is placed in a vertical list, no interline glue is placed above or below it. A good way to add interline glue around a rule is to place an empty box on both sides, thus '\null\hrule\null'.

2.16 Macros

This material is intended for users who rarely use macros but are otherwise experienced with TeX. A macro is a list of tokens that's been given a name, typically because it is used a lot in certain documents. For example, if the following is used many times in a document '\medskip$\bullet$\enskip', it can be given a name, such as \section, and defined as a macro by saying '\def\section{\medskip$\bullet$\enskip}'. After this definition, the macro can be *expanded* by simply saying '\section'. To expand a macro means to expand the tokens that constitute it. Those tokens are also called the *replacement text* of the macro, since they replace the macro name during expansion.

However, the feature that makes macros so useful and powerful is the use of parameters. By using parameters, each expansion of the same macro may be different. Our macro above may be extended by adding a parameter, such as the name of the section. After '\def\section#1{\medskip$\bullet$\enskip{\it#1}}', each expansion must supply some text that will be typeset as the name of the new section. Thus, e.g., '\section{Advanced Techniques}' will expand the tokens of the macro, and replace the parameter #1 by the argument 'Advanced Techniques'.

➠ Note the difference between *parameters*, which are formal and have no fixed value, and *actual arguments*, which are text and commands that TeX can process. The notion of parameters also generalizes the concept of a token. Two types of tokens have been mentioned earlier, a control sequence or a single character. From now on, a token can also be a macro parameter, something of the form '#x', where x is one of the digits $1, 2, \ldots, 9$. A macro can have up to nine parameters and, each time it is expanded, arguments must be supplied to replace each parameter.

Our first macro is now extended by adding another parameter, '#2', whose value is the section number. Defining '\def\section#1#2{\medskip$\bullet$\enskip{\bf#2.\thinspace}{\it#1}}', each expansion must have the form '\section{Advanced Techniques}6'. The second argument, '6', is not enclosed in braces because of the following rule: "The argument of a macro is a single token (the next token in the input stream, except that it cannot be any of the braces), unless it is delimited or enclosed in braces." The first argument 'Advanced Techniques' is long, so braces are used to indicate its boundaries. The second argument is a single character (a token), so no braces are necessary; but what is a delimited argument?

When a macro is defined, each of the parameters #1, #2 can be followed by any characters (except, of course, '#' and '{'), which are its delimiter. If this is done, then each argument in any of the expansions must be followed by the same characters. Our \section macro is now extended to include delimiters: \def\section#1;#2.{\medskip$\bullet$\enskip{\bf#2.\thinspace}{\it#1}} A typical expansion of the new macro is '\section Advanced Techniques;26.' The first argument is everything up to, but not including, the first ';'. The second argument is everything following the ';' up to the next period. The delimiters themselves are skipped and do not become part of the actual arguments.

A simple example of delimiters is '\def\test1#1#2. #3\end{...}'. The first parameter has no delimiter, so it must be either a single token or a brace. However, it must be *preceded* by a '1'. The second token is delimited by the two characters

'.␣' (not just a period), and the third one, by the characters \end. Since the delimiters are skipped, the \end is not expanded (what would happen if it were?). If we now expand '\test10123. 45\end', the parameters will be '0', '123', '45'. If the expansion does not provide the right delimiters—such as in '\test210123. 45\end', '\test10123.45\end'—TeX issues the error message
! Use of \test doesn't match its definition.

When a parameter is delimited, the argument may contain braces, but they must be balanced. Thus in

```
\def\x#1\end{\message{'#1'}}
\x 1{2\end3}4\end5
```

the argument is '1{2\end 3}4'.

Delimited parameters may get complex and confusing (see example on [203]), so the \message command may be used to find out the precise value assigned to each parameter:
\def\test1#1#2. #3\end{\message{arg1=#1; arg2=#2; arg3=#3}...}.
Another useful debugging tool is the \tracingmacros command, discussed among advanced macro features.

In addition to \def, macros can also be defined by \let and \chardef, and as active characters. It has already been mentioned that an active character is a control sequence whose name is limited to a single character, without even a '\', but whose replacement text can be of any length. The \chardef command [44] is, in a sense, the opposite of an active character. It defines a control sequence whose replacement text is limited to just one number (in the range 0–255), but whose name can be any valid control sequence name.

Thus \chardef\abc=98 or \chardef\abc='\b defines \abc as a macro whose replacement text is the character code of 'b'. It is equivalent to \def\abc{\char98}. Also \chardef\active=13 defines \active as a macro whose value is the number 13.

The \let command is still different. Its general form is
\let⟨control sequence⟩=⟨token⟩. It defines the control sequence as being identical to the token. Thus defining

```
\def\a{X}
\let\g=\a \def\k{\a}
\g\k
```

produces 'XX'. If we now redefine \a, the meaning of \k will change (since it was defined by \def) but \g will not change. Thus \def\a{*} \g\k produces 'X*'. The difference between \def and \let is now clear. \let\g=\a creates \g as an independent macro that does not change if \a is modifed. In contrast, \def\k{\a} Creates \k as a macro pointing to \a, so it is always dependent on \a.

Other ways to define macros are discussed in Chapter 5.

2.17 Formats

When TeX starts, there are no user-defined macros; the only commands available are the primitives, of which there are about 300. The primitive \def can be used to define macros, which are then added to the repertoire of available commands. When \def is executed in the stomach, it loads the replacement text of the new macro into a special table in memory.

When a large document, such as a book, is developed, many macros may have to be defined. Each time the document is typeset, all the definitions have to be prepared by the stomach and loaded in memory, which may be time consuming. In such a case, it is better to convert the macros into a *format*. A format is a collection of commands (macros and active characters), written in a special way on a file, to facilitate rapid loading.

A complete TeX system includes a program called INITEX [39] that is used to install TeX. INITEX is like TeX but can also prepare formats and hyphenation tables and perform other tasks. To prepare a format, INITEX should be run, the relevant macros should be defined, and the \dump command [283] should be executed. This command dumps the table containing the macros from memory onto a file. That file is called a format file, and it can later be loaded fast simply by copying its contents directly into memory. A format file is loaded, like any other file, by the \input command.

A very useful format, the plain format, comes with every TeX implementation. It is written on file plain.fmt, and it consists of about 600 macros that are described throughout *The TeXbook* and are useful for general-purpose typesetting. The plain format is also listed in [App. B]. Many commonly used commands, such as \bye and \smallskip, are part of this format.

When macros are developed for a specific document, they are normally used together with the plain format. It is easy to create a new format containing all the plain macros plus any user-defined ones. See [344] for directions.

Textures, a Macintosh TeX implementation, makes it particularly easy to create formats since it can execute the \dump command without any need for INITEX. It is also easy to load any existing format in Textures at the flick of the mouse, without having to type any commands.

> Do not ye yet understand,
> that whatsoever entereth in at the mouth
> goeth into the belly,
> and is cast out into the draught?
> — *The Bible, Matthew 15:17*

> ...It is that which at this instant, issuing out of a labyrinthine tangle
> of yeses and nos, makes my hand run along a certain path on
> the paper, mark it with these volutes that are signs: a double
> snap, up and down, between two levels of energy, guides this
> hand of mine to impress on the paper this dot, here, this one.
> — *Primo Levi, The Periodic Table (1984)*

3. Boxes and Glue

Boxes and glue are two of the main tools used by TeX. A box is an indivisible rectangular structure that can hold text and other items. Glue is simply *spacing*. It can be horizontal or vertical spacing, and it can be made as rigid or as flexible as desired.

Boxes are used by TeX for two purposes:

- To prepare a typeset page. This, of course, is the main task of TeX. Each character of text is a box. A word is a larger box made up of individual character boxes (Figure 3.3). A line is a long narrow box made up of word boxes strung horizontally and separated by (horizontal) glue. A complete page is also a box, made up of line boxes stacked vertically and separated by (interline) glue.
- To make it possible for the user to place text anywhere on the page. Without explicitly using boxes, the only output possible is a page, neatly set with lines of text, all of the same width. Usually, this is what we want but—if text has to be placed at certain positions (Figures 3.4 and 3.5), to achieve special effects—boxes should be used (see also example on [389]).

Many of the following examples show boxes surrounded by rules, to indicate their boundaries and layout. The rules, however, are not automatically supplied by TeX and have to be added explicitly.

Material placed in a box can be of the following types: text, other boxes, glue, kern, rules, penalties, and whatsits. All these types, except the last two, are discussed here.

Even though a box is a flat, two-dimensional structure, it has three dimensions: height, width, and depth (Figure 3.1). The reason is that boxes normally contain text, and a character of text has those three dimensions [63]. Figure 3.2 shows some text characters in their boxes. The dimensions of each character, and of its box, are determined by the font designer.

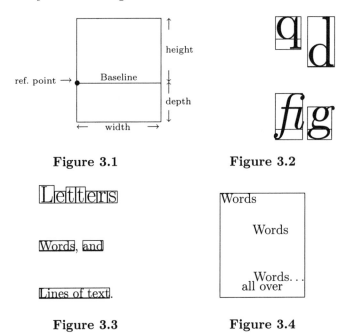

Figure 3.1

Figure 3.2

Figure 3.3

Figure 3.4

Figure 3.5 shows special effects possible with boxes and glue

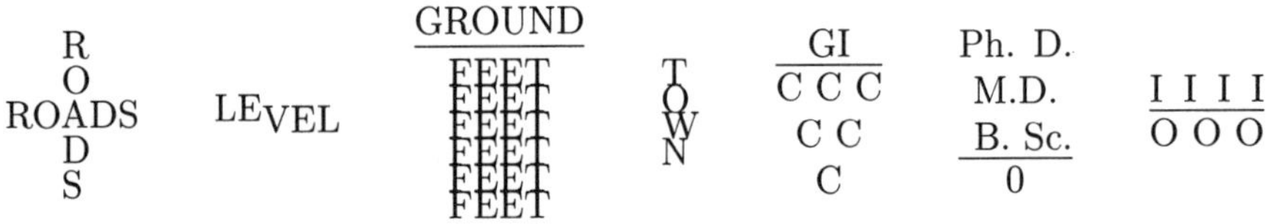

Figure 3.5

▸ **Exercise 3.1:** (Tongue in cheek.) Find good titles for the examples in Figure 3.5.

Boxes can be combined with rules to "box" text. Glue is used in Figure 3.6 below to separate the text and the rules (the diagram was created with macro `\Cboxit`, page 188).

I	bi	bid	bird	bride	bridge	bridged

abridged	dried bags	bridge ads	sad bird egg

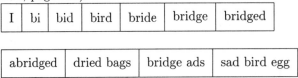

Figure 3.6

An important feature of TeX is that it does not know the shape of the characters that it typesets, just the dimensions of each character box. As a result, TeX does not complain when a character is narrower than its box, or when it sticks out of the box. This is an important property of boxes, used to achieve special effects. Things such as notes typeset on the margin, and characters typeset on top of each other, can be done using boxes with text sticking out of them.

A character of text is normally considered a box. However, it is not always treated as such. We show on page 42 that text placed in a \vbox is treated differently from boxes placed there. Also, the \lastbox command (section 3.28) distinguishes between text and boxes.

There are two types of boxes, horizontal (\hbox) and vertical (\vbox). The former is simpler and will be discussed first.

3.1 \hbox

An \hbox is created by the command \hbox{<material>}, and it has the following properties:

■ The material is placed in the box horizontally, from left to right; it becomes a *horizontal list*.

■ The box will not be broken across lines; it is an indivisible unit.

Examples:

1. 'A test of a box at the end of a \hbox{2.5 inch} line'. The string of text '2.5 inch' will be typeset with the rest of the text and become part of the current line. However, the line will not be broken inside this string. If the string ends up being placed around the middle of the line, this will cause no problem. If it is placed toward the end of the line, however, the line may end up being too long, which causes the well-known 'overfull box' error message. The reader should try the following test:

```
\hsize=2.5in
A test of a box at the end of a 2.5 inch line\par
A test of a box at the end of a \hbox{2.5 inch} line
```

The \hbox command has another form \hbox to<dimen>{<material>}. This creates a box whose width is the given ⟨dimen⟩. Thus \hbox to1in{<material>} will create a 1 inch wide box. However, we have to supply exactly 1 inch's worth of material to fill up the box; otherwise we end up with an error message. It is best to consider this form of the command as a *promise*; we promise TEX that we will supply just enough material to fill up the box. If we break our promise, TEX gets upset.

2. \hbox to37pt{material}\hbox to35pt{material}. The word "material", in font cmr10, is 36.13898pt wide. The first box is thus underfull, and the second one, overfull.

This form of the \hbox command should thus normally be used with flexible glues. We have said before that a glue is simply spacing. It turns out to be more useful than just spacing, since it can be flexible, and it can be made as flexible as desired. The most useful form of flexible glue is \hfil, a command that will skip horizontally any necessary distance. We can use \hfil to fill up any empty space in a box. Thus:

3. `\hbox to50pt{material\hfil}` creates $\boxed{\text{material}\quad}$. The text is left justi-fied in the box, and the glue is stretched, by the necessary amount, to fill up the empty space on the right. Since we happen to know the width of the text, it is easy to figure out that the `\hfil` causes a skip of $50 - 36.13898 = 13.86102$ pt, but this knowledge is normally unnecessary.

4. `\hbox to50pt{\hfil material}` creates $\boxed{\quad\text{material}}$. The text can easily be right justified. `\hbox to50pt{\hfil material\hfil}` will create $\boxed{\quad\text{material}\quad}$. The text will be centered in the box because the two pieces of `\hfil` have the same stretchability (infinite) and are therefore stretched by the same amount.

▶ **Exercise 3.2:** What is the result of `\hbox to50pt{\hfil\hfil material\hfil}`? As a reminder, these tests will work if there is room, on the current line, for 50 pt's worth of material. The most useful examples of this command are of the form:

5. `\hbox to\hsize{\hfil Heading\hfil}`. This will create

Heading

The box has the width of a line (`\hsize`), so the user should make sure that it occupies a new line by itself. Otherwise, the line it is on would be overfull. This is done by placing the box at the start of a new paragraph, or by terminating the preceding line with an `\hfil\break`. The result would be to center the word 'Heading' on a line by itself.

The special form `\hbox to\hsize` is so useful that it is given the short name `\line`. We can now write `\line{\hfil Heading\hfil}`, but remember to place this box at the start of a new line. The command `\line{\hfil material\hfil}` is so useful that it is also given a name, `\centerline`. We can thus write `\center-line{Heading}`.

6. The `\hfil` glue can also be used in the math modes. Expressions such as `$\hfill a\over b+c$` or `$b+c\over a\hfill$` are commonly used. They result in $\frac{a}{b+c}$ and $\frac{b+c}{a}$, respectively.

The examples above show that boxes placed in an `\hbox` can be shifted horizon-tally by means of globs of glue (or kern). It is also possible to move them vertically by means of the `\raise` and `\lower` commands. Thus

7. `\hbox{\kern4pt A\kern8pt BC\kern8pt D\kern4pt}` produces $\boxed{\text{A BC D}}$, but

```
\hbox{\kern4pt\raise4pt\hbox{A}
\kern8pt BC
\kern8pt\lower6pt\hbox{D}
\kern4pt}
```

results in $\boxed{^{\boxed{A}}\text{BC}\ _{\boxed{D}}}$.

An `\hbox` is constructed by placing its components side by side such that their baselines are aligned. When `\raise`, `\lower` are used, the baselines are no longer aligned, and the baseline of the `\hbox` is defined, in such a case, as the baseline shared by the components before any vertical movements. The previous example can now be redrawn as:

The box now has a depth, as a result of the lowering.

The last command that affects the shape of an \hbox is 'spread⟨dimen⟩', which spreads the box beyond its natural width. An \hbox spread12pt{⟨material⟩} makes the box 12 points wider than its natural size. If the material in the box has no flexibility, it cannot spread to fill up the additional space, resulting in an underfull box. This is why 'spread' is normally used with flexible glues. To demonstrate spread, we start with \hbox{A\hfil B\hfil\hfil C}, which results in ABC. The flexible glues are set to zero since there is no reason to stretch them to any other size. The box is now spread by \hbox spread 12pt{A\hfil B\hfil\hfil C}, which produces A B C. The 12pt spread is equally divided between three pieces of flexible glue, so each piece is stretched to 4pt. In contrast, the construct \hbox spread12pt{ABC} has no flexibility. It produces ABC and results in an 'underfull box' message.

▶ **Exercise 3.3:** What is the result of

```
\hbox spread-2pt{A B C}
\hbox{A B C}
\hbox spread7pt{A B C}
```

(Note the spaces between the letters.)

8. The TₑX logo is defined [356] as
\def\TeX{T\kern-.1667em \lower.5ex\hbox{E}\kern-.125em X}. Note how the 'E' is placed in a box, so \lower can be used to push it down. The letters are pushed closer by means of a negative kern.

9. When TₑX reads a source line, it converts the end-of-line character (which may be a ⟨return⟩, cr-lf, or anything else) into the value of parameter \endlinechar, which is normally character code 13 (⟨return⟩). This character behaves, in the box, like a space and may affect the contents of the box. Thus

```
\hbox{AB
C}
```

creates AB C but

```
\hbox{AB%
C}
```

creates ABC. This is why special attention should be paid to end-of-lines when constructing an \hbox.

10. \hbox spread-2pt{\kern0.5pt\hss} creates a strange box with a negative width. The natural width is that of the \kern (0.5pt), and the spread reduces that width by 2pt, to −1.5pt. The \hss has thus to shrink to −2pt to allow that.

The examples and diagrams above intuitively show how the dimensions of an \hbox are determined. The rules are as follows: The (natural) width of an \hbox is the sum of the widths of its components; however, the width can be explicitly specified by to⟨dimen⟩ or it can be modified by spread⟨dimen⟩.

The height of an \hbox is that of the highest component, and the depth, that of the deepest one. However, \raise and \lower commands can change that. When those commands are used, the baseline remains where it was before the vertical shifts, and the height and depth are determined [77] by the maximum distance between the baseline and the top and bottom of the shifted components.

An \hbox can have negative width, but nothing will be typeset in such a case. The height and depth of an \hbox can never be negative [77], as is demonstrated by later examples.

The only components of an \hbox that have dimensions are text, other boxes, rules, glue, and kern. Things such as penalties and whatsits don't have any dimensions so, placed in an \hbox, they don't affect its size (they can, however, affect the size of a \vtop; see example 25).

11. The \unhbox command opens a box. Its effect is to remove the boundaries of the box. Thus (see section 3.2 for the meaning of \box0, \box1)

```
\setbox0=\hbox{B}
\setbox1=\hbox{A\copy0C}
\setbox2=\hbox{A\unhbox0C}
```

creates \box1 of the form ABC, and \box2 of the form ABC.

Removing unnecessary box boundaries saves memory and also lets glues from adjacent boxes interact (see section 3.7).

The last example is different.

12. \hbox{} creates an empty \hbox. Such a box turns out to be so useful that it has a special name, \null.

▶ **Exercise 3.4:** What is the result of \hbox to24pt{}?

3.2 Box Registers

To help the user handle boxes, TEX features 256 box registers. Such a register is best thought of as a storage place where a box can be created and saved. The box can be used later and can either be erased or be left in the register and reused. Only one box can be placed in a register at any time, and placing a new box in a register erases any box that's already there. The registers are named \box0 through \box255 and are all identical, except that \box255 is reserved for special use by the OTR.

The command \setbox0=\hbox{⟨material⟩} places a box in register \box0. Note that this operation *does not typeset anything*; it just saves the box in the register. The register can later be typeset by saying \copy0; it can be typeset *and emptied* by the command \box0, and its contents can be listed in the log file by \showbox0. It can also be used in other commands, such as \vsplit.

A box is normally emptied when it is used. This is a good rule since boxes can get quite large and consume a lot of memory, and TEX cannot rely on the user to empty them and reclaim the memory. The \copy command makes it possible to use a copy of the box, while retaining its original contents. Note that when a box register is empty (either nothing has been placed in it, or it has been used and emptied), it is considered **void**. The special command \ifvoid can be used to test for empty box registers. Its syntax is

\ifvoid0 <code for box0 empty> \else < code for box0 not empty> \fi.

▶ **Exercise 3.5:** What is the result of \box0 \ifvoid0 Yes \else No \fi?

▶ **Exercise 3.6:** What's the result of \setbox0=\null \ifvoid0 Yes\else No\fi?

13. Saying '\setbox0=\hbox{A\hbox{B} C}\showbox0' creates, in the log file,

```
1  > \box0=
2  \hbox(6.83331+0.0)x25.13893
3  .\tenrm A
4  .\hbox(6.83331+0.0)x7.08336
5  ..\tenrm B
6  .\glue 3.33333 plus 1.66666 minus 1.11111
7  .\tenrm C
```

Lines 1–2 show the name of the box register and its dimensions (height+depth) x width.

The box has four top-level components, as indicated by the four lines with a single dot on the left. The second component is the box with the 'B', making the 'B' a level-two component (note the two dots on line 5). The third component is the space preceding the 'C', which has become a glob of glue. The special glue notation on line 6 is discussed in section 3.9.

Note the difference between lines 3, 5, and 7, listing text, and line 4, listing a box. This is another aspect of the difference between boxes and characters of text.

14. Since a box may have many components and may be nested by other boxes, the listing produced by \showbox can get very long, and can therefore be restricted. The following:

```
\setbox0=\hbox{A\hbox{B} C}
\setbox1=\hbox{X\box0\hskip1pt\vrule Y}
\showboxdepth=1
\showbox1
```

creates, in the log file,

```
> \box1=
\hbox(6.83331+0.0)x41.53896
.\tenrm X
.\hbox(6.83331+0.0)x25.13893 []
.\glue 1.0
.\rule(*+*)x0.4
.\tenrm Y
```

The listing of \box1 is now limited to top-level components, because of the \showboxdepth=1 command. Similarly, a \showboxdepth=2 would limit such listings to lines with at most two dots. The empty pair of square brackets '[]' following \hbox(6.83331+... means that there is unshown material (level-two components) in the \hbox.

The line `.\rule(*+*)x0.4` does not show the height and depth of the rule, since they depend on the other components of the box and are determined when the box is constructed. Rules are discussed in section 3.22.

15. There is a similar `\showboxbreadth` parameter, which limits box listings in a different way. Setting `\showboxbreadth=2` limits such a listing to the first two components of any level. Thus

```
\showboxbreadth=2
\showbox1
```

produces

```
> \box1=
\hbox(6.83331+0.0)x41.53896
.\tenrm X
.\hbox(6.83331+0.0)x25.13893
..\tenrm A
..\hbox(6.83331+0.0)x7.08336
...\tenrm B
..etc.
.etc.
```

The '`.etc.`' indicates that there are more, unshown, top level components; the '`..etc.`' says the same thing about the level-two components.

▸ **Exercise 3.7:** What output is produced by

```
\setbox0=\hbox{A\hbox{B} C}
\setbox1=\hbox{\box0\hskip1pt\vrule Y}
\showboxbreadth=1
\showbox1
```

☛   The `\showbox` command is a very useful debugging tool when working with boxes.

A box register can be used with other commands, such as `\vsplit` (section 3.27) and `\lastbox` (section 3.28).

There is a subtle point, involving `\box0` and the **plain** format, that users should keep in mind. The **plain** format says [361] `\setbox0=\hbox{\tenex B}`. This is used in the definition of macro `\bordermatrix`. As a result, `\box0` is not void at the start of a job. If a **plain** job starts with `\showbox0`, the result will be

```
> \box0=
\hbox(0.0+6.00006)x8.75002
.\tenex B
```

This could lead to errors if the document contains a construct such as `\setbox0=\hbox{..\box0..}`, or if it uses the test `\ifvoid0....`

When material is placed in a box register, the dimensions of the register are available in the parameters `\ht`, `\dp`, and `\wd`. They can be examined and also changed. Thus:

16. After setting `\setbox0=\hbox{tp}`, we can say `\showthe\ht0`, to get the height of `\box0` (6.15079pt) in the log file. We can also say `\wd0=0pt`, which will set the width of `\box0` to zero and will cause the contents to stick out to the right tp.

17. Suppose that some small diagrams need to be placed horizontally, side by side, and to be numbered. The diagrams have different widths and each should have its number centered below it.

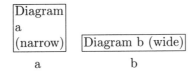

A simple approach is to place each diagram in a box register, typeset them with, for example, `\hbox{\copy0\kern12pt\copy1}`, and typeset the numbers by a command such as `\hbox{\foot0a\foot1b}`, where `\foot` is a macro such as `\def\foot#1#2{\hbox to\wd#1{\hss#2\hss}\kern12pt}`.

18. The construct `\setbox0=\hbox{}\wd0=24pt` creates an empty `\hbox` of height and depth zero, and width 24pt.

The quantities `\ht`, `\wd`, and `\dp` are called ⟨box dimension⟩ [271] and have ⟨dimen⟩ values. Even though their values can be modified, they behave differently from ⟨dimen⟩ registers. One can say, for example, '`\ht0=0pt`' or '`\ht0=2.1\ht0`' (the latter amounts to a multiplication by 2.1), but the command '`\advance\ht0 by ...`' is invalid because the `\advance` operation must be followed [276] by a ⟨dimen variable⟩ or an ⟨integer variable⟩, and the box dimensions are not any of those. (An integer variable is either a count register or an integer parameter. All the integer parameters are listed on [272–273].)

When developing a large document, using many box registers, a box register can be reserved for a special use by the `\newbox` command. Saying `\newbox\notes` will reserve the first nonused box register and assign its number to `\notes`. Commands such as `\setbox\notes=...`, `\copy\notes`, `\unhbox\notes` can later be used to manipulate the box register. This is better than using an explicit register number such as `\box0` or `\box1`, since the same number may be used by different macros in different parts of the document for conflicting purposes.

3.3 Modes

The concept of boxes is intimately tied up with that of the *mode* [Ch. 13]. At any given time, TEX can be in one of six modes. The main two modes are horizontal and vertical, to be denoted H and V, respectively. TEX is in H mode when it is reading the text of a paragraph; it is in V mode between paragraphs.

When TEX starts, it is in V mode. It starts reading the source file, executing any commands that are read. The source file may start with commands such as `\setbox...`, `\vskip`, `\hrule`, `\hbox`, `\vbox`, `\def`, and they are all executed in V mode. When TEX reads the first letter of a paragraph (or anything that's inherently horizontal; see below), it switches to H mode, reads to the end of the paragraph (or until it finds something that does not belong in a paragraph; see below), and switches back to V mode. It then typesets the paragraph, appends it to the page (actually, to the main vertical list), and reads the next thing on the source file.

What things do not belong inside a paragraph? Anything of a vertical nature such as \topglue, \vskip, \hrule (an hrule sounds like something horizontal, but this is exactly what you would place *between* paragraphs, in V mode, and not inside a paragraph, in H mode). Such things are called *vertical mode material* (see [222]), and they automatically terminate a paragraph and change the mode to V. There are also *inherently vertical commands* (see [286]).

TeX changes the mode to H when it sees the first character of a paragraph. What if the user wants to switch to H mode without typesetting anything? This can be done by the \leavevmode command or by any *inherently horizontal command* (see [283, 278]).

▶ **Exercise 3.8:** The \leavevmode command is not a primitive. Find ways to implement it.

The two internal modes are the *internal vertical* and the *restricted horizontal* modes (IV and RH). TeX is in one of those modes when it is building a box. When TeX sees a command to build an \hbox, it enters RH mode and builds the box by stringing things in the box from left to right. The \vbox command causes TeX to switch to IV mode, where it stacks components in the box, top to bottom.

Note that the user may specify a box (of either type) inside a paragraph. In such a case TeX switches temporarily to IV or RH mode, builds the box, and then returns to H mode to complete the paragraph. When text is placed in a \vbox (section 3.4), the mode is temporarily changed from IV to H, the text is set in paragraphs, and the mode is switched back to IV. The modes may thus be *nested*, and TeX may be in several modes at the same time. The example on [88] is an extreme case of mode nesting.

When TeX is in one of the math modes, it builds a math formula as a horizontal list, using the rules in [App. G]. In inline math mode, that list is eventually placed in the current line of text. In display math mode, the math list is placed between lines of text, as part of the current paragraph.

3.4 \vbox

The second type of box is a \vbox, created by \vbox{<material>}. In some respects such a box is the vertical opposite of an \hbox. It is constructed in IV mode, so material placed in it is stacked vertically such that the reference points of the components are vertically aligned (however, see \moveleft, \moveright on page 45). Also, a \vbox will not be broken across pages; it is indivisible. However, the two types of boxes are not simply the opposite of each other, and there are important differences between them.

The main difference has to do with text placed in a box. When text is placed in an \hbox, TeX stays in RH mode, the characters are laid left to right, and each becomes a component of the \hbox. On the other hand, when text (or a \vrule) is placed in a \vbox—at the top level, not inside any component—it is handled in H mode. The text is broken into lines that are stacked vertically, each becoming a component of the \vbox. This can also be expressed by saying that everything placed in an \hbox—characters, boxes, glue, etc.—is treated in the same way. In a \vbox, however, text is treated differently from boxes, glue, and other items.

19. \vbox{\hbox{A}} results in a \vbox whose only component is the \hbox with the 'A', thus A. The dimensions of the \vbox are those of the 'A'. In contrast, \vbox{A} produces a \vbox with a paragraph of text. The paragraph contains just a single line with the single letter 'A' (and the paragraph indentation), and that line is the only component of the box. The dimensions of the \vbox are those of the line of text. So the height and depth are those of the 'A', but the width, surprisingly, is \hsize. The idea is that text placed in a \vbox should be broken into lines of width \hsize. If a different size is desired, \hsize can be changed inside the box. Such a change is local and does not affect the value of \hsize outside the box. As a result, \vbox{A} has the shape

        A

(note the paragraph indentation), and \vbox{\hsize=.5in\noindent A} creates A .

Rule. The width of a \vbox is that of the widest component. However, if the box contains text or a \vrule at the top level, its width is \hsize.

20. \vbox{\hsize=1in\hbox{A}\hbox{BC}} creates A BC. Note that \hsize is ignored.

21. \vbox{A\par\hbox{B}\hbox{C}} The 'A' starts a paragraph and also switches the mode from IV to H. The \par terminates the paragraph and switches the mode back to IV. The two hboxes that follow are consequently handled in IV mode and are stacked vertically. Note that the mode is changed to RH inside each of the hboxes and changes back to IV following each box. The final result is

        A
    B
    C

Result. If a \vbox does not contain text, its layout is easy to figure out. If it contains text or a \vrule, then some experience is needed to predict its layout and dimensions.

22. \vbox{\kern4pt \hbox{M} \kern2pt \hbox{i}} creates M i. The width of the \vbox is that of the widest component, and there are no surprises. However, when text is placed in such a box, even mixed with other components, the box's width becomes \hsize.

23. \vbox{\hbox{A}Text} results in

    A
        Text

The box with the 'A' is placed in IV mode, so whatever follows it is placed below it. The 'T' of 'Text', however, changes the mode to H, which starts a new paragraph. This is why 'Text' is on a new line.

▶ **Exercise 3.9:** Guess the results of `\vbox{Text\hbox{A}}`.

▦▶ The second difference between an `\hbox` and a `\vbox` has to do with the depth. The depth of an `\hbox` is determined by the component whose depth is the largest. The depth of a `\vbox`, in contrast, is defined as the depth of the *bottom component* (however, see discussion of `\boxmaxdepth` in section 3.5). Thus a `\vbox{\hbox{q}\hbox{p}}` has the depth of the 'p' ⊡. However, the precise rules for the depth are more complex and should be reviewed on [80].

▦▶ The third difference between the two types of boxes is the way the components are separated. In an `\hbox` the individual components are not automatically separated. It's up to the user to insert glue and kern to separate them. In a `\vbox`, in contrast, the normal interline glue is automatically inserted between top-level components. The idea is that the components of a `\vbox` are normally lines of text and should be separated accordingly. Thus in the `\vbox{\hbox{q}\hbox{p}}` above, the two components are separated with `\baselineskip` ⊡.

24. Because of the interline glue, the dimensions of the three boxes below are affected by empty boxes:

```
\vbox{\hbox{}\hbox{a}}
\vtop{\hbox{}\hbox{a}}
\vtop{\hbox{\write16{abc}}\hbox{a}}
```

Because of the inserted `\baselineskip`, the first example has a height of 12pt and the other two, a depth of 12pt. Note that the `\write` in the third example is converted into a whatsit, which itself has no dimensions. As far as dimensions are concerned, the last two examples are identical.

```
\vbox(12.0+0.0)x5.00002              \vbox(0.0+12.0)x5.00002
.\hbox(0.0+0.0)x0.0                   .\hbox(0.0+0.0)x0.0
.\glue(\baselineskip) 7.69           .\glue(\baselineskip) 7.69
.\hbox(4.30554+0.0)x5                .\hbox(4.30554+0.0)x5
..\tenrm a                           ..\tenrm a
```

```
\vbox(0.0+12.0)x5.00002
.\hbox(0.0+0.0)x0.0
..\write*{abc}
.\glue(\baselineskip) 7.69
.\hbox(4.30554+0.0)x5
..\tenrm a
```

25. The box `\vtop{\hbox{p}}` has the dimensions of the 'p'. Its height is 4.30554pt, and its depth, 1.94444pt. We now place a penalty (which is a whatsit, and therefore dimensionless) at the top of the box '`\vtop{\penalty0\hbox{p}}`'. Note that its height becomes the height of the penalty, namely zero, and its depth becomes the height plus depth of the 'p', which is 6.24998pt. This shows how a dimensionless item can affect the dimensions of a box.

Another difference between the two types of boxes is in the way blank spaces are treated. They are important in an \hbox and sometimes have to be explicitly suppressed. In the vertical modes, in contrast, spaces at top level are ignored (see answer to [ex. 13.2]).

The height of a \vbox can be explicitly specified by a to⟨dimen⟩. The height can be increased or decreased by a spread⟨dimen⟩, similar to an \hbox. Also, when components are stacked in a \vbox, they can be shifted horizontally by the \moveleft, \moveright commands.

26. \vbox to.4in{\vfil\hbox{M}\nointerlineskip\hbox{i}\vfil} creates

. The two flexible glues are stretched by the same amount to vertically center the two boxes with 'M' and 'i'. The interline glue between the 'M' and 'i' boxes is eliminated by \nointerlineskip.

▶ **Exercise 3.10:** Guess the result of \vbox to12pt{}.

27. The interline glue can be suppressed for the entire box by the command \offinterlineskip

```
\vbox spread12pt{\offinterlineskip\vfil
 \hbox{M}
 \moveright5pt\hbox{i}
 \moveleft5pt\hbox{t}\vfil}
```

creates .

The two flexible glues are spread by 6pt each, for a total of 12pt.

28. Similar to the previous example, this shows how the width of a \vbox is determined when its components are shifted horizontally.

```
\vbox spread12pt{\offinterlineskip\vfil
 \hbox{M}
 \moveright10pt\hbox{i}
 \moveleft5pt\hbox{t}\vfil}
```

creates .

A \vbox is constructed by stacking its components such that their reference points are vertically aligned. The reference point of the entire box is located on this line. After shifting the components horizontally, their reference points are no longer aligned [80], but the reference point of the entire box is not affected. Boxes moved to the left do not widen the \vbox and may stick out on the left. Boxes moved to the right, however, may widen the entire \vbox. The width is determined [81] by starting at the reference point and moving right until reaching the farthest right edge of any component box. The width of a \vbox must be nonnegative, but the height and depth can be negative, in which case the box will not show up in print.

29.

```
\vbox{\hsize=10pt \hbox{A}\noindent B}
\vbox{\hsize=10pt \moveright10pt\hbox{A}\noindent B}
```

Both boxes contain the box with the 'A' and, below it, a paragraph of text with just the 'B'. The first box is 10pt wide, but, in the second box, the \moveright moves the box with the 'A' to the right boundary, thus widening that box by the width of the 'A' (its width becomes approximately 17.5pt).

30. A \vbox with \hbox{i} and \hbox{t} stacked. After shifting the 'i' to the left and the 't' to the right, the layout is. The 'i' sticks out on the left, while the 't' has widened the box.

Migration of vertical material. When a paragraph of text is typeset, either on the page or in a \vbox, lines of text are stacked vertically with interline glue preceding each line (except the first). If a line contains vertical mode material, placed there by the user, the material is taken out by TeX and is placed below the line. This process is called *migration of vertical material* [105]. The material associated with the \vadjust, \insert, and \mark commands falls under this category.

However, if the vertical material is buried inside a box, it may not be discovered by TeX and may never be executed. In an experiment such as

```
\setbox0=\vbox{\hsize=3in ...<text>}
\vadjust{\kern10pt} \hbox{\mark{XX}}
... <text>
\hbox{\vadjust{\kern11pt}} \mark{YY} ...<text>}
\showboxbreadth=500 \showboxdepth=1 \showbox0
```

the output will show \kern10pt and \mark{YY} placed between certain lines of text, but \mark{XX} and \kern11pt will remain locked in their boxes, not affecting the output.

Thus when a line of text is added to a vertical list, the following items are actually added [105] in this order: the interline glue, the line itself (as an \hbox), any migrated vertical items, and any interline penalty.

3.5 \boxmaxdepth

As mentioned earlier, the depth of a \vbox [80] is the depth of the bottom component. There is, however, another constraint. The depth of a \vbox is limited to the current value of parameter \boxmaxdepth. If \boxmaxdepth=1pt, then the depth of any \vbox created cannot exceed 1pt.

In the plain format, \boxmaxdepth=\maxdimen [348], so it has no effect on the depths of boxes. However, \boxmaxdepth can always be changed by the user.

31. To illustrate \boxmaxdepth we start with \setbox1=\vbox{\hbox{p}} \showbox1, which results in

```
> \box1=
\vbox(4.30554+1.94444)x5.55557
```

```
.\hbox(4.30554+1.94444)x5.55557
..\tenrm p
```

The dimensions of the \vbox are those of the 'p'. After setting \boxmaxdepth=1pt, however, the same construct creates

```
> \box1=
\vbox(5.24998+1.0)x5.55557
.\hbox(4.30554+1.94444)x5.55557
..\tenrm p
```

The depth of the letter 'p' in font cmr10 is 1.94444pt, which becomes the depth of the inner \hbox. The setting of \boxmaxdepth, however, limits the depth of the \vbox to 1pt. In order for the material to fit in the \vbox, its height should be incremented by the difference. The height is therefore increased from 4.30554pt to 5.24998pt, a difference of .94444pt. This is equivalent to lowering the reference point (or the baseline) of the \vbox by .94444pt. The diagram below (using the 'p' from cmr17) illustrates the box before and after the depth limitation. Note that the two baselines are aligned, which moves the second box higher.

▶ **Exercise 3.11:** Align the tops and bottoms of the two boxes.

▶ **Exercise 3.12:** What is the reason for the space between the two boxes, and how can it be eliminated?

The rules for determining the dimensions of a \vbox can now be summarized:
- The width of a \vbox is that of the widest component. If the box contains text or a \vrule at top level, its width is \hsize. Components shifted to the right can extend the width as shown above.
- The depth is that of the bottom component, but is limited to the current value of \boxmaxdepth. If the bottom component has no dimensions, the depth is zero.
- The height is the sum of the vertical dimensions (height plus depth) of all the top-level components, minus the depth of the box. The height may be increased if the depth is limited by \boxmaxdepth.

32. A \vbox can be combined with text and may appear anywhere inside a paragraph. Its baseline will simply be aligned with that of the current line. Since the box tends to be high, it may cause a large separation between its line and the line above it. The only problem is that the box is indivisible and, if it ends up toward the end of the line, that may cause it to be overfull. Similarly, a \vbox ending up at the bottom of a page may have to be moved to the following page, leaving the current page underfull.

3.6 \vtop

The rules above imply that a \vbox is normally high but has a small depth. TEX also supports something called a \vtop, which is a \vbox whose height is the height of the top component. The depth of a \vtop is the combined vertical sizes of all its components, minus its height (but see [81] for the precise rules). Two important differences between a \vtop and a \vbox are described later. One has to do with the meaning of the keywords to and spread; the other one concerns the behavior of the \topskip glue.

Here is text in a tiny vbox, to show the baselines in a vbox and a vtop. Note how they are aligned.

Here is text in a tiny vtop, to show the baselines in a vbox and a vtop. Note how they are aligned.

33. \vtop{\hrule height20pt depth28pt}. This box gets the height of its top component (20pt). Its depth is the size of the rest of the material (28pt). The case \vtop{\hrule height20pt depth28pt\vfil} is similar. We do not insist on any specific dimensions, so all the flexible glues are set at their natural values. The \vfil has a natural value of zero, which gives the \vtop a depth of 28pt.

34. \vtop{\hbox{A}\hbox{B}}. This box gets the height of the letter 'A' (6.83331pt). Since both 'A' and 'B' have zero depths, the rest of the \vtop (from the bottom of the 'A' to the bottom of the 'B') has size \baselineskip. The depth of this \vtop is thus 12.0pt A̲B̲.

35. \vtop{AB}. The height is, as before, 6.83331pt. The depth is zero, since both letters have zero depth. The width of this box is \hsize, since it contains text. This is in contrast with the previous example, where the box contained other boxes.

36. \vtop{\hrule AB}. The height is that of the rule (0.4pt; see section 3.22). The depth is the size of the rest (6.83331pt). Note that there is no interline space between the rule and the text, but there is the usual \parskip glue (0pt plus 1pt) preceding the paragraph 'AB'.

▶ **Exercise 3.13:** What is the width of the \vtop above?

37. \vtop{\hrule height2pt depth3pt AB}. Similar to the previous example, the height is 2.0pt and the depth, 9.83331pt.

▶ **Exercise 3.14:** A \vbox normally has a large height and a small depth. The opposite is true for a \vtop. How can one create a box with a large height and a large depth?

The topskip glue. (And its relation to a \vbox.) When \box255 is constructed, a special glue is inserted above the top line, to keep its baseline exactly 10pt from the top of \box255. This creates a "flush top" effect, which adds to the uniform appearance of the pages. This glue is called \topskip [114] and has a plain value of 10pt. Note that TeX does not simply insert \topskip at the top of the page. It measures the height of the top line, and inserts the difference between \topskip and that height (a negative difference is changed to zero). If the top line is a \vbox, it usually has a large height, which causes \topskip to become zero. A \vtop, with a small height, behaves differently in this respect.

The following examples show how a \vtop is treated differently from a \vbox when either of the keywords to or spread is used [81]. When we say \vtop to20pt{...} we promise TeX that we will supply exactly 20 point's worth of material for the box. Since we rarely know in advance the exact size of our material, we usually use flexible glues with to or spread. A \vtop normally has a small height and a large depth, so the flexible glues typically are part of the material that contributes to the depth. This is why it is appropriate for the 20pt to become the depth, rather than the height, of the \vtop. This is the opposite of the situation in a \vbox.

38. \vtop to20pt{\hrule AB\vfil}. The \vtop is constructed in two steps. The first step is to treat it as a \vbox. Its components become the rule (height 0.4pt), the \parskip glue (0pt plus 1pt), a paragraph consisting of a single line of text with height 6.83331pt and depth zero, and the \vfil. We require the height of this \vbox to be 20pt, so the \vfil is set to $20 - (0.4 + 6.83331) = 12.76669\text{pt}$. The depth of this \vbox is the depth of the bottom component (\vfil, which has zero depth).

The second step is to convert this \vbox to a \vtop by moving the reference point such that the height of the \vtop becomes that of the top component (0.4pt). The depth thus turns out to be $20 - 0.4 = 19.6\text{pt}$.

39. \vtop to20pt{\hrule AB\vfil\hbox{\strut}}. Similar to the above. In step 1, a \vbox with the following components is created:
- a rule of height 0.4pt;
- the \parskip glue (0pt plus 1pt);
- the line of text, with height 6.83331pt and depth zero;
- the \vfil;
- a \baselineskip glue, which is set to 3.5pt, to give the usual separation between lines;
- the \hbox with the strut (height 8.5pt, depth 3.5pt).

The depth of this \vbox becomes the depth of the bottom component (the box with the strut, 3.5pt), and the height should be 20pt. The \vfil is thus set to $20 - (0.4 + 6.83331 + 3.5 + 8.5) = 0.76669\text{pt}$. We end up with a \vbox of total vertical size $20 + 3.5 = 23.5\text{pt}$.

In step 2, this is converted to a \vtop by sliding the reference point. The height of the \vtop should be 0.4pt, so its depth ends up being $23.5 - 0.4 = 23.1\text{pt}$.

▶ **Exercise 3.15:** What is the result of \vtop to20pt{\vfil\hbox{\strut}}?

40. \vtop to20pt{\boxmaxdepth=3pt\vfil\hbox{\strut}}. Here we try to limit the depth of the \vtop to 3pt, but this does not work! In step 1, the command \boxmaxdepth=3pt creates a \vbox with depth 3pt. The extra 0.5pt of the strut depth is added to the height, but the height, of course, is required to be 20pt. The resulting \vbox thus has a total vertical size of 23pt, and the \vfil is flexed accordingly. In step 2, the reference point is moved up to create the \vtop with zero height, leaving a depth of 23pt.

41. \vtop to20pt{\vfil\hbox{\strut}\null}. \null is an empty \hbox. We do not expect it to affect the layout of the \vtop much, but it does! In step 1, a \vbox is created with the following: the \vfil, the \hbox with the strut (8.5 + 3.5pt), a \baselineskip glue of 8.5pt, and the \null. The depth of the \vbox is zero (the depth of the \null), and its height should be 20pt. However, there is 8.5 + 3.5 + 8.5 = 20.5pt worth of material in the box, so it is declared overfull. In step 2 the reference point is moved up, ending with a \vtop of height 0pt and depth 20pt.

▸ **Exercise 3.16:** What are the dimensions of \vtop to20pt{\null\vfil\hbox{\strut}}?

▸ **Exercise 3.17:** What is the result of \vtop to3pt{}?

42. \vtop spread20pt{\hrule AB\vfil}. In step 1, a \vbox is constructed to its natural size (i.e., the \vfil is set to zero), and then the \vfil is stretched to 20pt. The box thus contains the \hrule (0.4pt), a \parskip (0pt), the text (6.83331 + 0pt), and a \vfil of 20pt. The depth is zero and the height, 0.4 + 6.83331 + 20 = 27.23331pt.

In step 2, the reference point is moved to set the height of the \vtop to 0.4pt. The depth is thus the remaining 26.83331pt.

43. \vtop spread20pt{\hrule AB\vfil\hbox{\strut}}. Again, in step 1, the \vfil is stretched to 20pt, and the \vbox contains the rule, the text, the \vfil, a \baselineskip of 3.5pt, and the strut box. The depth is that of the strut (3.5pt), and the height is the rest, 0.4 + 6.83331 + 20 + 3.5 + 8.5 = 39.23331pt. The total vertical size is 3.5 + 39.23331 = 42.73331pt. In step 2, the height becomes 0.4pt, and the depth is thus the remaining 42.33331pt.

It is highly recommended that the reader experiment with each of the preceding examples by saying \setbox0=\vtop{...} \showbox0.

44. A \vtop can be combined with text and may appear anywhere inside a paragraph, like a \vbox. Its baseline will be aligned with that of the current line and, since a \vtop tends to have large depth, it will normally stick below its line. The only vtop in a problem is that the box is indivisible and, if it ends up toward paragraph the end of the line, it may cause it to be overfull.

3.7 Opening Boxes

The \unhbox command has already been mentioned. There is a similar \unvbox command, and both are discussed here in some detail.

The \unhbox command is intrinsically horizontal [283] and causes TeX to start a new paragraph. When used inside a \vbox, as in

```
\setbox1=\hbox{Text}
\setbox2=\vbox{\unhbox1}
```

it changes the mode inside the \vbox from IV to H. As a result, the \vbox gets width \hsize. It's like typesetting text in a \vbox. When used inside an \hbox:

```
\setbox1=\hbox{text}
\setbox2=\hbox{\unhbox1}
```

the \unhbox simply contributes text to \box2. The result is identical to the assignment \setbox2=\hbox{text}.

The \unvbox command is the opposite. It is intrinsically vertical [286] and, when it appears in H mode, it causes TeX to terminate the current paragraph. When \unvbox appears in one of the vertical modes, as in

```
\setbox1=\vbox{\hsize=2in text}
\setbox2=\vbox{\unvbox1}
```

it does not have much effect. \box2 gets a width of 2 inches, and its only component is a short, single-line paragraph. The line is placed in \box2 as an \hbox. However, "the appearance of a vertical command in RH mode is forbidden" [286]; so an \unvbox inside an \hbox (as in \setbox2=\hbox{\unvbox1}) will cause an error.

The following shows another feature of \unvbox.

```
\setbox0=\vbox{\hsize=4in ..text..}
\setbox1=\vbox{\hsize=5in\unvbox0}
```

The \unvbox does not change the original line breaks! It creates \box1 with a width of 4 inches, not 5 inches, and with the original line breaks from \box0. The explanation is that the first command typesets the text in \box0 as a set of hboxes, each 4 inches wide. The \unvbox unpacks these hboxes into \box1 but does not change their width.

A common example where \hbox{\unvbox...} is needed is the case where we have text in \box0 and \box1, each perhaps 2 inches wide, and we want to typeset them side by side.

This is easy to do with \line{\box0\hfil\box1}. However, boxes are indivisible and, if we are close to the bottom of the page, the entire block of text would have to be moved to the next page, leaving the current page underfull. We would thus like to open the boxes by saying

```
\setbox2=\line{\unvbox0\hfil\unvbox1}
\unhbox2
```

but this is impossible because of the previous restriction. The reason for the restriction is now apparent. If we open \box0 and \box1, there would be no way to construct \box2 as an \hbox, with components placed left to right. This is a simple problem with no simple solution.

There are also \unhcopy and \unvcopy commands that are similar to \unhbox and \unvbox, but only open copies; the original boxes remain intact.

3.8 Simple Box Examples

45. The first example is a simple "read between the lines," | R | E | A | D |, done by

```
\def\Vrule{\vrule height10pt depth2pt width.8pt}
\hbox{\Vrule\ R \Vrule\ E \Vrule\ A \Vrule\ D \Vrule}
```

46. Here is an \hbox with seven components, four pieces of \kern and three inner boxes. The first of those is a small \hbox with the word "Text." Its dimensions are also implicitly determined. The third one is a \vbox with two small \hboxes vertically centered. The height is specified as 0.4 inches, which should be taken as a promise; we promise TEX that we will provide exactly 0.4 inches' worth of material for the box. This is why flexible glues are almost always used in such a case. In our case, two pieces of \vfil are used to vertically center the words "Text" and "Mext."

The second inner box is a \vbox with text. Its height is determined by the height of the text, and its width is the current value of \hsize. This is why \hsize was set, inside the box, to 0.3 inches.

```
\hbox{%
\kern20pt
\hbox{Text}
\kern7pt
\vbox{\hsize=.3in\noindent Text}
\kern5pt
\vbox to.4in{\vfil\hbox{Text}\hbox{Mext}\vfil}
\kern2pt
}
```

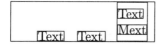

The vertical gap between the "Text" and "Mext" boxes is the usual interlineskip.

47. This is identical to the previous example except that the components are enclosed in a \vbox, so they are stacked vertically, the \kern command now means *vertical* rigid glue, and the dimensions of the outer \vbox are implicitly determined. Its width is that of the widest component, since all the components are boxes or glue (there is no text at the upper level of the \vbox). Its height is the sum of the heights of its components, and its depth—which is zero—is that of the bottom component (the \kern2pt).

```
\vbox{
\kern20pt
\hbox{Text}
\kern7pt
\vbox{\hsize=.3in\noindent Text}
\kern5pt
\vbox to.4in{\vfil\hbox{Text}\hbox{Mext}\vfil}
\kern2pt
}
```

▶ **Exercise 3.18:** Use \showbox to display the contents of these examples in your log file.

      A deeper understanding of boxes can be gained by modifying the previous examples in different ways (changing or eliminating glues, or using to or spread). The reader should use \showbox to see the effects of each modification.

3.9 Glue

Even though the concept of glue is central to TEX, and is as important as boxes, glue does not have a good definition. We mentioned earlier that glue is spacing. Glue can also be visualized as globs of flexible stuff, connecting boxes together either vertically or horizontally. Glue has two attributes, namely, its natural size and its flexibility. The natural size of a glue is a dimension, such as 24pt or 1.56in, and the flexibility can range from zero to infinity.

The flexibility itself is described by two dimensions, the stretchability and the shrinkability, so a glob of glue can be described as the triplet $< w, y, z >$, where w is the natural size, and y, z, the stretchability and shrinkability, respectively. In a TEX program, the notation 12pt plus 5pt minus 3pt is used for the triplet $< 12, 5, 3 >$. Such a glue can stretch up to $12 + 5 = 17$ pt and shrink down to $12 - 3 = 9$ pt. Either of the plus or minus components may be missing, but the plus must precede the minus. This may lead to subtle errors, such as in

```
\def\xskip{\hskip 12pt plus5pt}
... glues have plus and \xskip minus components ...
```

After the macro is expanded, the text has the form
```
... glues have plus and \hskip 12pt plus5pt minus components ...
```
Naturally TEX considers the word minus to be the shrink of the glue, and expects it to be followed by a dimension. The solution is to define \def\xskip{\hskip 12pt plus5pt \relax}. The \relax serves to take TEX out of the context of the glue.

▶ **Exercise 3.19:** What is the result of ... `glues have \xskip plus and minus components ...`, (assuming that no `\relax` is used)?

Glues are used in many cases. When a box such as `\hbox to2in{<material>}` is constructed, the user has to supply exactly 2 inches' worth of material to fill up the box and avoid an error message. However, the exact width of text and boxes is usually unknown, which is why flexible glues are normally used in such a case. The glue will be flexed by an amount depending on the size of the material.

The following simple experiment should be performed to clarify this point. `\setbox0=\hbox to100pt{\hskip 0pt plus 100pt abc} \showbox0`. The glue has a natural size of zero but is allowed to stretch up to the entire width of the box. Since the box also contains text, the glue will stretch to fill up that part of the box that's not taken up by the text. Running this example produces, in the log file,

```
> \box0=
\hbox(6.94444+0.0)x100.0, glue set 0.84723
.\glue 0.0 plus 100.0
.\tenrm a
.\tenrm b
.\kern0.27779
.\tenrm c
```

The message '`glue set 0.84723`' means the glue was stretched to 84.723% of its maximum stretchability. It was thus stretched to 84.723pt, which means that the width of the text "abc" must have been $100 - 84.723 = 15.277$pt. Now that we know the width of the text, we perform the similar experiment `\setbox0=\hbox to100pt{\hskip100pt minus50pt abc} \showbox0`, where the glue will have to shrink by the same amount (15.277pt) to leave room for the text. Since 15.277pt is now 30.556% of its maximum shrinkability, we end up with the message '`glue set - 0.30556`' in the log file (the negative sign indicates shrinking of the glue).

The test `\hbox to100pt{\hskip0pt plus100pt abc\hskip0pt plus100pt}` is a simple extension of the previous ones. The total stretch necessary is, as before, 84.723pt. Since the two globs of flexible glue have the same stretchability, they are stretched by the same amount, so each is stretched by 42.361%. The glue set ratio is thus 0.42361. The result is that the text is centered in the box, thus
⎡_____abc_____⎤.

▶ **Exercise 3.20:** What is the result of
`\hbox to100pt{\hskip0pt plus100pt abc\hskip0pt minus100pt}`?

48. This example uses glues with different flexibilities `\hbox to100pt{\hskip 0pt plus 50pt abc\hskip 0pt plus 100pt}`. The total amount of stretch is the same 84.723pt. However, since the second glob of glue has twice the stretchability of the first one, it is also stretched twice as much. The quantity 84.723pt is thus divided in three; the first glue is stretched by $84.723/3 = 28.241$pt, and the second glue, by $84.723 \times 2/3 = 56.482$pt. The glue set ratio is thus $28.241/50 = 56.482/100 = 0.56482$, and the result is ⎡____abc_____⎤. The text is typeset one third of the way from the left edge of the box.

▶ **Exercise 3.21:** What is the precise meaning of the previous sentence?

3.10 Infinitely Flexible Glues

The familiar command `\hbox to100pt{\hfil abc\hfil}` is short for `\hbox to100pt{\hskip0pt plus100pt abc\hskip0pt plus100pt}`. The meaning of `\hfil` is a horizontal skip by 0pt, but with infinite stretchability.

It is possible to define glues with stretch and shrink components that have infinite flexibility. Something like `12pt plus 1fil minus 10pt` can shrink down to 2pt but can stretch by any required size. More common is the glue value '`0pt plus1fil`', which can take any nonnegative size. In fact, the command `\hfil` is equivalent to `\hskip0pt plus1fil`. The keyword `fil` represents one unit of infinite flexibility and, each time a `fil` is used, TeX has to calculate the amount by which the glue should be flexed.

Note that a `fil` can only appear in the stretch and shrink components of a glue, not in its natural size. Thus '`\hskip 1fil plus...`' is invalid.

Another interesting point is that any multiples of a `fil` can be used. Thus the quantities `2fil` and `3.71fil` are valid. They still have infinite stretchability but, when used together with a `fil`, will stretch twice as much, or 3.71 times as much.

49. An illustrative example is

```
\newcount\coumt
\loop
\hbox to300pt{\hskip0pt plus\coumt fil Text\hskip0pt plus1fil}
\advance\coumt by1
\ifnum\coumt<10\repeat
```

which produces

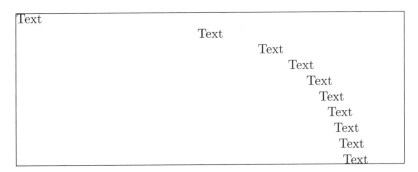

In the first iteration, `\coumt` equals zero, so there are zero units of `fil` on the left, and one unit on the right, of the word "Text." The word is thus typeset left justified.

In the last iteration, there are nine units of `fil` on the left, and the same single unit on the right of the word. The word itself has a width of 20pt, so the difference $300 - 20 = 280$pt has to be divided between 10 units of `fil`. Each such unit is thus set to 28pt. The first `\hskip` is expanded to $9 \times 28 = 252$pt, and the second one to 28pt.

▸ **Exercise 3.22:** What is the glue distribution in the third line of the preceding example?

▸ **Exercise 3.23:** Imagine a smooth curve connecting the reference points of the 'T's in the ten iterations of the word "Text." What is the equation of the curve, assuming that the y-coordinate varies from 0, on the top line, to 9 on the bottom one?

In the expression \coumt fil, the value of \coumt should be less than 16,384 [72]. In general, the total number of fil units in a box should be less than 32,768. Thus the box

```
\hbox to100pt{\hskip0pt plus1fil\hskip0pt plus16383fil
x\hskip0pt plus16383fil}
```

creates . The 'x' is centered. Adding one more unit of fil, however, brings the total to 32,768, and the result of

```
\hbox to100pt{\hskip0pt plus2fil\hskip0pt plus16383fil
x\hskip0pt plus16383fil}
```

is

x   ⸺⸺⸺⸺⸺ (unpredictable).

▸ **Exercise 3.24:** Write a macro \oval#1#2 that should create an oval, similar to the ones below, using '#1' as a building block, and '#2', as the width of the oval. The height is always 21 lines. The expansions \oval.{50} \oval O{200} create the ovals:

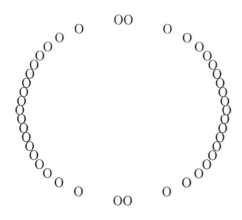

3.11 Negative Multiples of fil

It is possible to use glues of the form 0pt plus-1fil or 0pt minus-1fil, and the following examples attempt to show how those glues behave.

TEX executes each of the examples in two steps. In the first step it counts the number of fil units and calculates the amount by which a fil should be stretched. In the second step the glues are actually set. A fil set to a positive value has positive width and causes a skip to the right. Similarly, a fil set to a negative value has negative width and thus causes a skip to the left. Thus the glob of glue 0pt plus1fil normally causes a skip to the right but, in certain cases, can cause a skip to the left.

50. The first example is \hbox to100pt{\hskip0pt plus2fil x\hskip0pt plus-1fil}.

The width of the x is 5.2778pt and the total width of the box is 100pt. The two pieces of flexible glue should, therefore, be stretched to fill up the difference of 94.7222pt. In step 1, TEX counts $2 + (-1) = 1$ unit of fil. Each fil is thus stretched to 94.7222pt. The first \hskip thus skips 2×94.7222pt to the right and the second \hskip, 94.7222pt to the left. The result is

x

51. A similar example is \hbox to100pt{\hskip0pt plus3fil x\hskip0pt plus-1fil}. Here there are $3 + (-1) = 2$ units of fil. Each fil is thus stretched to $94.7222/2 = 47.3611$pt. The first \hskip skips 3×47.3611pt to the right and the second one, 47.3611pt to the left. The result is

x

▶ **Exercise 3.25:** Use the previous examples to figure out what happens in \hbox to100pt{\hskip0pt plus4fil x\hskip0pt plus-1fil}.

52. A similar example is
\hbox to100pt{\hskip0pt plus1fil x\hskip0pt plus-2fil}.
There are $1+(-2) = -1$ units of fil, so a -fil should be stretched to 94.7222pt. As a result, a fil is stretched to -94.7222pt. The first \hskip thus skips 94.7222pt *to the left*, the x is typeset (outside the box), and the second \hskip skips 2×94.7222pt to the right. The result is

x

This example can also be written as \hbox to100pt{\hfil x\hfilneg\hfilneg} (see section 3.12 for \hfilneg).

53. The shrink component of a glue can also have negative values. The following example can easily be analyzed:

```
\hbox to100pt{\hskip50pt minus-1fil x\hskip50pt minus2fil}
\hbox to100pt{\hskip50pt minus-1fil x\hskip50pt minus3fil}
\hbox to100pt{\hskip50pt minus-1fil x\hskip50pt minus4fil}
```

In the first line, the \hbox has a natural size of $50+5.2778+50 = 105.2778$pt, so the glues have to shrink by 5.2778pt. This amount is divided between the $-1+2 = 1$ unit of fil, so each fil is set to 5.2778pt. The first glue shrinks by -5.2778pt (i.e., it expands) to 55.2778pt. The second glue shrinks by 2×5.2778 to 39.4444pt. In the second line there are 2 units of `fil`, so each is set to 2.6389pt. In the third one, each `fil` is set to 1.75926pt. The results are

3.12 The filneg Glues

The primitive command `\hfilneg` is equivalent to `\hskip 0pt plus-1fil`. It is thus the opposite of `\hfil` and can cancel it. There is also a `\vfilneg` command.

54. Perhaps the simplest case is `\line{\hfil...\hfilneg}` (see [ex. 12.17]). This is equivalent to `\line{\hskip0pt plus1fil...\hskip0pt plus-1fil}` (compare with the examples above). The total number of `fil` units is $1 - 1 = 0$, so the two flexible glues cancel each other. The material placed in the box will have to fill the box exactly, resulting almost always in an over- or underfull box.

55. A practical example of the use of `\hfilneg` is centering the last line of a paragraph.

The last line of a paragraph is usually shorter than the rest (it has the glue `\parfillskip` ready to stretch on the right-hand side), and the following settings will cause it to be centered:

```
\leftskip=0pt plus1fil
\rightskip=0pt plus-1fil
\parfillskip=0pt plus2fil
```

The last line of the paragraph ends with a `\parfillskip` glue, followed by `\rightskip`. The sum of these glues is now `0pt plus1fil`, which equals `\leftskip`, so the line is centered. On all the other lines, the `\leftskip` and `\rightskip` glues cancel each other. A different solution to this problem, using `\lastbox`, is shown in section 3.28.

▶ **Exercise 3.26:** Modify the above to typeset the last line of a paragraph right justified.

56. Two more practical examples of the use of `\vfilneg` are the `\filbreak` macro, described on [111], and the `\beginsection` macro, on [355]. This macro is expanded at the start of each section of a document. It starts with `\vskip 0pt plus.3\vsize \penalty-250 \vskip 0pt plus-.3\vsize...`.

TeX can stretch the first `\vskip` up to 30% of the page height. If it decides to break the page at the penalty, the new section will start at the top of the next page (where the second skip will be discarded). If the penalty point is not close to the bottom of the page, TeX will not break the page at that point, and the two glues will cancel each other. The new section will start on the same page.

The effect is that if 30% or more of the page remains at the end of a section, the next section will start on the same page; otherwise, the rest of the page will be left blank, and the next section will start on the next page.

3.13 Glues with Negative Dimensions

The glue `0pt plus1fil minus1fil` is especially interesting. It has a natural size of zero but it can stretch to infinity and also shrink to minus infinity (bear in mind that, in TEX, infinity is the value of `\maxdimen`). This glue can thus have any dimension, positive or negative. Since it is very useful, there is a TEX primitive, called `\hss`, whose meaning is equivalent to `\hskip 0pt plus1fil minus1fil`. There is also a corresponding `\vss`.

57. The first example is simple: `\hbox to0pt{x\hss}`. We construct a box with a width of zero. The total width of the material in the box should thus be zero. The material consists of two things, the letter 'x' (whose width is 5.2778pt) and the `\hss` glue. The glue is thus shrunk to -5.2778pt.

When the box is typeset, the letter 'x' is typeset, moving the reference point to the right, as usual, by the width of the letter. The `\hss` is then executed, moving the reference point to the left (since it is negative) by the same amount. The result is that the letter 'x' is typeset *without moving the reference point*. The effect of typesetting `\hbox to0pt{x\hss}` can be described graphically as ɪx where the thick bar represents a box of width zero. This illustrates the fact that the contents of a box is not limited to the boundaries of the box; it can stick out of the box without upsetting TEX. In our example, the 'x' sticks out of the right side of the box and will overlap anything that follows the box.

58. A more useful example is '`\hbox to0pt{/\hss}=`', which, when typeset in H mode, produces '$\neq$'. This construct is so important that it is defined [82] as a `plain` format macro: `\def\rlap#1{\hbox to0pt{#1\hss}}`.

Similarly, the construct `\hbox to0pt{\hss x}` produces xɪ. The letter now sticks out of the left edge of the box and will overlap anything typeset to the left of the box. This operation is also defined as a macro:
`\def\llap#1{\hbox to0pt{\hss #1}`.

▸ **Exercise 3.27:** Why do the two constructs `=\llap{/}` and `\rlap{=}/` produce the different results '$\neq$' and '$\neq$'.

One use of these operations is to place things on the margins of the page. The code `\line{\llap{\sevenrm Left}Text\hfil Text\rlap{\sevenrm Right}}` produces

_{Left}Text   Text_{Right}

Of course, as with many other things in TEX, this can be done in other ways. The rule on the left margin of this paragraph was placed by means of the command `\line{\kern-1em\vrule\hfil\vbox{Of course,...}}`

The concept of overlapping can easily be generalized. It is possible to overlap things above and below the baseline by macros such as

`\long\def\ulap#1{\vbox to0pt{\vss #1}}`
`\long\def\dlap#1{\vbox to0pt{#1\vss}}`

It is also possible to typeset material centered on the reference point and occupying no space. Assuming the following definitions:

`\def\xlap#1{\hbox to0pt{\hss #1\hss}}`

```
\long\def\ylap#1{\vbox to0pt{\vss #1\vss}}
\long\def\zlap#1{\ylap{\xlap{#1}}}
```

a construct such as 'ab-\zlap{+}-ijk' results in ab$\mp$ijk. The '+' is centered on the baseline, between the 'b' and 'i', without occupying any space.

The plain macro \centerline is defined as \line{\hss<material>\hss}. This way, the material can be wider than \hsize and no error message will be issued.

3.14 Predefined Fills

The plain format defines, on [357], the useful macros \dotfill, \hrulefill, \leftarrowfill, and \rightarrowfill. They can be used to fill up a horizontal space with dots, rules, or arrows.

59. Here are two simple examples:
\hbox to100pt{A\dotfill B\hrulefill C}   produces   A........ B_____ C
\hbox to100pt{\leftarrowfill\ width \rightarrowfill} results in
←——— width ———→

These macros use the infinite quantity fill (with two l's; see section 3.15) and also \leaders. Other predefined fills can be created using the same techniques.

▶ **Exercise 3.28:** What is the result of \hbox{A\dotfill B}?

3.15 Different Orders of Infinity

The code \hbox to100pt{\hskip0pt plus100pt x\hskip0pt plus1fil} creates x_____. The glue on the left is not stretched at all! We say that the two glues have *different orders of flexibility*. A single fil has infinitely more flexibility than the plus100pt. Similarly, the fill and filll keywords represent infinite flexibilities of higher orders. They are infinitely greater than fil. Thus the two tests
\hbox to100pt{\hskip0pt plus1fil x\hskip0pt plus1fill}
\hbox to100pt{\hskip0pt plus1fill x\hskip0pt plus1filll}
produce the same result as the previous one. The glues on the left are not stretched at all, because the glues on the right have a higher order of infinity.

The reader will recall that \hfil is equivalent to \hskip 0pt plus 1fil. Similarly, \hfill is equivalent to \hskip 0pt plus 1fill. The primitives \vfil, \vfill are also available, but no \hfilll or \vfilll commands are available in the plain format (they can, of course, easily be defined).

These higher orders are useful when a macro with a fil is used and we want to nullify the effect of the fil. If we cannot change the macro definition, we can use fill outside the macro. An example is the predefined fills, such as \dotfill, mentioned earlier. They use fill and, as a result: (1) Using fil in connection with them has no effect; (2) using fill, it is possible to modify their effect; (3) using filll, their effect can be nullified. The following constructs demonstrate this:

```
\def\hfilll{\hskip0pt plus1filll}
\hbox to100pt{A\dotfill B\hfil C}
\hbox to100pt{A\dotfill B\hfill C}
```

```
\hbox to100pt{A\dotfill B\hfilll C}
```

produce

A................BC
A........B   C
AB   C

▶ **Exercise 3.29:** What are the relative strengths of `1fill` and `16383fil`?

Infinite glue components can participate in addition (and subtraction). The commands

```
\skip0=3pt plus1fill minus2fill
\advance\skip0 by 1pt plus2fil minus3filll
```

will set the value of `\skip0` to 4pt `plus1fill minus3filll`. The higher-order infinities completely displace the lower-order ones.

3.16 The Glue Set Ratio

When a box contains too much or too little material, we think of it as overfull or underfull. There are, however, four classes of "bad" boxes. A box may be "underfull," "loose," "tight," or "overfull," depending on the relationship between its badness and the values of the `\hbadness` and `\hfuzz` (or `\vbadness` and `\vfuzz`) parameters. To fully understand "bad" boxes, we introduce the concepts of glue set ratio and badness.

The concept of setting glue is best demonstrated by considering a line of text. Such a line contains word boxes, with globs of glue separating them. The glue has to be flexed to bring the size of the line to the desired size, which we denote l (normally `\hsize`). Each glob is a triplet $< w_i, y_i, z_i >$, and flexing the glue is done by calculating the total amount of stretch and stretching each glob of glue in proportion to its y_i component. This is similar in the case of shrinking.

To do this, TeX

1. calculates the three sums $W = \sum w_i$, $Y = \sum y_i$, $Z = \sum z_i$;
2. calculates the natural size L of the line. This is the sum of the widths of all the word boxes, plus W.
3. calculates the amount of variation Δ of the line by $\Delta = L - l$;
4. calculates the *glue set ratio* r by

$$r = \begin{cases} 0, & \text{if } \Delta = 0 \text{ (ideal);} \\ \Delta/Z, & \text{if } \Delta > 0 \text{ (shrink);} \\ -\Delta/Y, & \text{if } \Delta < 0 \text{ (stretch).} \end{cases}$$

Notice that r is nonnegative.

5. changes the size of each glob of glue from w_i to either $w_i + ry_i$ or $w_i - rz_i$, depending on the sign of Δ;
6. in the case of a stretch ($\Delta < 0$), stretches the line by any necessary amount. Such a box may eventually become loose or even underfull. It may also remain "good" if the amount of stretch is small (see section 3.18).
7. in the case of a shrink ($\Delta > 0$), shrinks each glob to at most $w_i - z_i$ but not less (spaces less than that size are considered unreadable). Such a box may eventually

become tight or even overfull, or it may remain "good;"

8. after the breakpoint has been determined, discards any glue following the last word in the line. Discardable items [95] are glue, kern, penalty and, in an \hbox, also "math-on, math-off" items.

9. modifies the above operations, of course, if the line contains infinities like \hfil, \hfill, or \hss. The highest order of infinity is always used, and the lower ones are ignored.

▶ **Exercise 3.30:** What is the glue set ratio in

```
\hbox spread4pt{\hbox to5pt{\hfil}%
  \hskip0pt plus10pt\hbox to5pt{\hfil}}
```

3.17 The Definition of Badness

When a box is constructed, TEX calculates a quantity called the *badness* of the box [97]. This quantity is used to determine whether the box is "good" or "bad," and to classify it (section 3.18). If the box contains a line of text, its badness is used by the line-break algorithm. If the box contains a page of text, its badness is used by the page-break algorithm.

Intuitively, the badness measures how much the glues in the box had to be flexed so that the material would fit in the box. With this in mind, it seems that the glue set ratio r is a good basis for the definition of the badness. It is natural to define the badness as proportional to Δ. It should, however, also be inversely proportional to Y (or Z). Imagine two boxes flexed by the same amount, say 10pt, where one has a maximum flexibility of 20pt and the other, only 5pt. The more flexible box should not be considered bad at all, whereas the other one, flexed to twice its permitted maximum, should be judged very bad.

The actual definition of badness involves r^3 instead of r. This makes sense since, for $r < 1$, r^3 is not much different from r but, for $r > 1$, r^3 is greater than r and grows faster than r. As a result, small changes in r (say from 3 to 4) cause large changes in r^3 (from 27 to 64). Thus r^3 is a more sensitive measure than r. The badness itself is defined as the integer nearest $100 \times r^3$ and is limited to the range [0,10000]. A badness of more than 10,000 is considered infinite and is truncated to 10,000.

The badness is accordingly defined as the integer nearest $\min(100 \times r^3, 10000)$. Note that any infinite flexibility (like \hfil, \hfill) causes Y or Z to become infinite, so the badness in such cases is zero. To speed things up, an efficient method is used that does not always yield the exact value of r^3. So, when $r = 3$, the badness comes out as 2698 instead of 2700.

60. Examples
- $\Delta = 9$pt and $Y = 10$pt $\Rightarrow$ badness = 72.9
- $\Delta = 2Y$ or $r = 2 \Rightarrow$ badness = 800
- $2\Delta = Y$ or $r = 50\% \Rightarrow$ badness = 13

▶ **Exercise 3.31:** What is the badness of the box in the previous exercise?

▶ **Exercise 3.32:** What does a badness of 100 tell us about the relative sizes of Δ, Y, and Z?

3.18 Classifying "Bad" Boxes

The badness of a box is calculated when the box is constructed, and it serves to classify the box as either "underfull," "loose," "good," "tight," or "overfull." The badness can also be displayed in the log file by \showthe\badness, it can be typeset by \the\badness, and it can be tested by, for example, \ifnum\badness<500. The parameter \badness is set to the badness of the most recent box constructed [229], and is "read only", it cannot be changed.

The class of an \hbox is determined [302] by comparing its badness to the value of parameter \hbadness, whose plain value [29] is 1000. The ⟨dimen⟩ parameter \hfuzz is also used in this process. The rules are as follows:

An empty \hbox is always "good." A nonempty one is "underfull" or "loose" if its glue has to be stretched and its badness is greater than \hbadness. In such a case, the box is "underfull" if its badness > 100, and "loose" if ≤ 100.

Similarly, a nonempty \hbox is either "overfull" or "tight" if its glue has to shrink and its badness exceeds \hbadness. As mentioned above, glue does not shrink below its maximum shrinkability. If, after shrinking a box to the maximum, the difference between its width and the desired size is greater than \hfuzz (or if \hbadness < 100), then it is "overfull." The badness of an overfull box is set [229] to 1,000,000 (also see below). The plain value of \hfuzz [30] is 0.1pt.

The paragraph builder uses a different classification [97] for the purpose of comparing adjacent lines in a paragraph. It uses the four classes: very loose, loose, decent, and tight. This classification is discussed in Chapter 14.

61. We set \hbadness=50. It is now easy to create a "loose" box. When the box \hbox to2pt{\hskip0pt plus1pt\vrule width1pt} is constructed, the glue has to stretch to its maximum, implying $r = 1$, so the badness becomes 100. Since \hbadness < 100, the box is loose.

62. When \hbadness is restored to 1000, even the box \hbox to3pt{\hskip0pt plus1pt\vrule width1pt}, which is worse than the previous one, comes out good. The glue has to stretch to twice its maximum, so the glue set ratio r becomes 2, and the badness becomes $100 \times 2^3 = 800$. Since the badness is still less than \hbadness, the box is good.

63. We stretch more, \hbox to4pt{\hskip0pt plus1pt\vrule width1pt}. The glue set ratio is 3, making the badness equal 2698 (the precise value should be 2700). The box is now underfull.

64. We now turn to cases where glue has to shrink. We start with '\hbox to1pt{\hskip0pt minus1pt\vrule width2pt}'. The glue is shrunk by its maximum permitted shrinkability, resulting in $r = 1$ and badness = 100. The box is good.

65. The same box, with badness 100, becomes tight if we set \hbadness=50.

66. We return \hbadness to 1000, and squeeze our box a little more, \hbox to0.9pt{\hskip0pt minus1pt\vrule width2pt}. The material inside cannot shrink to less than $2 - 1 = 1$pt, so the width of the box remains 1pt. However, the difference between the actual and desired widths is $1 - 0.9 = 0.1$pt, which does not exceed \hfuzz. Our box is still good, but if we squeeze it even more, it will become overfull. Also, if we limit \hbadness to below 100, this case will yield an overfull box.

The badness of an \hbox is set to 1,000,000 if, after maximum shrinking, it is still too wide. Note that the box may not even be overfull, because of the leniency of \hfuzz.

67. Setting \hbadness=10000 implies that no \hbox would be considered underfull. Setting \hbadness=10000 and \hfuzz=\maxdimen will also prevent such boxes from becoming overfull.

A constructed \vbox is classified in a similar way, using \vbadness, but \vfuzz is used instead of \hfuzz.

▶ **Exercise 3.33:** Construct a \vbox with a layout and dimensions equivalent to \hbox to0.9pt{\hskip0pt minus1pt\vrule width2pt}, and verify that it behaves in a similar way.

3.19 The Interline Glue

TeX uses the three parameters a, b, and c to determine the vertical spacing between lines (or, in general, to determine vertical glue). To illustrate the rules, let's consider two lines, the current one and the next one. Each has a height, a depth, and, of course, a baseline. The rules are as follows:

1. The next line is set such that the distance between the baselines is made equal to a.

2. TeX now looks at the distance between the bottom of the current line and the top of the next one. If that distance $< b$, the next line is moved down such that the distance is set equal to c.

Rule 1 is used most of the time. Parameter a is called \baselineskip; it is glue (it can have flexibility) and is normally set to 12.0pt (rigid). It can, however, be changed at any time.

Rule 2 is used if the next line has a tall character (such as a first letter set in a large size). Such a character increases the height of the line box, and setting the next line according to rule 1 would cause a situation like the one below:

Rule 2 discovers that the distance between the bottom and the top $< b$ (in fact, it is negative) and moves the next line down to make that distance equal to c.

Parameter b is called \lineskiplimit, it is a dimension, and its **plain** value is 0.0pt. Parameter c is called \lineskip, it is glue, and its **plain** value is 1.0pt (rigid). The merits of adding flexibility to parameters a and c are discussed on [78].

3.20 Suppressing the Interline Glue

The normal interline glue in a vertical list can be suppressed by two commands. A \nointerlineskip placed between two components suppresses the glue between them. An \offinterlineskip placed anywhere in a vertical list suppresses the interline glue for the rest of the list. The effect of the latter can be global if it is placed at the start of the document. It can be localized by placing it in a group.

To understand the commands, a short discussion of interline glue is necessary (see complete algorithm on [80, 282]). When a vertical list is being constructed, interline glue is placed between most components to keep the distance between successive baselines equal to \baselineskip. When a component is appended to the list, the size of the glue placed above it depends on its height and on the depth of the previous component. The size should thus be set to $b - p - h$, where h is the height of the current component, p is the depth of the previous one, and b is the value of \baselineskip (we ignore the flexibility of b).

This is why the depth of the previous component has to be saved when a vertical list is constructed. It is saved in parameter \prevdepth, which is updated each time a component is appended to the list. At the start of the list, when there is no previous component, \prevdepth is set to -1000pt. When \prevdepth has this value, the interline glue is suppressed. When the next component is a rule, TeX sets \prevdepth to -1000pt to suppress interline glue below the rule. This suggests a way to suppress the interline glue at any desired point; just set \prevdepth to -1000pt. This is how \nointerlineskip is implemented [352].

▶ **Exercise 3.34:** How is the interline glue suppressed above a rule?

\prevdepth can easily be "shown", for example, by an experiment such as

```
\def\P{\showthe\prevdepth}
\vbox{\P\hbox{\strut}\P\hbox{,}\P\hrule\P\hbox{y}\P}
```

As mentioned earlier, the \offinterlineskip command (which is a plain macro) can be used to suppress the interline glue for an entire vertical list, or even globally, for all or part of a document. The first thing that comes to mind, when trying to implement such a command, is to set to zero the three quantities that affect the interline glue, namely \baselineskip, \lineskip, and \lineskiplimit. This usually works but is not general enough. In fact, it is one of the typical mistakes so often made by the notorious B. L. User (see [ex. 12.10 and 21.1]). Such settings do not work if the components of the vertical list have negative depths or heights.

A better approach is to start by setting \baselineskip to some large negative value. Trying to separate consecutive baselines by \baselineskip will always bring the bottom line above the top one (even if the heights and depths are negative). Next, set \lineskiplimit to a large positive value. This would always make the distance between the top of the current line and the bottom of the previous line less than \lineskiplimit, so TeX would have to use \lineskip as the interline glue. We set \lineskip to 0pt.

The actual definition of \offinterlineskip [352] is

```
\def\offinterlineskip{\baselineskip=-1000pt
  \lineskip=0pt \lineskiplimit=\maxdimen}
```

A good example of the use of \offinterlineskip to suppress interline glue is in tables that should have vertical rules. Several such examples are shown on [245–247].

3.21 \kern

A \kern is similar to glue [75], with two differences: (1) \kern is rigid; (2) \kern specifies a point where a line, or a page, should not be broken. Since a box is indivisible anyway, \kern is used in a box to indicate rigid spacing. It is interesting to note that the same command, \kern, indicates horizontal spacing when used in an \hbox and indicates vertical spacing when used in a \vbox.

There is a special kern called \mkern [168, 280]. It is only allowed in math mode and will not be discussed here. There are also such things as implicit and explicit kerns. They are only used by the hyphenation algorithm [454] and will not be described here either.

3.22 Rules

Rules, both horizontal and vertical, are traditionally used in typesetting. In TEX, a rule does not necessarily have to be long and thin; it has three dimensions, like a box, and can have any rectangular shape.

There are two types of rules, \hrule and \vrule. The general command to create an \hrule is \hrule height<dimen> width<dimen> depth<dimen>, and similarly for a \vrule. If the dimensions are not explicitly given, they are calculated from the context.

68. The command '\hrule height10pt width5pt depth0pt' creates ▌ (looks vertical). Similarly '\vrule height5pt width10pt' creates ▬ (looks horizontal).

▸ **Exercise 3.35:** Both rules above should be $10 \times 5\,$pt rectangles, yet they look different. Is this an optical illusion?

The shape of the rule does not depend on whether it is H or V, and the difference between the two types is in the context in which they can be used, not in their shapes. An \hrule is considered vertical material and can be part of a vertical list. A \vrule is the opposite and can only appear in horizontal lists. The reason for this convention is that a horizontal rule is a good separator between items stacked vertically, whereas a vertical rule is a natural separator for items laid horizontally, from left to right.

As a result, a \vrule should be used inside a paragraph, such as this |, or in an \hbox. An \hrule should be used between paragraphs or in a \vbox.

Any unspecified dimensions of a rule are determined [221] by these defaults:
- The height of an \hrule is 0.4pt, and the depth is 0pt.
- The width of a \vrule is 0.4pt.
- Other dimensions are determined by extending the rule to the size of the smallest box containing it. An example of this rule is the \vrule above. Its depth is set equal to the depth of the line it happens to be on.

69. \hbox{p\vrule*} results in ꒰p꒱. The rule is extended to the height and depth of the box.

70. \vbox{\hsize=24pt\parindent0pt p\hrule*} creates . The rule is extended to the width (24pt) of the box, but this example shows two more important properties of rules:

■ No interline glues are automatically inserted on either side of a rule placed in a \vbox.

■ The 'p' starts a paragraph. The \hrule, being vertical material, terminates the paragraph. The '*' starts a new paragraph.

▸ **Exercise 3.36:** What's a good way to add the normal interline glue on both sides of a rule?

71. To create something like 'Name_____' we have to use a \vrule. An \hrule is vertical material and would terminate the paragraph (try it). The construct above was created by Name\vrule height-2pt depth2.4pt width1in. The combination of negative height and large depth pushes the rule down. An identical result is obtained by Name\vbox{\hrule height-2pt depth2.4pt width1in}.

▸ **Exercise 3.37:** How does one get Name_____?

72. \vbox{\hsize=.5in\vrule height5pt} creates a box, .5 inches wide, with the paragraph indentation at the top, followed by the rule, followed by the end-of-paragraph (\parfillskip) glue. The \vrule is H mode material and is treated as text.

73. \vbox{\hsize=.5in\hrule} creates a somewhat surprising result. The height and depth of the \hrule are 0.4pt and 0pt, respectively. Its width is extended to the width of the surrounding box, which is zero! The \hsize=.5in specification is not obeyed in this case, because the box does not contain text. Text would have been set in lines of width 0.5 inches, but without text, the width of a \vbox is that of its widest component, regardless of \hsize.

The only component of our box is the rule, whose width should be that of the box! As a result, both widths are set to zero, and nothing shows up in print.

74. Adding a small box to the previous example gives us \vbox{\hsize=.5in \hbox{X}\hrule} and results in X̲. The width of the \vbox is that of the 'X', and the rule is extended accordingly.

75. Adding text, however, forces the width of the box to be precisely 0.5 inches, and extends the rule to that size. Thus \vbox{\hsize=.5in X\hrule} results in X̲.

76. \setbox0=\vbox{\hrule \hbox to1in{...}} The width of the rule is unspecified, so it is always extended to the width of \box0. Originally it is 1 inch, but if we later say \wd0=0pt, the rule will shrink to zero width.

77. Rules can be used to create boxes with negative dimensions. The command \hbox{\vrule height-1pt depth-1pt width-1pt} creates an \hbox with negative width. The height and depth of an \hbox, however, are never negative, so they are set to zero. Similarly, \vbox{\hrule height-1pt depth-1pt width-1pt} creates a box with negative height and depth, but width zero.

78. The word *strut* has already been mentioned. It refers to a \vrule with width zero. A standard strut is part of the plain format and is defined, on [353], as \vrule height8.5pt depth3.5pt width0pt (the actual definition is slightly more complicated and takes into account the current mode). Such a rule does not show up in print and is used to open up boxes. Inexperienced users find it hard to believe that such a rule can be useful, but a glance at [478] shows that it is one of the most frequently mentioned terms in *The TEXbook*.

A horizontal strut can also be defined. It is an \hrule with height and depth of zero. Surprisingly, such a thing is rarely used (but see discussion of \hphantom in section 3.24).

79. Rules can be used to typeset rulers. The ruler below follows the example on [58].

4 in

It is created by this compact set of macros:

```
\def\1{\vrule height 0pt depth 2pt}
\def\2{\vrule height 0pt depth 4pt}
\def\3{\vrule height 0pt depth 6pt}
\def\4{\vrule height 0pt depth 8pt}
\def\ruler#1#2#3{\leftline{$\vcenter{%
  \hrule\hbox{\4#1}}\,\,\rm#2\,{#3}$}}
\def\\#1{\hbox to .125in{\hfil#1}}
\def\8{\\\1\\\2\\\1\\\3\\\1\\\2\\\1\\\4}
\ruler{\8\8\8\8}4{in}
```

3.23 Placing Accents, an Example

Accents (or rather *diacritical marks*) are popular in many "common" languages (such as Scandinavian, Slavic, Vietnamese) as well as in "exotic" languages (such as Vedic, Sanskrit), which are often transliterated into Latin characters for the benefit of most readers. TEX provides many accents, and more can be developed with METAFONT. The plain format has several macros that can be used or modified to handle newly developed accents. It is relatively easy to place an accent above or below a character (an accentee) by using a simple construction involving boxes and glue.

For simple cases, the primitive \accent [286] produces good results. It should be followed by an ⟨8-bit number⟩ and a character. The ⟨8-bit number⟩ is the position of the accent in the current font, and that accent is placed above or below the character (see examples of the use of \accent on [356]). Thus \accent'30a \accent"7Ef \accent"7FP produces ą f̃ P̈. When no character follows, \accent just produces the accent. Thus \accent'27{}... produces ˚.

There may be an assignment between the ⟨8-bit number⟩ and the character, normally for a change of font. Thus \accent"5E\tt P produces P̂.

However, \accent does not always produce the desired results. \accent'30g, for example, results in g, clearly not what we wanted. Also multiple accents, such as in è or P̈, are sometimes called for, which again cannot be handled by \accent. This is why general methods should be developed for handling accents.

Since an accent is a character in a font, a general approach to accents is to use METAFONT to design accents that are exactly bounded by their boxes. The height and width of such a box should be those of the accent, and the depth should be zero. A simple box construction can then be used to place the accent properly either above or below any character. A few examples of such 'accents' are shown here with their boxes (a little space has been placed between the character and its box, for improved readability).

⌣ ∼ ⌢ ∧

Such an accent can easily be placed above and/or below an accentee with a macro such as

```
\font\acc=filename
\def\ifnnull#1{\def\inner{#1}\ifx\inner\empty\else}
\def\genaccent#1#2#3{\leavevmode\setbox0=\hbox{#3}%
\vbox{\offinterlineskip
\ifnnull{#1}\hbox to\wd0{\hss\acc\char#1\hss}\kern.2ex\fi
\vbox to\ht0{\copy0\vss}
\vtop{\null\vbox to\dp0{\vss}
\ifnnull{#2}\kern.2ex\hbox to\wd0{\hss\acc\char#2\hss}\fi}}}
```

where \acc is the name of the accent font. Macro \genaccent places its #1 parameter above, and #2 below, the character specified by #3. Either accent may be empty. Note that the value of ex is taken from the current font, not from \acc. The accent font does not even need to have a defined x height. The string H̃B̰ŏx̂ is created by

```
{\genaccent{21}{}H\genaccent{}{20}B%
\genaccent{22}{}o\genaccent{23}{}x}
```

Multiple accents are produced when the third parameter of \genaccent is itself an accented character. For example, P̰̃ is created by
\genaccent{20}{21}{\genaccent{22}{23}P}.

Notice that the accent has to fill its box vertically but not horizontally. It may be wider or narrower than its box, so long as it is centered in the box. This is because the accent box is always centered, by \genaccent, with respect to the character, and the reference point is moved to the left by the width of the character, not that of the accent.

The macro can be further generalized by:

■ making the accent font name a parameter, with a default value of, perhaps, \cacc (computer accent). With 256 characters per font, it's hard to imagine a situation where the two accents would have to come from different accent fonts.

■ making the gap between the accentee and the accent a parameter, with a default value of .2ex.

■ adding a parameter or two to specify whether the accents should be left, center, or right justified.

If METAFONT is not available, existing accents from the cm fonts can be used. The problem is that the bounding box of such an accent is normally higher than the accent and may also have a depth. Some common examples are: ⌑ ⌑ ⌑.

Simply placing such an accent above a character results in something like P; totally unacceptable! Macro `\supaccent` below has been developed to handle such cases. It places an accent with a tall bounding box above a given accentee.

To illustrate the development of the macro, we create P̊ in several steps. We start by placing the accent and the accentee side by side, ' ̊P'; we then raise the accent by −1ex, which drops it close to the baseline ' ̥P'. Now we raise the accent an amount equal to the height of the accentee, resulting in ' ̊P'. Finally, we overlap the accent and the accentee by means of an `\rlap` to get P̊. Slightly better results are obtained by raising the accent a little less (compare this discussion with the plain macro `\AA` on [356]).

The next step is to generalize our construction into a macro `\supaccent#1#2`, where `#1` is the character number of the accent in the current font, and `#2` is the accentee.

```
\def\supaccent#1#2{\leavevmode
\setbox0=\hbox{#2}
\dimen0=\ht0 \advance\dimen0 by-1ex
\rlap{\raise.67\dimen0\hbox{\char#1}}\box0{}}
```

Applying our macro to an 'l', however, produces 'l̊'. The accent is wider than the character. The solution is to center the accent horizontally in an `\hbox` as wide as the character. The next version of our macro is

```
\def\supaccent#1#2{\leavevmode
\setbox0=\hbox{#2}\dimen0=\ht0
\advance\dimen0 by-1ex \rlap{\raise.67\dimen0
\hbox to\wd0{\hss\char#1\hss}}\box0{}}
```

The tests

```
\supaccent{'27}P \supaccent{"7D}a \supaccent{"7E}l
\supaccent{"7F}{Pa} \supaccent{'26}{\hbox{a\kern-.1em l}}
\supaccent{'25}{Pl}
```

result in P̊ ắ l̃ P̃a ãl P̌l.

The macro can also be used in math mode. The two constructs

```
$x^{\sevenrm\supaccent{"7E}a}$
$x^{\fiverm\supaccent{"7E}u}_{\fiverm\supaccent{"7F}w}$
```

produce $x^{ã}$ $x^{ũ}_{w̃}$.

The double accent P̊ is obtained by `\supaccent{'27}{\supaccent{"16}P}`, but better results are obtained by raising the accent a little more. We thus generalize the definition of `\supaccent`

```
\def\supaccent#1#2#3{\leavevmode
\setbox0=\hbox{#2}\dimen0=\ht0
```

```
\advance\dimen0 by-1ex \rlap{\raise#3\dimen0
\hbox to\wd0{\hss\char#1\hss}}\box0{}}
```

and define a macro \dbaccent for double accents

```
\def\dbaccent#1#2#3{\supaccent{#1}{\supaccent{#2}{#3}{.8}}{.8}}
```

The results ắ ĕ ĩ ŏ ũ, seem satisfactory.

▸ **Exercise 3.38:** Develop a macro \subaccent#1#2 where #1 and #2 are any characters. The macro places #1, as an accent, below #2.

In principle, accents can also be handled by \halign or by \matrix. The plain format \oalign and \ooalign macros [356] are examples of the former, and [ex. 18.46] of the latter. In practice, however, boxes seem the best tool for working with accents.

3.24 Phantoms and \smash

The control sequence \phantom<material> [178] behaves like infinitely flexible glue and expands to occupy the space normally taken by the material. Thus \hbox{} produces the empty box ⬚, whose dimensions are determined by the string 'AB'. Similarly, \hbox{p} creates a box ⬚ᵖ containing just the 'p', which should be compared with ⁀AᴮB̲p̲.

Similarly, a \vphantom<material> expands vertically to the height and depth of the material, but with width zero, like a strut. \hbox{\vphantom{AB}p} creates a box containing just the 'p' ⎸p⎸, but with the height of the 'AB'. The control sequence \mathstrut is defined, on [360], as a '\vphantom('.

There is also an \hphantom<material>, which expands horizontally, to fill up the width of the material. \vbox{\hphantom{AB}\nointerlineskip\hbox{p}} results in p̲⬚, which has the height and depth of 'p' and the width of 'AB'. The \hphantom always has height and depth of zero, so it is equivalent to a horizontal strut.

The \smash<material> control sequence is different. It typesets the material and moves the reference point to the right by its width. However, it makes the height and depth zero. Thus \hbox{\smash{One Two}} creates O̶n̶e̶ ̶T̶w̶o̶ and \vbox{\smash{\hbox{Three}}\hbox{Four}} results in ⎡Three⎤ Four⎦.

80. The special effect ════════════ Heading ════════════ is a simple example of \phantom and is achieved by (see [412])

```
\vbox{\hsize=2.9in
 \line{\hrulefill\phantom{ Heading }\hrulefill}
 \kern-6pt
 \line{\hrulefill\hbox{ Heading }\hrulefill}
}
```

Both \smash and the phantoms are used mostly in math mode. Underlining is a simple example of their use outside math mode (compare with [ex. 18.26]).

81. Underlining text. This is discouraged in TEX (see answer to [ex. 18.26]). There is an `\underline` primitive, but it can only be used in math mode, and its use [178] is limited. There is also an `\underbar` macro [353], which uses `\underline` to underline text in an `\hbox`.

Here we show how to underline text by placing a rule just below it. The construct `\vtop{\hbox{gap}\hrule}` generates <u>gap</u>, but this does <u>not always</u> work properly. One improvement uses `\smash`. `\vbox{\smash{\hbox{gap}}\hrule}` results in <u>gap</u>, the rule is placed at the baseline, crossing any descenders. Another way is to place the rule always at the same depth, below the descenders, by a box such as `\vtop{\hbox{\vphantom{y}<material>}\hrule}`. The `\vphantom` acts as a strut, causing the rule to be placed at that depth.

▸ **Exercise 3.39:** Write a macro `\triMax#1#2` where each argument is a character or a box (and thus has height, depth, and width). The macro is to create a box whose height is the maximum of the heights of the two arguments, and similarly for its depth and width.

82. Sometimes the book designer decides to start certain paragraphs with a big letter, as shown below. This is easy to get by placing the letter in a smashed `\hbox to0pt`. All the dimensions of the box are zero, and space is reserved for it by hanging indentation. The following paragraph (compare with [v])

G ENTLE READER: This is an example of the use of **smash**, one of the more esoteric control sequences. It reserves the space on the left of the paragraph using hanging indentation, and places the big 'G' in a smashed, lowered hbox.

was created by

```
\noindent\hangindent=40pt\hangafter-2
\smash{\lower12pt\hbox to 0pt{\kern-\hangindent\beeg G\hss}}%
\hskip-16pt{\sc ENTLE} R{\sc EADER}: \strut This is an example
. . .
. . . smashed, lowered hbox.
```

The best way to understand this construct is to experiment with it and see what happens if the `\smash`, `\lower12pt`, `\hskip-16pt`, and `\strut` are eliminated.

83. In a table such as the one below, a '`\phantom`)' can be used to align the two x's in the left column. Compare the following:

```
\halign{$\hfil#$ = &#&#\cr
x&0&1\cr
f(x)&1&2\cr}
```
Produces
$$x = 01$$
$$f(x) = 12$$

```
\halign{$\hfil#$ = &#&#\cr
x\phantom)&0&1\cr
f(x)&1&2\cr}
```
Produces
$$x = 01$$
$$f(x) = 12$$

84. Sometimes, a compact diagram, such as a below, is needed. It is easy to create by:

```
\vbox{\offinterlineskip\tabskip0pt
\halign{#&#&#&#\cr
x& &x&x\cr x\cr
x&x& &x\cr &x& &x\cr}}
```

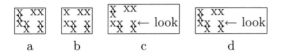

    a   b   c   d

If the diagram has a missing row, such as in b, a \phantom can be used to open up the right space:

```
\halign{#&#&#&#\cr
x& &x&x\cr\phantom{x}\cr
x&x& &x\cr &x& &x\cr}
```

Now we want to add the word "look" with an arrow, as in c. Since the 'l' and 'k' are higher than the 'x', unwanted vertical space shows up in the diagram. The solution is to use \smash to nullify the height of 'look'. Diagram d was obtained by

```
\halign{#&#&#&#\cr
x& &x&x\cr x\cr
x&x& &x\smash{$\leftarrow$ look}\cr
 &x& &x\cr}
```

▸ **Exercise 3.40:** Use \smash and phantoms to create the pattern:
    Co$\bigparallel$een
McCu$\bigparallel$ough

85. If a paragraph contains a character that's a little taller than normal, such as :, and we don't want it to affect the interline spacing, its height can be suppressed by \smash{$\vdots$} (note how close the dots ⋮ are to the line above). The next example is similar.

86. A small diagram such as $\mathbf{{}^{*\cdots}_{**}{}^{**}$ is needed. It is slightly taller than a normal line so we want to (1) drop it so it is centered on the current line; (2) smash it so it does not affect the interline space. The basic diagram is not hard to create. The asterisk is placed in a box '\setbox10=\hbox{$*$}', and the bottom line is \hbox{$\cdots$\copy10}. The line above it uses an \hphantom and a \kern to place the asterisks at the proper place:
'\hbox{\hphantom{$\cdots$}\kern\wd10\copy10\copy10}'. The entire construct is

```
\setbox10=\hbox{$*$}*
\def\room{\hphantom{$\cdots$}}
```

```
\smash{\hbox{$\vcenter{\offinterlineskip
\hbox{\room\kern5\wd10\copy10$\cdots$}
\hbox{\room\kern3\wd10\copy10\copy10}
\hbox{\room\kern \wd10\copy10\copy10}
\hbox{$\cdots$\copy10}}$}}
```

The first of the three diagrams below shows the final result. The second one shows the effect of removing the \smash; the interline glue has been stretched to accommodate the diagram. The third diagram shows the effect of replacing the \vcenter with a \vbox.

3.25 More Advanced Examples

87. Here is an \hbox, 3 inches wide, with four components. The first is a \vbox, 0.5 inches high. Even though we specified \hsize=30pt, its width is zero, since it is empty. The second component is an \hbox, 0.5 inches wide, whose height and depth are those of a strut (8.5pt high and 3.5pt deep).

The third one is a \vbox whose height is explicitly specified and whose width is that of its only component, an \hbox with an asterisk. The last component is a \vtop. The specification \hsize=30pt is obeyed since this box contains text (the single character '+'). Note that the text is indented. Also note the kerns and \hfil.

```
\hbox to3in{\kern10pt%
\vbox to.5in{\hsize=30pt}\kern20pt
\hbox to.5in{\strut\hfil}\kern30pt
\vbox to.25in{\vfil\hbox{*}\vfil}
\hfil
\vtop to.2in{\hsize=30pt +\vfil}\kern10pt
}
```

88. The next example is a \vbox, none of whose dimensions is specified. In such a case, its height becomes the total height of its components, and its width is that of the widest component (because it does not contain text at upper level).

Note how the \vrule is extended to the vertical size (height plus depth) of the larger of the two vboxes surrounding it. Also the \hrule is extended to the width (3 inches) of the box immediately containing it.

Here is the listing for this example,

```
\vbox{
 \hbox to3in{\hfil\bf A Heading\hfil}
 \hrule depth2pt
 \hbox{\vbox{\hsize=1in textg}%
 \kern2pt\vrule\kern2pt
 \vbox{\hsize=1in\noindent textf}}
}
```

A Heading	
textg	textf

89. This example involves lines of text, of unknown sizes, that have to be typeset one above the other, each centered on the page. If the lines are "These lines make," "an," "example," then we want to typeset them as:

These lines make

an

example

The problem is the width of the rule. It should be that of the longest line of text, but we don't know how long the lines will be. The solution uses two steps. The first is to stack the text lines vertically in \box0 and measure its width. The second step is to typeset our lines the way we want (centered), followed by a rule of width \wd0.

```
\def\lina{These lines make} % the text can also be
\def\linb{an}               % read from a file or
\def\linc{example}          % from the keyboard
\setbox0=\vbox{
\hbox{\lina}\hbox{\linb}\hbox{\linc}
            }
\vbox{
\centerline{\lina}\centerline{\linb}\centerline{\linc}
\nointerlineskip\centerline{\vrule width\wd0 depth.4pt}
      }
```

▸ **Exercise 3.41:** Modify the above example such that the width of the rule is that of the shortest line.

▐▶ **90.** The many examples shown here use macro \boxit below to surround a box with four rules. Its first parameter is the material to be "boxed". This can be either straight text or a box of any type. The second parameter is the amount of space, in points, between the rules and the material. Note the \long definition, which is necessary since #1 can be more than one paragraph of text.

```
\long\def\boxit#1#2{\vbox{\hrule
 \hbox{\vrule\kern#2pt%
  \vbox{\kern#2pt #1\kern#2pt}%
 \kern#2pt\vrule}\hrule}}
```

The macro creates a \vbox with three components, two hrules at the top and bottom, and an \hbox in between. The \hbox also has three components, two vrules on the left and right, and the material in between. The spacing is achieved by pieces of \kern, and the most interesting point is that all the dimensions of the rules are implicitly determined.

▶ **Exercise 3.42:** Modify \boxit to produce an \hbox instead of a \vbox.

Sometimes it is necessary to "box" just a small amount of material, such as a single word, in H mode. Below is a version of \boxit, called \ppboxit, that uses a \vphantom as a strut, so that all results have the same height and depth (why are 'Pp' used?). It also places thin spaces on both sides of the argument.

```
\def\ppboxit#1{\leavevmode\hbox{\vrule\vbox{\hrule%
  \hbox{\vphantom{Pp}\thinspace#1\thinspace}\hrule}\vrule}}
```

▶ **Exercise 3.43:** Macro \boxit creates a box with depth zero (why?), so the code p\boxit{\hbox{p}}0 results in 'pp0'. Define a macro \depthit#1 similar to \boxit, but creating a box with the depth of #1. A typical use of \depthit is A{\depthit{p}}o{\depthit{l}}o{\depthit{g}}y, resulting in 'Apology'.   Use two different methods!

91. Macro \lbx below is similar but also draws the baseline. Its only parameter is the material to be boxed. It proceeds in three steps. In step 1, it places the text in \box6. In step 2, it builds three boxes, in box registers 7, 8, and 9, with the necessary rules. In step 3, the three boxes are combined, and the text is added, with zero dimensions, just above the baseline. Note that the macro works in either H or V mode, but does not work for an empty argument.

```
\def\lbx#1{\setbox6=\hbox{#1}%
\setbox7=\vbox{\hrule\hbox to\wd6{\vrule height\ht6\hfil\vrule}}%
\setbox8=\hbox to\wd6{\vrule\hrulefill\vrule}%
\setbox9=\vtop{\hbox to\wd6{\vrule depth\dp6\hfil\vrule}\hrule}%
\vbox{\offinterlineskip\box7\ht6=0pt\wd6=0pt\dp6=0pt\box6\box8\box9}}
```

The interested reader should also look into macro \makeblankbox, which is part of the manmac format (even though it is not mentioned in [App. E]).

92. The diagram below is similar to the one on p. 111 of *The* METAFONT*book*. It can serve as the basis of a TeX program that typesets a maze.

It was created by

```
\def\\#1{\hbox to 11pt{\hss$#1$\hss}}
\def\verti{\rlap{\vrule height8.5pt depth3pt}}
\def\under{\smash{\lower3.5pt\rlap{\vrule width11pt height0.4pt}}}
\def\over{\smash{\raise8.5pt\rlap{\vrule width11pt height0.4pt}}}
\halign{\indent#\cr
```

```
\verti\over\\a\\b\verti\under\\c\verti\\d\over\\e\verti\cr
\\e\verti\over\under\\a\\b\\c\\d\cr
\\d\\e\verti\over\\a\over\\b\verti\\c\cr
\\c\\d\\e\under\\a\verti\\b\cr
\verti\under\\b\under\\c\under\verti\\d\\e\over\under\\a\verti\cr
}
```

which is, surprisingly, very readable. The table itself is created as a single-column
\halign. The rules are placed in dimensionless boxes, so they don't occupy any
space.

3.26 Exercises

▶ **Exercise 3.44:** Typeset the example below using the following specifications: The
value of \hsize is 5in. The whole thing is set in a \vbox to3in that has a \vfil
at the bottom. There is a \vskip 0.15in above the \hrule, and .1in below it.
Both columns are 2-inch-high vboxes with vfils at the bottom. The width of the
right-hand column is 3.25in, and the width of the left-hand one is automatically
determined by the width of the text boxes in it. Also, the spaces around the
\vrule are equal and are automatically computed as: 6.5in − 3.25in − 0.4pt −
(size of left-hand column). Note the parts that are boldface and slanted. The right-
hand column has no indentations, and there is a paragraph separation of 12pt.
There are a \vskip0.1in above the words "Technical Support" and a space of
0.2in for the president's signature.

Lonely Publishers
1234 Box and Glue Blvd., Suite 56
Stretch Village, Ill 56567

Caleb Summers **President**	Dear Customer
	Thank you for buying *Lonely*™, and welcome to the growing family of **Lonely** customers!
(123) 555-1212	Sincerely yours,
Technical Support (800) 555-1313	Caleb Summers President

▶ **Exercise 3.45:** Change the top of the previous exercise to:

Preferred Publishers
1234 Moriah Woods Blvd., Suite 56
Elk Grove Village, Ill 56567

where the width of the \hrule is automatically determined by the widest line of the title.

▸ **Exercise 3.46:** Use the \boxit macro above to typeset the following in a \vbox, 3 inches wide.

<u>**Your Comments Count**</u>

1. What is your age group?

   1–20 ☐ 21–31 ☐ 30– ☐

2. Which of the following have provided you with the most helpful information about our company on your visit?

Employee ☐   Advertising   ☐
Brochure ☐   Agent   ☐
Display   ☐   Other _____ ☐

▸ **Exercise 3.47:** Use boxes, glue, and the \boxit macro to typeset the diagram below. The height is 48pt, the depth, 30pt, and the width, 65pt.

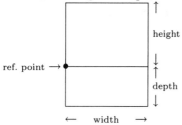

▸ **Exercise 3.48:** Define a macro \varrowfill#1 similar to \leftarrowfill and \rightarrowfill, such that the expansion \varrowfill{20pt} will create ↕.

▸ **Exercise 3.49:** Use boxes, glue, arrows, and rules to draw the following diagram:

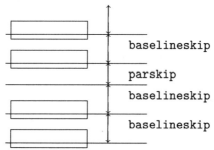

Each of the smaller boxes has a width of 60pt, a height of 10pt, and a depth of 4pt. The size of the arrows is $10 + 4 + 2 \times 4 = 22$pt, except the parskip arrow, whose size is $2 \times 4 = 8$pt.

▶ **Exercise 3.50:** Write a macro `\leadbox#1#2` that creates a box surrounded by rules on all four sides. The height and width of the box are #2, and the thickness of the rules is determined by the parameter `\newdimen\thick`. Parameter #1 is typeset centered in the box. The test `\leadbox{\hrule height4pt width4pt}{10pt}` should create ▣, and the test `\leadbox{\hbox{$\bullet$}}{10pt}` should, similarly, create ⊡ (assuming that `thick=1pt`).

▶ **Exercise 3.51:** Define a macro `\stagbox#1#2` such that `\stagbox{10pt}{9pt}` will typeset ⌐⌐. Parameter #1 is the height and width of the box created; #2 is the length of each rule (the thickness is again `\thick`). Each rule is moved by an amount equal to #1-#2.

▶ **Exercise 3.52:** Define a macro `\crossrules#1` such that `crossrules{10pt}` will typeset ✚. The macro always creates three rules, two have dimensions #1 by 0.2#1, and the third is a square, 0.6#1 units on the side.

▶ **Exercise 3.53:** Define a macro `\fourdiamonds` to create ◈. The macro uses the `\diamond` symbol (in math mode) and positions this symbol in a box, 10.33333pt on the side. A vertical kern of −1.5pt is used to move the diamonds closer together.

3.27 \vsplit

The `\vsplit` command is used to split a `\vbox` in two. It can be used to split any `\vbox`, even one without any text (perhaps with just rules, boxes, glue, etc.), but its intended use is to split a `\vbox` with lines of text. One practical use of `\vsplit` [417] is to split a long and narrow page into two or more equal parts, and typeset them side by side, for a double- or multi-column page. Another example, on [395], shows how `\vsplit` can be used to typeset short footnotes in a triple-column format.

The general form is `\vsplit<number> to<dimen>`, where ⟨number⟩ is the number of a box register that should contain a `\vbox`. The command splits the specified amount of material off the top of the register and makes it available. Thus `\vsplit0 to.5in` will split 0.5 inches' worth of material off the top of `\box0` and typeset it. Similarly, `\setbox1=\vsplit0 to.5in` will save the split material, as a `\vbox`, in `\box1`.

Two parameters, `\splittopskip` and `\splitmaxdepth` [124], are involved in the splitting. In the example above, the depth of the new box, `\box1`, is limited by the current value of `\splitmaxdepth` (instead of the usual `\boxmaxdepth`). Also, the top component of `\box0` becomes a new glob of glue, adjusted to set the baseline of the top line of text a distance of `\splittopskip` from the top. The `plain` value of `\splittopskip` [348] is 10pt (same as `\topskip`), and that of `\splitmaxdepth` is `\maxdimen` (same as `\boxmaxdepth`).

⏵ Normally, the box to be split contains lines of text. The split, obviously, can only occur between lines, not inside a line. Thus in the above example, the amount of material removed from `\box0` is not necessarily 0.5 inches. However, `\box1` is always set to a height of 0.5 inches, which means that it may come out under- or overfull. In fact, the exact point of split is determined by penalties and badnesses in the same way as a page break. A `\vsplit`, then, is similar to a page break except that no insertions are involved.

After the material is removed from \box0, any discardable items remaining at the top are discarded. The \splittopskip glue is then added to the box, but only if it is not void.

These are the rules governing a box split. They are illustrated by the following examples.

93. A \vbox without text \setbox0=\vbox{\hrule height2pt} is split by \setbox1=\vsplit0 to1pt in the middle. The height of \box1 is set to 1pt. The split removes the (indivisible) rule from \box0 and places it in \box1, which becomes overfull. Nothing is left in \box0, so it becomes void.

For most of the examples below, we use \setbox2=\hbox{\vrule height7pt depth2pt} and \setbox0=\vbox{\copy2\copy2}. This creates

```
> \box0=
\vbox(19.0+2.0)x0.4
.\hbox(7.0+2.0)x0.4 []
.\glue(\baselineskip) 3.0
.\hbox(7.0+2.0)x0.4 []
```

\box0 now looks as if it contains two lines of text, making it easy to analyze the results.

94. \box0 is split to a small size by \setbox1=\vsplit0 to2pt. The split is inconveniently located inside the top line of text and the box ends up being split between the two lines. The top line is placed in \box1, whose height is 2pt, so it is overfull (5pt too high). Its contents is

```
> \box1=
\vbox(2.0+2.0)x0.4
.\hbox(7.0+2.0)x0.4 []
```

The remaining line in \box0 gets \splittopskip glue above it, to bring its baseline 10pt below the top.

```
> \box0=
\vbox(10.0+2.0)x0.4
.\glue(\splittopskip) 3.0
.\hbox(7.0+2.0)x0.4 []
```

95. The split is done to the same size, but with a depth limitation of 1pt:

```
\splitmaxdepth=1pt
\setbox1=\vsplit0 to2pt
```

The only difference is that \box1, whose depth is now 1pt, is even more overfull (6pt too high). Normally, when the depth of a \vbox is limited, its height is increased by the difference. The height of \box1, however, is explicitly set to 2pt, so it cannot be changed.

96. A split `to13pt` occurs inside the second line of text. The page break algorithm decides that the best splitting point is between the lines. Only one line is removed and placed in `\box1`, which is now underfull. To suppress the error message, just use `{\vbadness=10000 \global\setbox1=\vsplit0 to13pt}`. The same situation will occur for any split size less than 19pt. When we split `to19pt`, both lines are placed in `\box1`, and `\box0` remains empty.

▸ **Exercise 3.54:** Eliminating the error message may not be enough. Show how to restructure `\box1` to eliminate the underfull condition.

97. A penalty of 10,000 is used to tie the two lines together:

```
\setbox0=\vbox{\copy2\penalty10000\copy2}
\setbox1=\vsplit0 to12pt
```

Even though the split is only `to12pt`, not enough for two lines, both lines are split and placed in `\box1`, which becomes overfull as a result.

98. Flexible glue is added at the top of `\box0`. It ends up in `\box1` and can prevent over/underfull situations.
We set `\setbox0=\vbox{\vskip0pt plus1pt\copy2\copy2}`, and split `to7.5pt`. The result is

```
> \box1=
\vbox(7.5+2.0)x0.4, glue set 0.5
.\glue 0.0 plus 1.0
.\hbox(7.0+2.0)x0.4 []
```

The glue is stretched by 0.5pt to fill up `\box1`.

▸ **Exercise 3.55:** If we now split `to8.5pt`, the glue will have to stretch by 1.5pt, so the glue set ratio would be 1.5. Why don't we get an 'underfull box' message in this case?

99. A `\vbox` can only be split at certain points [110] such as at a glue. Using `\nointerlineskip` to suppress the interline glue, such as in `\vbox{\copy2\nointerlineskip\copy2}`, will make the box unsplittable. Any attempt to split it will remove its entire contents and leave it empty.
In contrast, the box `\vbox{\copy2\offinterlineskip\copy2}` has `\lineskip` between its components, and can be split there.

100. Consider the general problem of splitting a `\vbox` in two equal parts. The first idea that comes to mind is to calculate half the height of the box and split to that size.

```
\newdimen\half
\setbox0=\vbox{<lines of text>}
\half=\ht0 \divide\half by2
\setbox1=\vsplit0 to\half
```

The following diagrams show why this does not generally work. We assume boxes with lines of text separated by `\baselineskip` glue. If the box has an even number of lines, a split to half its height will occur either inside the glue in the middle of the box, which is good, or inside the line preceding that glue, which is too early. For an odd number of lines, a split to half the height will normally occur inside the middle line, which will cause the split part to be one line shorter than the remaining part.

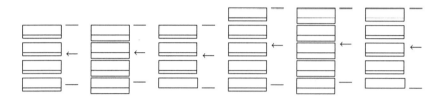

A general solution involves a loop. We start with half the height, split the box tentatively, and compare the heights of the resulting parts. If the top part is smaller than the bottom one, the split size is increased by one point, and the process is repeated. The process stops when the height of the top part becomes ≥ that of the bottom part. An example is shown on [417], in connection with double-column pages. This problem is also treated in detail in (Ref. 13, p. 81).

3.28 \lastbox

It is possible to examine the contents of a box and also break it up into its components, by means of the \last*xx* commands. There are four of them: \lastbox, \lastskip, \lastkern, and \lastpenalty [271]. There are also the three commands \unskip, \unkern, and \unpenalty. The commands can be either placed in the box when it is constructed, or applied to it after it's been opened (by an \unhbox or \unvbox). If the last component of a box is a glue, its value will be reflected in \lastskip. Two things can be done at this point: (1) \skip0=\lastskip; (2) \unskip. The first saves the glue value for future use, and the second removes it [280]. This is similar for \lastkern and \lastpenalty. If the last item is a box, the command \setbox0=\lastbox will both set \box0 *and* remove the last box.

The '\lastbox' command cannot be used in V mode to remove the last box of the current page. When used in the math modes, it always produces a void box. It should, therefore, only be used in H, RH, and IV modes.

If we place \setbox0=\lastbox after the first character of a paragraph, we get a void \box0. A character of text is not recognized as a box by \lastbox. If we leave V mode by means of \leavevmode, TEX starts a new paragraph and inserts the paragraph indentation, which is an \hbox of width 20pt. The simple test \leavevmode\setbox0=\lastbox therefore sets \box0 to an \hbox(0.0+0.0)x20.0.

The construct \setbox0=\hbox{\hbox{abc}}\unhbox0 opens \box0 and starts a paragraph with the \hbox{abc} as the only text. TEX, however, creates the paragraph indentation and inserts it preceding that box. If we say \setbox1=\lastbox \setbox1=\lastbox, the first command will set \box1 to \hbox{abc}, and the second one, to the indentation box. It is, therefore, a good idea to use \lastbox inside a box. Thus \setbox2=\hbox{\unhbox0 \setbox1=\lastbox....

101. Four components are placed in \box1. The box is then opened inside \box2, and the last three components are removed. The first component (the 'A') remains in \box2.

```
\newdimen\Tkern
\setbox1=\hbox{A\kern1pt\hbox{B}\quad}
\setbox2=\hbox{\unhbox1
\skip0=\lastskip \unskip \showthe\skip0
\setbox0=\lastbox \showbox0
\Tkern=\lastkern \unkern \showthe\Tkern
} \showbox2
```

will display in the log file

```
> 10.00002pt.          skip0
> \box0=
\hbox(6.83331+0.0)x7.08336
.\tenrm B
> 1.0pt.               Tkern
> \box2=
\hbox(6.83331+0.0)x7.50002
.\tenrm A
```

As an alternative, \box1 is simply opened, and the last three components are removed. The remaining component (the 'A') is simply typeset.

```
\newdimen\Tkern
\setbox1=\hbox{A\kern1pt\hbox{B}\quad}
\unhbox1
\skip0=\lastskip \unskip \showthe\skip0
\setbox0=\lastbox \showbox0
\Tkern=\lastkern \unkern \showthe\Tkern
```

102. This sets \box2 to the reverse of \box1.

```
\newdimen\Tkern
\setbox1=\hbox{\kern1pt\hbox{B}\quad}
\unhbox1
\skip0=\lastskip \unskip
\setbox0=\lastbox
\Tkern=\lastkern \unkern
\setbox2=\hbox{\hskip\skip0\box0\kern\Tkern}
```

▶ **Exercise 3.56:** Generalize the above example to the general case where \box1 may have any number of components.
Consider the commands
\setbox1=\vbox{\hbox{A}\hbox{B}\setbox0=\lastbox\copy0}. There are three components in \box1, the two boxes with the 'A' and 'B', and the \baselineskip glue between them. The \setbox0=\lastbox removes the bottom box (with the 'B') and places it in \box0. The \copy0 appends the box back to the bottom of

\box1, but since TeX always inserts the usual interline glues in a vertical mode, another \baselineskip glue will automatically be inserted before the box is appended. The result will be a \box1 with four components, the two boxes and, between them, two pieces of interline glue. To avoid the extra glue, just say \setbox1=\vbox{\hbox{A}\hbox{B}\setbox0=\lastbox\nointerlineskip\copy0}.

Note that, even though we said \copy0, the contents of \box0 is not available outside of \box1. To make it globally available, just say \global\setbox0=....

103. The \lastbox and \unskip commands are used, in this example, to measure the widths of all the columns of an alignment (compare this with macro \endvrulealign on [392]). We start by building an alignment in \box0:

```
\setbox0=\vbox{\tabskip=1pt\halign{#\tabskip=2pt
&#\tabskip=3pt\cr a&b\cr C&DE\cr}}
```

We now open \box0, remove its last box (the bottom alignment row), place it in \box2, and copy \box2 back into \box0. The code is \setbox0=\vbox{\unvbox0 \global\setbox2=\lastbox \copy2}. The next step is to loop, remove the tabskip glue with an \unskip, and remove and place the last column from \box2 in \box3, whose width can be displayed or used in any way.

```
\unhbox2
\loop
\unskip
\setbox3=\lastbox
\ifhbox3 \message{\the\wd3;}%
\repeat
```

▶ **Exercise 3.57:** The loop above uses \ifhbox3 to decide when to terminate the loop. It would be more elegant to test for a void \box3. The code is simple: \ifvoid3 \else\message{...}\repeat. Why is this wrong?

104. The last line of a paragraph is usually shorter than the rest (it has a \parfillskip glue ready to stretch on the right-hand side). How can a paragraph be typeset with the last line centered? A simple solution is to use \lastbox. The paragraph is set in a \vbox, and the \lastbox command is used to break up the bottom line of text, which can then easily be centered.

The centering is done by opening the line box and placing it between two \hfills (note, two l's), to neutralize the infinity of the \parfillskip.

```
\def\lastCent#1{\setbox0=\vbox{#1} \unvbox0 \setbox2=\lastbox
\setbox3=\line{\hfill\unhbox2\hfill} \nointerlineskip\box3}
```

As a reminder, a more elegant method, using flexible glues, is described in the section on \filneg.

It should be emphasized that \lastbox is a horizontal command. A user trying to use \lastbox in macro (especially inside a loop) should, therefore, be aware of the following points:

1. Any spurious spaces created by the macro will be appended to the current list, following the boxes being isolated by \lastbox. As a result, the next \lastbox would see a space and return nothing.

2. When the macro is expanded at the start of a paragraph, the paragraph indentation (which is a box) may directly precede the list of boxes being isolated and may be gobbled by a **\lastbox**.

The **\last**xx commands are powerful and can be used to manipulate boxes in sophisticated ways. They have two limitations:

■ When removing components from a box, the user should know what the next component is (a box, glue, kern, or penalty) and should use the right command. If the next component is glue, and **\lastkern** is executed, the resulting kern would be zero.

It is possible to write a loop where all four commands are tried at each step, to see what the next component is. Chapter 17 discusses this technique in detail.

■ A box may also contain rules and whatsits. Unfortunately, the TeX primitives **\lastrule**, **\lastwhatsit** do not exist. Also, the **\lastbox** command does not recognize a character of text as a box. As a result, the **\last**xx commands cannot always be used to completely break up a box.

He had forty-two boxes, all carefully packed,
With his name painted clearly on each.
But, since he omitted to mention the fact,
They were all left behind on the beach.
— *Lewis Carroll, The Hunting of the Snark*

Her romantic mind was like the tiny boxes, one within
the other, that come from the puzzling East, however
many you discover there is always one more
— *James Matthew Barrie, Peter Pan*

For in and out, above, about, below,
'tis nothing but a magic shadow-show,
play'd in a box whose candle is the sun,
round which we phantom figures come and go.
— *Omar Khayyam, Rubaiyat*

4. Paragraphs

To understand paragraphs, the reader should be familiar with the following short summary of TeX's main operations and modes (see [Ch. 13] for a more detailed discussion of modes).

When TeX starts, it is in vertical mode. It reads the source, and when it sees the first character to be typeset, it switches to horizontal mode. This starts the first paragraph, which is fully read into memory. TeX then switches to vertical mode, where it sets the paragraph and takes care of page breaking. Typesetting the paragraph is done by breaking it into lines that are appended—each as an \hbox— to the *main vertical list* (MVL). Following that, TeX determines whether there is enough material for a full page. If there is, the page-breaking algorithm is invoked to decide where to break the page. It moves the page material from the MVL to \box255 and invokes the output routine. Some material is usually left in the main vertical list, eventually to appear at the top of the next page. The output routine can modify \box255, can place new material in it, or can return some material from it to the MVL. Eventually, the output routine should execute a \shipout to append the next page to the .dvi file. TeX then stays in v mode and continues reading the source file.

Figure 4.1 shows the structure of a typical paragraph and the main parameters that control it. As usual in TeX, all the parameters can be modified, making it easy to control the paragraph shape and the spaces inside and between paragraphs. The following parameters are shown in Figure 4.1, are summarized on [274], and are explained elsewhere in this chapter.

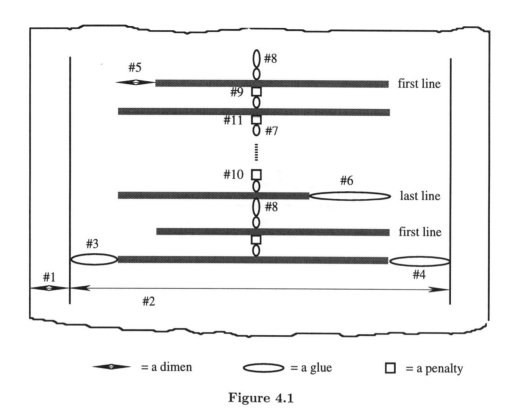

Figure 4.1

#1-\hoffset, [251], a ⟨dimen⟩ specifying how much the left margin should be offset from its normal value of 1in. The plain value is 0pt.

#2-\hsize, [26–27], a ⟨dimen⟩ specifying the horizontal size of the printed area. Its plain value is 6.5in.

#3-\leftskip, [100], a ⟨glue⟩ placed on the left of every line. It is used to create space on the left of a paragraph. Its plain value is 0pt, and it can only be changed between paragraphs.

#4-\rightskip, similar to \leftskip but inserted on the right.

#5-\parindent, [86], a ⟨dimen⟩ used to indent the first line of a paragraph. Its plain value is 20pt.

#6-\parfillskip, [100], a ⟨glue⟩ used to fill up (on the right) the last line of a paragraph. Its plain value is 0pt plus1fil.

#7-\baselineskip or \lineskip, [78], ⟨glue⟩ parameters controlling the interline spacing. They are discussed in sections 3.19 and 4.5.

#8-\parskip, [104], a ⟨glue⟩ inserted *above* a paragraph. It comes in addition to the normal interline space, and its plain value is 0pt plus 1pt.

In addition to the parameters above, there are also *penalties*, both inside lines and between lines, that help control line and page breaks. They will be mentioned here in brief and are fully discussed in other chapters.

Penalties controlling page breaks are placed between lines of text. The most important ones [104] are the following:

#9-\clubpenalty, to discourage a page break after the first line of a paragraph. The plain value is 150.

#10-\widowpenalty, similarly to discourage a page break before the last line of a paragraph. The plain value is the same.

#11-\interlinepenalty, placed between every pair of lines and used to control (encourage/discourage, force/prevent) page breaks between individual lines. The plain value is 0.

\brokenpenalty is placed below any line that ends with a hyphen. This discourages a page break after such a line. The plain value is 100.

Penalty values are arithmetically added when appropriate. Thus the penalty following the first paragraph line is clubpenalty+interlinepenalty. If that line ends with a hyphen, \brokenpenalty is also added. There are a few more vertical penalties—and many horizontal penalties, controlling line breaks—which are not shown in Figure 4.1.

▶ **Exercise 4.1:** What penalty is inserted between the lines of a two-line paragraph?

Math formulas are normally part of a paragraph, and the \vadjust command (section 4.10) makes it possible to temporarily leave the paragraph, do something of a vertical nature, and return.

To break a line in the middle of a paragraph, just say \break, or (to avoid stretching the line) \hfil\break.

4.1 Paragraph Start and End

A paragraph starts when TEX is in V mode and it reads a character of text or a command that's horizontal in nature. Examples are \indent, \noindent, \hskip, and a math-shift (a single '$'). See [283] for the full list of horizontal commands. At that point, the steps are as follows:

- The \parskip glue is inserted above the new paragraph.
- The mode is switched to H and a new horizontal list is started.
- If indentation was not prohibited, an empty '\hbox to\parindent' is inserted at the start of the new list.
- The contents of \everypar is inserted into the list.
- The page builder is exercised.
- Whatever caused the paragraph to start is now read again and executed.

A paragraph is terminated by \par, by a blank line, or by a vertical command, such as \vskip or \medskip (see [286] for the complete list). Note that an empty line is converted to a \par in the mouth. In IV mode, inside a \vbox, a paragraph can also be terminated by the '}' that closes the \vbox.

Incidentally, the \par command is ignored in vertical mode. Thus if you say \par\par, the second \par (which is already read in V mode) is ignored. Also, the \par command is illegal in any math mode and in arguments of macros (unless the macro is defined as \long). The steps at the end of a paragraph are as follows:

- An '\unskip \penalty10000 \hskip\parfillskip' is inserted. The \unskip removes any unnecessary glue that happens to be present at the end of the paragraph. The \parfillskip glue serves to fill the right end of the last line.

■ The paragraph is set (i.e., all line breaks are determined).

■ Parameters \hangindent and \hangafter (section 4.9) are reset to their default values of 0pt and 1, respectively.

■ The mode is switched to V and the page builder is exercised. No glue is added below the paragraph, just above the next one, when it starts.

4.2 \everypar

• The \everypar command can be used to do something, or to typeset something, at the beginning of **every** paragraph. This paragraph is preceded by the command '\everypar{$\bullet\,$}'. An \everypar{} can be used to nullify the effect of \everypar.

A common use of \everypar is to apply certain parameters to several paragraphs. This makes sense for parameters such as \looseness, \parshape, \hangindent, and \hangafter, which are reset after every paragraph.

4.3 Horizontal Lists

A horizontal list is something created by TEX when it is in H mode—reading material for a paragraph—or in RH mode—collecting material for an \hbox. Such a list contains mostly text to be typeset and interword glue. Following is a complete list of items that may appear in a horizontal list [95].

■ A box. This can be a single character, a word, a line, an \hbox or a \vbox explicitly inserted by the user, or a box inserted by a macro. Note that a ligature is essentially a single character box but is different from a normal character box since it may be replaced back by the original characters, if TEX decides to break a line at that point.

■ A glob of glue. This can be implicit—such as interword glue—or explicit, such as \hskip. Note that the value of the interword glue (the size of spaces between words) depends on the "space factor" (page 92). Also included in this category are \leaders.

■ \kern. Kern is rigid glue, with the difference that a line cannot be broken at a kern. Note that if a line is broken at a glue, the glue is discarded. This is also true for page breaks.

■ A penalty. Such an item expresses the desirability of breaking a line at a certain point. A positive penalty discourages TEX from breaking a line, while a negative one encourages it to break a line. Penalties can be inserted explicitly, by the user (using \penalty [97], or \break [94]), or automatically by TEX. Automatic penalties are inserted after a hyphen (\hyphenpenalty [96]), after a math binary operator (\binoppenalty [174, 446]), after a math relation (\relpenalty [174, 446 and ex. 18.20]), and around a displayed math formula (\predisplaypenalty \postdisplaypenalty [189]).

■ A \vrule.

■ A whatsit [226]. This is a token list whose expansion is deferred until the next \shipout. TEX generates whatsits for a \special or for anything that has to do with non \immediate writing to files [227]. A macro placed in such a token list will only be expanded when the \shipout routine opens the whatsit and places its contents in the .dvi file.

- A discretionary break [95]. This is like specifying hyphenations manually, but is very general.
- "math-on" and "math-off" characters.

A line break can only occur [96] at a glue, a penalty, a "math-off," or a discretionary break. Remember, glue and penalty are discardable! They are discarded at a line break (and also at a page break).

The \showbox command, which has already been mentioned, is most useful when trying to understand horizontal mode and line breaks. Using it extensively can shed light on many mysteries. The concept of a math list should also be mentioned. This is what TeX prepares in math mode. Math lists are explained on [157] and may be "shown" with a \showlists command [158–159].

4.4 The Interword Glue

The interword glue [75–76] depends on the font and is specified by the font designer. It is placed between words when a horizontal list is built. However, fine typesetting calls for sophisticated interword spacing, where punctuation marks are normally followed by more space than normal.

In TeX, periods, '?', '!', and ':' are followed by 1.5 times the normal interword space [73–74]. To eliminate this extra space, use a tilde '˜' or a control space '\␣' (what's the difference?). Example:

Rev. Dr. Thos. X. Johnson, Esq. married Rev. Dr. Eliz. Y. Jackson, Ph.D.

Rev. Dr. Thos. X. Johnson, Esq. married Rev. Dr. Eliz. Y. Jackson, Ph.D.

The first example uses ties to suppress the extra spacings, and it looks better. Bibliographies tend to contain many items in short notation, such as "Proc. Amer. Inst. Elect. & Electr. Eng.," so \frenchspacing (page 92) should be used to eliminate all extra spaces.

However, in TeX, the characters '.', '?', '!', and ':' are not followed by an extra space if they are preceded by an uppercase letter [74]. This works nicely in "J. Johnson," where the period does not signal the end of sentence. However, in "I work at JPL. We do work for NASA. It is mostly CAD/CAM." the sentences are not followed by extra spaces, which looks wrong, compared with the rest of the document. To add that space, simply say [ex. 12.6] "I work at JPL\hbox{}." or "I work at JPL\null.".

Consecutive spaces in the source file count as one space. Nonconsecutive spaces, however, sometimes accumulate and show up in print. This occurs when TeX is in H or RH mode, and can be the reason for many mysterious problems. Notice that many of the macros in the TeX literature have % signs placed at strategic points to suppress such unwanted spaces. An example is '\hbox{. \setbox0=\hbox{x} \setbox0=\hbox{x}}'. It creates a box with a period followed by three spaces (this example is analyzed on page 95).

Following is a detailed discussion of the way the interword glue is determined.

When a horizontal list is being constructed, either as part of the current paragraph or as an \hbox, spaces between words become globs of glue. The size of each glob depends on the current font and is determined by the font designer. Each font used for normal typesetting must contain four parameters that specify the interword glue. They are the natural size of the glue, its stretch and shrink components, and its extra space.

▸ **Exercise 4.2:** How can one "show" the interword glue components of a given font?

As is usually the case with TEX, the rules for interword spacing are not built in, but are special settings of a general, flexible, mechanism. That mechanism is based on "space factor codes," which are assigned to each character, and on a parameter called the "space factor" (denoted f). The sfcodes and f can be used to implement different spacing rules.

The main idea is that, following a word, we usually want the regular interword glue specified by the font designer. Following certain punctuation marks, such as semicolon and comma, we want spacing with more stretchability and less shrinkability than normal. At the end of a sentence—i.e., after a period, question mark, exclamation mark, or a colon—we want the same thing and also extra space. However, if any of the special punctuation marks is preceded by an uppercase letter, we want it to be followed by regular spacing.

The space factor is the main tool used to implement this idea. It is denoted f, and it is a ⟨number⟩. It is maintained while a horizontal list is constructed, and it is updated each time a box is appended to the list. Initially, f is set to the neutral value of 1000. When a noncharacter box (or a math formula) is appended to the list, f is again set to 1000. However, when a character is appended, f is set to the \sfcode of the character (which we denote g). The value of f can be "shown" by \showthe\spacefactor; it can be changed by \spacefactor=<number>. The interword glue following the box is determined by the current value of f as follows:

■ The stretchability of the glue is multiplied by $f/1000$. Thus if we want more stretchability after certain characters, their sfcodes should be set to more than 1000.

■ The shrinkability is multiplied by $1000/f$. Such characters will be followed by interword glue with less shrinkability.

■ If $f \geq 2000$, the extra space is added to the interword glue. This is useful for strong, end-of-sentence, punctuation marks.

The following experiment demonstrates the rules above and discusses a few other ones. To fully understand it, we need to know the sfcodes (g values) assigned to the different characters.

When TEX starts, all sfcodes are set to 1000 except the ones of the 26 upper case letters, which are set to 999 [345] (this has to do with an upper case letter preceding a period). The plain format sets the \sfcode of right parenthesis, right quote, and right square bracket to zero [345], to make those characters "transparent" to the space factor (see page 93).

The plain macros \frenchspacing and \nofrenchspacing [351] are opposites. The latter starts sophisticated spacing by assigning the special sfcodes below to certain characters. The former changes all those sfcodes back to the neutral 1000. The plain format starts by expanding \nofrenchspacing on [364].

Char.:	.	?	!	:	;	,
sfcode:	3000	3000	3000	2000	1500	1250

Example. A horizontal list is constructed by \hbox{a␣B␣c.␣)␣C.␣:␣;␣'␣,␣]␣} and is "shown" in the log file. The result is

```
> \box0=
\hbox(7.5+2.5)x83.75008
.\tenrm a
.\glue 3.33333 plus 1.66666 minus 1.11111
.\tenrm B
.\glue 3.33333 plus 1.66498 minus 1.11221
.\tenrm c
.\tenrm .
.\glue 4.44444 plus 4.99997 minus 0.37036
.\tenrm )
.\glue 4.44444 plus 4.99997 minus 0.37036
.\tenrm C
.\tenrm .
.\glue 3.33333 plus 1.66666 minus 1.11111
.\tenrm :
.\glue 4.44444 plus 3.33331 minus 0.55556
.\tenrm ;
.\glue 3.33333 plus 2.49998 minus 0.74074
.\tenrm '
.\glue 3.33333 plus 2.49998 minus 0.74074
.\tenrm ,
.\glue 3.33333 plus 2.08331 minus 0.88889
.\tenrm ]
.\glue 3.33333 plus 2.08331 minus 0.88889
```

The lowercase 'a' is followed by the normal interword glue of font cmr10. The uppercase 'B' has an \sfcode of 999, so it is followed by glue with a little less stretchability ($1.66666 \times (999/1000)$) and a little more shrinkability ($1.11111 \times (1000/999)$). A period has $g = 3000$, so it is followed by more stretchability and less shrinkability. Also, the extra space parameter of cmr10 (1.11111pt) is added to the natural size of 3.33333pt, making it 4.44444pt.

The ')' is followed by the same glue, since its \sfcode is zero. This is a result of an additional rule that says that a character with $g = 0$ does not affect f (it is "transparent" to the space factor).

Next comes a surprise. The 'C.' is followed by normal spacing. This is the result of yet another rule that says that $f < 1000$ and $g > 1000$ implies that f is set to 1000. The 'C' sets f to 999, and the period has $g = 3000$, so the two together combine to set f to 1000. This is how an uppercase letter followed by a period is followed by the normal interword space.

The colon has $g = 2000$, so it behaves like a period, but the flexibility is modified less. The semicolon has $g = 1500$, so no extra space is added, but the flexibility is modified somewhat. The right quote does not change f. The comma has $g = 1250$, so no extra space is added and the flexibilities are only changed a little. Finally, the ']' is "transparent" to the space factor (end of experiment).

▶ **Exercise 4.3:** Perform and analyze the experiment:
`\hbox{a,\showthe\spacefactor␣\spacefactor=1.\showthe\spacefactor}`.

To summarize, $f = 1000$ is neutral and does not affect the interword glue. It is appropriate for most characters. Values of $f > 1000$ increase the stretchability and decrease the shrinkability of the glue that follows. They are appropriate for punctuation marks inside a sentence. Values ≥ 2000 behave similarly and also add the extra space (`\fontdimen7`) to the glue. They are appropriate for end-of-sentence punctuation marks. Values < 1000 decrease the stretchability and increase the shrinkability. They may be used in rare cases where a line is overfull and should be shrunk a bit more than usual. Finally, a character with $g = 0$ does not change f.

This mechanism is complicated and the rules are involved because precise control of the interword spacing is crucial for high-quality typesetting, and high quality typesetting is what TEX is all about. This is also the reason why we are not yet finished! There are two more parameters that need to be discussed.

They are the ⟨glue⟩ parameters `\spaceskip` and `\xspaceskip` [76], which can be used to override the rules above. If `\spaceskip` is nonzero, it is used as the interword glue when $f < 2000$ (i.e., between words). Also, its stretch and shrink components are multiplied by $f/1000$ and $1000/f$, respectively. If `\xspaceskip` is nonzero, it is used as the interword glue when $f \geq 2000$ (i.e., between sentences). Its stretch and shrink components are not affected by f.

The `plain` macro `\raggedright` [356] simply sets `\spaceskip=.3333em` and `\xspaceskip=.5em` to obtain rigid interword glues. The same convention is used in some of the special paragraph examples in section 4.7.

▶ **Exercise 4.4:** The definition of `\raggedright` (on [356]) is

```
\def\raggedright{\rightskip=0pt plus2em
 \spaceskip=.3333em\xspaceskip=.5em}
```

Define a similar `\raggedleft`.

The following example illustrates the way `\raggedright` is used. Suppose that each chapter in a certain document should start with the chapter name typeset in large type. Since the name may be long, it may have to be broken on several lines, and it makes sense to use `\raggedright` in such a case. We may define a macro such as:

```
\def\chapter#1{\topglue1.5in
 \vbox{\raggedright\huge #1}
 \vskip.5in}}
```

where `\huge` is the name of the font used. This seems simple but is, in fact, wrong. The reason is that `\raggedright` sets both `\spaceskip` and `\xspaceskip` to values that depend on the current font. The way `\chapter` is defined, when `\raggedright` is expanded, the current font is not `\huge`. The correct definition is thus `\vbox{\huge\raggedright...}`.

▶ **Exercise 4.5:** What else should be added to the `\vbox` above for better appearance?

▶ **Exercise 4.6:** The box '\hbox{a b c}' has two pieces of normal interword glue, each with shrinkability of 1.11111pt, for a total shrinkability of 2.22222pt. We can shrink it by '\hbox spread-2.22222pt{a b c}' without it's being considered overfull. How can we use sfcodes to shrink it even more without making it overfull?

Finally, here is an experiment that demonstrates another, subtle aspect of the interword glue. When \hbox{.␣\␣\␣} is "shown," the result is predictable:

```
\hbox(1.05554+0.0)x13.88889
.\tenrm .
.\glue 4.44444 plus 4.99997 minus 0.37036
.\glue 3.33333 plus 1.66666 minus 1.11111
.\glue 3.33333 plus 1.66666 minus 1.11111
```

The period is followed by an end-of-sentence glue, as usual, but each control space sets f to 1000 and thus becomes a normal interword glue. However, when \hbox{.␣{}␣{}␣} is "shown," the result is surprising.

```
\hbox(1.05554+0.0)x16.11111
.\tenrm .
.\glue 4.44444 plus 4.99997 minus 0.37036
.\glue 4.44444 plus 4.99997 minus 0.37036
.\glue 4.44444 plus 4.99997 minus 0.37036
```

Here the two extra spaces, caused by empty groups, are identical to the glue following the period. Evidently, f is not changed when certain commands are executed in H or RH mode. Even in \hbox{.␣\setbox0=\hbox{x}␣\setbox0=\hbox{x}␣} the spaces are the same as above.

4.5 The Interline Glue

A knowledge of interline spacing [78–80] is necessary in order to control such spacings inside a paragraph.

When characters are set side by side to form a word, their boxes butt together (there are no spaces between characters in a word), which results in a uniform appearance of the words on the page. However, when lines are set vertically, one below the other, there should be some vertical space between them. The space should be as uniform as possible, but in practice, since lines contain characters of different sizes, the spacing may have to change.

TeX uses four parameters [78–80] to control line spacing: 1. \baselineskip—a ⟨glue⟩ parameter that we denote b. Its plain value is 12pt and it determines the normal space between consecutive *baselines*, not between the bottom of a line and the top of the line below. 2. \lineskiplimit—a ⟨dimen⟩ parameter that we denote l. Its plain value is 0pt. 3. \lineskip—a ⟨glue⟩ parameter to be called s. Its plain value is 1pt.

The spacing between the current line and the previous one is determined by the following rules:

    a. Normally, the distance between consecutive baselines is b.

    b. However, if the top of the current line and the bottom of the previous one are closer than l (perhaps because of some high character in the current line or a deep character in the previous one), that distance is set to s.

4. \prevdepth—a ⟨dimen⟩ parameter denoted p. It is normally set to the depth of the previous line, except at the start of a page, where it is set to -1000pt.

As with any parameters, it is easy to change the values of b, l, s, and p.

The two rules above are just an approximation. The actual rules used by TeX to determine line spacing are (assuming the current box has height h): a. If $p = -1000$pt, no interline glue is added. The two line boxes butt together. b. Otherwise, we calculate $x = b - p - h$ and compare. If $x \geq l$, which is the usual case, the interline glue x is appended just above the new box. Otherwise, glue s is used. c. \prevdepth is set to the depth of the new box.

Normally, $b = 12$pt, $h = 8.5$pt, and $p = 3.5$pt. Thus x is (approx.) zero. Also l is zero. However, if p or h is larger than usual (because of tall characters), x will become negative, in which case interline glue of s is used (normally 1pt).

Since b is glue, it may have flexibility, which, in turn, is transferred to x. See the bottom of [78] for a discussion on the flexibility of the interline glue.

There are two macros that control line spacings [352]. \nointerlineskip eliminates spacing between the current line and the next one simply by setting \prevdepth=-1000pt. \offinterlineskip eliminates spacing permanently (or until changed). This can easily be done by setting l to some large value (like \maxdimen [58, 347–348]). Now every pair of lines are closer than l, so the spacing is set to s, which is zero.

The \indent macro creates an indentation (an \hbox to\parindent) in the amount of \parindent (normally 20pt). Notice that \indent is a command whereas \parindent is a parameter. The \noindent command suppresses indentation of the current paragraph. \parskip is glue inserted *before* a new paragraph, not after the previous one.

4.6 Controlling Line Spacing

      The following experiments change the settings of the above-mentioned parameters and should be performed.

■ Setting \baselineskip=12pt, \parskip=0pt, produces text with 12pt between baselines and no extra vertical space between paragraphs. This creates pages with rigid vertical glues (except around displayed math formulas, if present) and does not leave TeX any flexibility in choosing page breaks.

■ \baselineskip=12pt, \parskip=12pt plus 2pt minus 1pt, results in 12pt of space between lines (normal single spacing) and (flexible) 24pt vertical distance between paragraphs.

■ \baselineskip=18pt, \parindent=0pt, generates nonindented, one-and-a-half spaced text.

■ \baselineskip=18pt, \parskip=18pt plus 2pt minus 1pt. Lines will have 1.5 spacing and there will be $18 + 18 = 36$pt of (flexible) space above each paragraph.

■ \baselineskip=24pt plus 2pt minus 1pt, \parskip=0pt \lineskip=2pt
Here the interline glue is double space and has flexibility. Vertical spacing between paragraphs is just the 24pt, and large characters cause a space of 2pt between the top and bottom of consecutive lines.

Since the last line of a paragraph is rarely full, it has to be filled with a glue. This glue should have a natural size of zero (in case the line is full) but infinite stretchability. So there is a parameter \parfillskip with value 0pt plus 1fil. Some people like to set paragraphs that are perfect rectangles, and this can now be achieved by \noindent \parfillskip=0pt (as in this paragraph). Of course, the paragraph should be long enough (I hope that this one is, otherwise garbage would have to be added) to provide TeX with enough flexibility in selecting line breaks. Otherwise, the last line will be an "underfull box." Another example of \parfillskip=0pt appears on page 98, where a centered paragraph with ragged margins is demonstrated.

The previous paragraph was typeset by (see [100] for why the \par is important here):

```
{\noindent\parfillskip=0pt Since the last line ...
... ragged margins is demonstrated. \par}
```

4.7 Different Paragraph Shapes

Two glue parameters, \leftskip and \rightskip [100], are inserted on the left and right of every line of the paragraph. They can be used to create narrow paragraphs, to indent paragraphs left or right, and to achieve other effects. This paragraph is a typical example, created by:

```
{\noindent\leftskip=.1in \rightskip=.25in
 Two glue parameters, ... created by:
\par}
```

This one is even more noticeable. In fact, we had to increase the tolerance in order to avoid "overfull boxes." In general, TeX permits easy control of paragraph shapes, which is perhaps one of its main advantages. Word processors also allow such effects but not with such precise control.

```
{\leftskip=.2in\rightskip=.3in\tolerance=1500
\noindent This one is ...
... but not with such precise control. \par}
```

Long quotations are traditionally done in a paragraph slightly narrower than normal. This paragraph was set with the \narrower macro, which works by advancing both \leftskip and \rightskip by \parindent. Using \advance instead of just setting the parameters has the advantage that such a paragraph can be used inside another—indented—one.

```
{\smallskip\narrower \noindent Long quotations are ...
... inside another---indented---one. \par}
```

Observe the settings used in this paragraph. It is indented on both sides with the right margin made flexible (this will reduce the number of necessary hyphenations). Then the two parameters \spaceskip and \xspaceskip [76, 356] are made rigid. The effect is to have fixed space between words and a ragged right margin. Here and there hyphenation is necessary but the total effect is similar to a typewritten paragraph. The same idea can be used for ragged left.

```
{\noindent\leftskip=.5in
 \rightskip=.5in plus 6em \spaceskip=.3333em
 \xspaceskip=.5em Observe the settings used ...
... for ragged left.\par}
```

> Now we have a centered paragraph. The settings in this case
> call for identical, flexible indentations on both sides. This,
> of course, centers the entire paragraph, generates ragged
> left and right margins, and eliminates most hyphenations.
> In addition, the interword glue was made rigid. It is easy
> to define a \raggedcenter macro to do just that. The
> reader should also experiment by removing the \noindent
> and the \parfillskip=0pt and comparing the results.

```
{\noindent\rightskip=.5in plus 6em \leftskip=\rightskip
\spaceskip=.3333em \xspaceskip=.5em \parfillskip=0pt
Now we have a centered ... comparing the results.\par}
```

▸ **Exercise 4.7:** Analyze the behavior of \narrower combined with \raggedright.

4.8 The Paragraph Indentation

When starting a paragraph, the first item is the indentation, which is an \hbox to\parindent. Note that the indentation is a box, and not a skip. Once it has been inserted, it can be removed by means of \lastbox, rather than by \unskip.

4.9 The Hanging Indentation

Many times only part of the paragraph should be indented. A common example is indenting the first two or three lines on the left, to reserve room for an illustration. TeX provides two parameters [102] that make hanging indentation easy. '\hangafter= n' specifies how many lines are to be indented (for negative n, indentation is applied to the first n lines; for positive n, it is applied to lines $n+1$, $n+2$, ...). '\hangindent= x' specifies a dimension x for the amount of indentation. A positive x means left indent and a negative one, right indent. This paragraph was typeset by:

```
\hangafter=-2 \hangindent=40pt \noindent Many times only ...
... was typeset by:
```

An important use of hanging indentation is to leave room in the paragraph for a small illustration, or to start it with a large letter, perhaps two or three times taller than normal. It is easy to leave that room at any of the four corners of the paragraph, but what if we need to reserve a small area elsewhere in the paragraph? The concept of *concatenating paragraphs*, discussed in section 4.12, can be used to leave room on the left or the right sides, but not at the top or bottom. The important command \parshape (section 4.16) can be used to create a rectangular hole anywhere inside a paragraph. This paragraph was typeset by:

```
\hangafter=2 \hangindent=-40pt \noindent An important use ...
```

```
... paragraph was typset by:
```

It is important to realize that the values of `\hangafter` and `\hangindent` are reset at the end of a paragraph. They apply to one paragraph only! To apply them to several paragraphs, `\everypar` should be used.

1. Many times, when the document contains a list of items, the writer wants each item to be typeset as a paragraph where all the lines are indented, and the first line has a short label.

2. A macro, `\item#1`, can be developed to achieve this. The first step is `\par`, to make sure that an expansion of the macro terminates the current paragraph.

3. The next step is to say `\hangafter=1 \hangindent=\parindent`. This guarantees that all the lines, except the first one, will be indented by `\parindent`.

4. Now should come an `\indent`, or a `\leavevmode`. This starts the new paragraph and places the indentation on the first line.

5. The label is now placed in an `\llap{#1\enspace}`. The label overlaps the paragraph indentation on its left and is separated from the first letter of the paragraph with an `\enspace`.

6. The last thing necessary is an `\ignorespaces`, to suppress any spaces between the macro expansion and the first letter of the paragraph.

The result is

```
\def\item#1{\par\hangafter=1\hangindent=\parindent%
\leavevmode\llap{#1\enspace}\ignorespaces}
```

which should be compared to the actual definition of the plain macro `\item` on [355].

▶ **Exercise 4.8:** Write a macro `\rItem#1` that places a short label on the right-hand side of a paragraph.

The same problem can be solved by a macro such as:
```
\def\listItem#1{\leavevmode\rlap{\kern\hsize\kern-\parindent#1}}
```
which is similar to the LaTeX macro `\@hangfrom`, and is easy to read.

The setting `\hangindent=3in \hangafter=0` will indent all the lines of the following paragraph by 3 inches on the left. The setting `\leftskip=3in` will do the same thing, but there are two differences:

■ The two hanging indentation parameters are reset at the end of the paragraph, while `\leftskip` has to be reset explicitly.

■ Hanging indentation is done by changing `\hsize` for the indented lines. In contrast, `\leftskip` and `\rightskip` just place glue on the left (or right) of each line, and do not change `\hsize`. This means that a displayed math formula will be placed differently in the two cases. In the former case, it will be centered with respect to the new `\hsize`; in the latter, it will be centered with respect to the original `\hsize`.

Example of hanging indentation

Macro `\specPar#1` typesets its argument as a heading, centered, and in boldface, above the following paragraph. It uses hanging indentation to indent the first line of the paragraph by an amount equal to the space on the left of the heading.

```
\def\specPar#1 {\par\setbox0=\hbox{\bf#1}
\hangindent=\hsize \advance\hangindent by-\wd0
  \divide\hangindent by2 \hangafter=-1
\centerline{\box0}\noindent}
```

The macro is straightforward, except for the following points:

1. It uses a space as a parameter delimiter. This absorbs the end-of-line following '#1' which otherwise would have been added to the indentation of the first paragraph line.

2. The expansion \specPar{...} should be followed by the text of the paragraph, either on the same line or on the following one. A blank line between them would reset the values of \hangindent and \hangafter.

" Here is a short passage set in small type to demo hanging indentation on both ends of a paragraph. This is done by splitting a paragraph in two, indenting the first part at the top and the second part at the bottom, then adding the big quotes by means of llap and rlap. The rest is just details of moving boxes and getting the spacing just right. **"**
The effect shown here is used a lot in books where high-quality typesetting is important. It has been achieved by setting certain paragraph parameters to special values, and by using boxes. The spaces for two big quotes are reserved by means of hanging indentation. The text in small type is actually made of two paragraphs, one indented at the top left and the other at the bottom right. The paragraphs thus look like one paragraph and are placed in box0. Eight lines of hanging indentation are used in this paragraph, and box0 is placed in this space simply by resetting its three dimensions to zero. The whole thing was done by:

```
\setbox0=\vbox{\hsize=2in\tolerance=9000\hbadness=10000\parskip=0pt
\parfillskip=0pt\parindent=0pt\baselineskip=9pt
\font\big=Helvetica at 50pt \font\small=cmr7 \small
\hangindent=24pt \hangafter=-2
\leavevmode\smash{\lower32pt\llap{\big''\small\quad}}%
Here is a short passage ... indenting the first part at the\par

\hangindent=-24pt \hangafter=3
top and the ... spacing just right.%
\smash{\lower24pt\rlap{\small\quad\big''}}}}
\wd0=0pt\ht0=0pt\dp0=0pt\box0
\nointerlineskip
\hangindent=2.2in \hangafter=-8
{\hsize=4.2in\noindent The effect shown here ... was done by:\par}
```

4.10 \vadjust

This command [105] is useful for controlling the shape of a paragraph. Its form is \vadjust{<vertical mode material>}. When TeX encounters this command at a certain point in a paragraph, it marks that point, saves the vertical material, and proceeds with setting the paragraph. When all line breaks have been determined, the lines are appended to the main vertical list, each as an \hbox. After the line with the marked point is appended, the vertical material from the \vadjust is also appended, followed by the rest of the paragraph. This command is commonly used to open up a space in the paragraph for an illustration e.g., \vadjust{\vskip 3in}. However, if at this point there are not 3 available inches on the page, TeX will scream 'overfull page'. The command \vadjust{\vfill\eject} will eject a page after the current line. Now look carefully at this paragraph. There is a \vadjust hidden in it. Can you detect its effect?

Another use of \vadjust is to place material on the margins of the page, from within a paragraph. Macro \rNote#1#2 below places a short, two-line note on the right margin. It is expanded at any point in the paragraph, and it uses a \vadjust short to temporarily escape from the paragraph. The short note here was created by note \rNote{short}{note} placed after '...right margin.'.

```
\def\rNote#1#2{\vadjust{\rlap{\kern\hsize\thinspace\vbox to0pt{
\sevenrm\hbox{#1}\hbox{#2}\vss}}}}
```

The height of the \vbox is explicitly set to zero. Its depth is zero because its bottom component is glue. As a result, the box does not affect the interline spacing in the paragraph.

▶ **Exercise 4.9:** Create a similar macro \lNote#1#2 for placing a two-line note on the left margin.

Another useful example of \vadjust is boxing a line of text. Macro \boxit, (page 75) can be used to surround any text with four rules, creating "boxed" material. How can we use \boxit to "box" one line in a paragraph? The problem is that we don't know the line breaks in advance, so we cannot tell \boxit what material to box. The solution is to use \vadjust to create an empty box as wide and high as a line of text, then superimpose it on the line. We first say: '\boxit{\line{\hfil}}0', then place this in a \vbox to0pt, to suppress its height (so it doesn't affect the placement of the lines of text), and finally use \vadjust to place it around the desired line. The box in this paragraph was created by saying:
\vadjust{\vbox to0pt{\vss\boxit{\line{\hfil}}0}}
after the word "placement."

See the Index for other examples of the use of \vadjust.

The author's personal experience shows that \vadjust is a potentially useful command. Many typesetting problems that initially appear formidable, or even hopeless, can easily be solved with a simple use of \vadjust.

4.11 \obeylines

This is a useful command that works [352] by changing the carriage-return character '^^M' into \par. As a result, each source line becomes a paragraph and is indented by \parindent. To eliminate this indentation, simply set \parindent=0pt. Compare the two rhymes below:

Violets are red
    Roses are blue
I use parindent
    Which, you know, is just glue

Violets are red
  Roses are blue
I'm still in bed
  And where, pray, are you?

In practice, \obeylines should be used inside a group, since we only want to use it temporarily. However, it is not enough to use a pair of braces to limit the effect of \obeylines. TeX must be under the effect of this command when it typesets the paragraph, and thus the paragraph has to be terminated before TeX runs into the right brace (remember, a right brace does not terminate a paragraph). This is why it is necessary to say \par, \smallskip, or something similar, just before (or just after) the right brace.

The control sequence \obeyspaces can also be extremely useful in certain situations. When it is in effect, consecutive spaces, which are normally ignored, are typeset. Thus saying

```
{\tt␣A%
␣␣first␣␣␣line\par
␣␣second␣␣␣line\par}
```

or, equivalently,

```
{\tt␣A\obeylines%
␣␣first␣␣␣line
␣␣second␣␣␣line\par}
```

results in

```
Afirst line
second line
```

Note how the three spaces between first and line were truncated to one space (and the same thing for the second line. These spaces can be preserved by saying \obeyspaces.

Also note that the two spaces at the start of both lines have been ignored. This happens because they were read in V mode, where all spaces are ignored. The \obeylines converts the end of each line to a \par, which changes the mode to V. One solution is to place a tilde at the start of each line, to change the mode to H. This is sometimes cumbersome, and a better solution is to change the definition of \obeylines to convert the end-of-line to a \par and also to switch back to H mode, to start the next paragraph. The original definition of \obeylines is \let^^M=\par, and what is needed is '\let\^^M=\par\noindent'. This can be done by

```
\def\parnoindent{\par\noindent}
\let^^M=\parnoindent
```

To see how this works, the reader should try a test of the form:

```
\def\parnoindent{\par\noindent}
\noindent A
{\tt\obeyspaces\obeylines%
\let^^M=\parnoindent
 first  line
   second  line\par}
```

4.12 Concatenating Paragraphs

Consider the following problem. We want to concatenate two consecutive paragraphs, such that TeX will treat them as one paragraph, but in the final document they will look like two distinct ones. There will be the normal \parskip glue between them, and the second one will start with the usual indentation.

On reading this, readers tend to ask *why* before they ask *how*. This effect can be useful if we want the two paragraphs to have a special shape. We may, for example, want to leave room for a diagram using hanging indentation, but the diagram may be taller than the current paragraph; or we may want to use \parshape to shape the two paragraphs as one unit.

Whatever the reason, the way to achieve this effect is to eliminate the usual \par (or the blank line) between the paragraphs and, instead, place commands to do the following:

- Fill up the last line of the current paragraph with flexible glue (\hfil).
- Place a \parskip glue below that line (\vadjust{\vskip\parskip}).
- Break the line (\break).
- Start the next line with the normal paragraph indentation (\indent).

The result \hfil\vadjust{\vskip\parskip}\break\indent (see [ex. 14.23]) is placed between the paragraphs. This method should be used sparingly since TeX handles concatenated paragraphs as one paragraph, which may exceed its memory capacity.

4.13 \par

This is a primitive that signifies the *end* of the current paragraph, not the beginning of the next one. It can be considered the opposite of \leavevmode, since it causes TeX to leave H mode. Normally there is no need to worry about it. The blank line between paragraphs is automatically converted to a \par, and \bye is a macro defined as \par\vfill\supereject, so it terminates the last paragraph.

However, in advanced applications, \par can be redefined to achieve special effects, and it is important to keep in mind that it terminates the current paragraph. Before the next one starts, all kinds of vertical commands (such as \vskip, \hrule) may be executed.

To redefine \par, just say \def\par{...}. However, the original meaning of \par should normally be saved, so it can be restored if necessary. This can be done by doing the whole thing in a group, or by something like:

```
\let\Par=\par
\def\par{...}
```

```
.
\let\par=\Par
```

What if we want the new definition of `\par` to be an extension of the original one, that is, to terminate the current paragraph, and then do other things? We cannot write `\def\par{\par...}`, so there is a plain macro, `\endgraf` [286], whose meaning is identical to `\par`. We can now say

```
\let\Par=\par
\def\par{\endgraf...}
.
.
\let\par=\Par
```

▶ **Exercise 4.10:** If we want something done at the end of every paragraph, we can redefine `\par`. How can we have something done at the start of every paragraph?

Here is an example of `\endgraf`. Suppose we want to start the first line of certain paragraphs in the left margin (*outdent* the paragraph). This is easy to do with a macro such as

```
\def\outdent#1 {\leftskip=-#1pt \hangindent=#1pt \hangafter=1%
\noindent}
```

however, parameter `\leftskip` will have to be reset after the paragraph (`\hangindent` and `\hangafter` are automatically reset). This can be done by changing the definition of `\outdent` to:

```
\def\outdent#1 {\bgroup\leftskip=-#1pt \hangindent=#1pt \hangafter=1
\def\par{\endgraf\egroup}\noindent}
```

Macro `\outdent` starts a group, and `\par` terminates it. Note that `\par` gets redefined when `\outdent` is expanded, but at that point the group has already started. The redefinition of `\par` is thus local to the group.

4.14 Example

Here is an application of `\everypar` that also involves redefining `\par` and demonstrates the nesting of modes. We want to place each paragraph, as it is being created, in a box (perhaps for later typesetting or analysis). This is easy to do by placing '`\setbox0=\vbox{`' before, and '`}`' after, each paragraph. To do it automatically, we try

```
\everypar={\setbox0=\vbox\bgroup
\def\par{\egroup\endgraf}}
```

This simple definition does not work, since it turns out to be recursive without limit. TEX executes the tokens in `\everypar` when it starts the paragraph, which causes it to open `\box0`. Inside the box, it starts the paragraph again, which causes it to execute `\everypar` again, which causes it to open `\box0` again, and so forth. The details are shown below. The solution is to disable the special definition of `\everypar` inside the box by

```
\everypar={\setbox0=\vbox\bgroup\everypar={}%
\def\par{\egroup\endgraf}}
```

Advanced readers can gain insight into the operations of TeX by carefully following the steps below. Suppose that TeX is in v mode and it reads the letter 'X'. It realizes that a new paragarph has started, and it goes through the following steps:

■ It backspaces over the 'X' (to read it again later).

■ It places \parskip glue below the previous paragraph.

■ It switches to H mode and starts a horizontal list that should eventually become the paragraph. Initially, the paragraph indentation is placed in the list.

■ It executes the tokens in \everypar [105]. This will open \box0 and, consequently, switch the mode from H to IV.

■ It reads the 'X' again. This will switch the mode to H inside the box (this H mode is now nested in IV, which itself is nested in another H mode). Another paragraph indentation will be placed in the box, followed by \everypar (the new, empty one) and by the text of the paragraph. TeX has started two paragraphs, one in the normal way (to be eventually typeset) and the other in \box0. Eventually both paragraphs would have to be terminated.

■ When a paragraph terminator is found (a blank line, a \par, or vertical mode material), TeX executes the definition of \par, which involves two steps.

■ 1. The \egroup is executed. This involves two substeps:

■ 1.1. The paragraph in \box0 is terminated by placing \unskip, \penalty10000, and \hskip\parfillskip [286] in the box. TeX determines the line breaks for the paragraph, places \leftskip and \rightskip glues in each line [100], and switches from the current H mode to the IV mode enclosing it. The box is still open.

■ 1.2. The box is closed, and the mode is switched from IV to the H mode enclosing it. TeX is now back in the original paragraph (the one that has only got the indentation so far).

■ 2. The \endgraf is executed. This terminates the original paragraph. TeX places an \unskip, a \penalty10000, and an \hskip\parfillskip in the horizontal list. TeX does the line breaking, and adds \leftskip and \rightskip to every line (there is only one line). The mode is switched to v.

The result is that each paragraph is built in \box0 but is also typeset as one blank line containing the indentation on the left and the flexible glues on the left and right. The normal vertical glues are placed between paragraphs, and a blank page is eventually typeset. The nesting and switching of modes can be followed with a \tracingcommands=1. The blank paragraphs can be seen by saying \output{\showbox255}. A typical layout is

```
> \box255=
\vbox(643.20255+0.0)x469.75499, glue set 621.20251fill
.\glue(\topskip) 10.0
.\hbox(0.0+0.0)x469.75499, glue set 449.75497fil
..\glue(\leftskip) 0.0
..\hbox(0.0+0.0)x20.0
..\penalty 10000
```

```
..\glue(\parfillskip) 0.0 plus 1.0fil
..\glue(\rightskip) 0.0
.\glue(\parskip) 0.0 plus 1.0
.\glue(\baselineskip) 12.0
.\hbox(0.0+0.0)x469.75499, glue set 449.75497fil
..\glue(\leftskip) 0.0
..\hbox(0.0+0.0)x20.0
..\penalty 10000
..\glue(\parfillskip) 0.0 plus 1.0fil
..\glue(\rightskip) 0.0
.\glue 0.0 plus 1.0fill
```

(The \leftskip is normally not shown if it is zero.) If we also want to typeset the text in the normal way, we can say '\box0' between paragraphs.

4.15 \prevgraf

This primitive [103] contains the number of the current paragraph line. It is set to zero at the beginning of each paragraph and is incremented each time a line is broken and is appended to the MVL. Its value can be shown by either \the\prevgraf, \number\prevgraf or \showthe\prevgraf. However, this can only be done in v mode, between paragraphs. Placing \the\prevgraf inside a paragraph will typeset a zero, since \prevgraf is only updated after the paragraph has been fully read and is being typeset.

The following commands illustrate how \prevgraf can be used. They also show that \par signifies the end of a paragraph, rather than the start.

```
\let\Par=\par
\newcount\linesofar
\def\par{\endgraf \linesofar=\prevgraf}
\everypar={\prevgraf=\linesofar}
```

The new definition of \par does two things:
■ \endgraf terminates the current paragraph and causes it to be typeset. This increments \prevgraf until it contains the total number of lines in the paragraph.
■ The assignment now saves this number in \linesofar.

The \everypar is executed at the beginning of the next paragraph (after TEX has set \prevgraf to zero), and it sets \prevgraf to \linesofar, the total number of lines in the previous paragraph(s).

All this is possible because \par signifies the end of a paragraph, rather than the beginning.

▶ **Exercise 4.11:** What's a good application for counting the number of lines in several consecutive paragraphs?

4.16 \parshape

This is the most general command that TEX provides to control the shape of a paragraph. Its general form [101] is \parshape= n $i_1\, l_1\, i_2\, l_2 \ldots i_n\, l_n$, and it creates a paragraph whose first n lines have lengths l_1, l_2, $\ldots$, l_n and are indented by i_1, i_2, $\ldots$, i_n from the left margin. If the paragraph has fewer than n lines, the extra specs will be ignored. If it has more than n lines, then $i_n\, l_n$ will be used repeatedly for the extra lines. The following properties of \parshape are important:

■ It only affects the next paragraph. This means that the \parshape command must be *immediately* followed by the next paragraph's text. An intervening blank line will nullify the \parshape.

■ TEX uses the current value of \prevgraf as a pointer to the next $i_k\, l_k$ pair of specs. Initially, \prevgraf=0, so $i_1\, l_1$ are used to typeset the first line. After that line, \prevgraf=1, so $i_2\, l_2$ are used for the second line, and so on. If \prevgraf is artificially changed to 5 at the start of the paragraph, TEX will typeset the first line using pair $i_6\, l_6$.

■ The effect of a \parshape can be nullified by placing \parshape=0 anywhere in the paragraph. As an example, if we apply \parshape to every paragraph (by, e.g., \everypar{\parshape...}), we can cancel this for a specific paragraph by placing \parshape=0 at its end.

We start with the simple example

```
\parshape 6 0pt\hsize 0pt.75\hsize 0pt.75\hsize 0pt.75\hsize
  0pt.75\hsize 0pt\hsize
```

This will typeset a paragraph with a normal first line, followed by four lines of length .75\hsize, followed by any number of normal lines. Such a paragraph has a four-line 'hole' in the right margin, perhaps for pasting in a picture later.

This simple example will not work if the current paragraph is too short (if it has less than five lines), and two approaches are shown below that make it possible to apply a single \parshape to two or more paragraphs.

■ The paragraphs are concatenated and one \parshape is applied to the total text. To do this, just place \hfil\vadjust{\vskip\parskip}\break\indent between the paragraphs. This has the disadvantage of concatenation, namely that the result is a long paragraph and may be too large for TEX to handle.

■ Keep the paragraphs separate, and use \linesofar, as above, to count the total number of lines in all the previous paragraphs. This way, TEX will always pick up the right \parshape specs.

The problem with this approach is that \parshape only "lasts" for one paragraph, so we have to make the \par temporarily invisible. Technically speaking, \par should be made invisible to the "mouth" and visible to the "stomach." This way, the mouth will read all the special paragraphs as one paragraph and will send the text, again as one paragraph, to the stomach *with the* \parshape *specifications*. This magical result is achieved by:

```
\newcount\linesofar
\def\par{{\endgraf \global\linesofar=\prevgraf}}
\everypar={\prevgraf=\linesofar}
```

To understand exactly how this works, one should be very familiar with the internal workings of TEX. It is thus no wonder that this approach was suggested by the author of TEX himself (Ref. 14). Another advantage of this approach is that \par does not have to be saved and restored.

The \parshape command can be very useful because its specifications can be created "on the fly," by a TEX program; they do not have to be entered manually. This is illustrated in [ex. 14.18] and was also used by A. Hocnig (Rcf. 15) to reshape a paragraph by creating a hole in it, for pasting a diagram later.

> Intuitively, a paragraph is a sequence of input lines that's ended by a blank line, by a \par command, or by an intrinsically vertical command....
>
> — *Paul Abrahams, TEX for the Impatient*

> Shall we buy treason, and indent with fears?
>
> — *William Shakespeare, Henry IV-1*

5. Macros

A command starting with a '\' is called a *control sequence*. If it consists of letters only (catcode 11), it is called a *control word*; if it consists of a nonletter (any catcode $\neq$ 11), it is called a *control symbol*. New control sequences can be defined by the user by means of \def and are called *macros*. A macro is thus either a control word or a control symbol defined by the user.

A simple example is \def\vec{$A_1,A_2,\ldots,A_n$}. This defines a macro whose name is \vec and whose value (or *replacement text*) is the string of tokens '$A_1,A_2,\ldots,A_n$'. When the macro is defined, the replacement text is simply saved in a table, for future use. Note that the replacement text should be balanced with respect to braces [275].

To use a macro, simply type its name. The macro is then *expanded*, which means that a copy of the replacement text is brought from the table, and is executed like any other string of tokens. Thus saying \vec, generates a copy of the token string '$A_1,A_2,\ldots,A_n$', and executes it.

The first step toward a full understanding of macros is to understand the difference between defining a macro and expanding it. Recall that a token is either a single character or a control sequence. When a macro is defined using \def, its definition is converted to tokens and is saved in memory as a list of tokens. That list is the replacement text of the macro. Nothing gets executed at definition time. When the macro is expanded, the tokens of the replacement text get executed; character tokens are typeset, macros are expanded, and primitives are executed.

A simple example is '\def\abc{1 \x a}'. Its replacement text is the 5-token list 1_{12}, ⊔10, \x, ⊔10, a_{11}. A better example is '\def\pq{\catcode'\\=12 \every-par{*}}'. This macro converts the '\' to a regular character (a somewhat dangerous thing to do) and sets \everypar to typeset an asterisk at the start of every paragraph. A common question is: Since the '\' is now a regular character, how does TEX recognize \everypar as a control sequence? The answer is—at definition time the \catcode and the \everypar are recognized as control sequences and are stored, together with the rest of the definition, as the replacement text of \pq. Nothing gets executed. Specifically, the '\catcode'\\=12' is not executed. When the macro is expanded, the tokens of the replacement text are executed and the catcode of '\' is actually changed. By that time, however, the \everypar has long been recognized as a control sequence, so it can be executed.

▶ **Exercise 5.1:** What are the tokens of the replacement text of \pq?

A macro, therefore, can save typing time. The biggest advantage of macros, however, is the flexibility made possible by the use of *parameters*. When a macro depends on parameters, each expansion may be different. Macro \vec above can be extended by \def\vec#1#2{$#1_1,#1_2,\ldots,#1_{#2}$}. It has two parameters, the first indicates the math variable (such as A, x) and the second, the last index of the vector (n, m, 100, etc.). Thus the expansion '\vec An' creates $A_1, A_2, \ldots, A_n$, and the expansion '\vec B{10}' creates $B_1, B_2, \ldots, B_{10}$. Note that if the argument is longer than one token, it must be enclosed in braces (but see section 5.1 for exceptions).

A word about terminology. When a macro \vec#1#2{...} is defined, it is customary to call the quantities #1, #2 *formal parameters* (or, in short, the parameters) of the macro. When the macro is expanded, as in '\vec B{10}', the quantities B and 10 are called the *actual arguments* (or, in short, the arguments) of the macro.

Also, up until now, a token meant either a single character or a control sequence. When using macros, however, the parameters (#1, #2) are also tokens.

▶ **Exercise 5.2:** What are the tokens in \def\abc#1{#1}?

▶ **Exercise 5.3:** What is the result of '\vec A{}'?

▶ **Exercise 5.4:** A two-parameter macro, such as \vec, should be expanded with two arguments. How can we create a (wrong) expansion with just one argument?

5.1 Delimiting Parameters

Defining a macro \def\test#1#2#3{\message{'#1',⊔'#2',⊔'#3';}}, the expansion '\test⊔abc⊔def⊔ghk' will display "'a',⊔'b',⊔'c';" since each parameter is simply the next token in the input stream. The text 'def⊔ghk' will simply be typeset. If we want the parameters to be 'abc', 'def', 'ghk', we either have to enclose each in braces {abc}{def} {ghk}, or use *delimiters*. The definition \def\test#1⊔#2⊔#3⊔{...} requires each argument to be followed by a space. Thus the expansion '\test⊔abc⊔def⊔ghk⊔' will consider 'abc' to be the first argument.

The concept of delimiters has already been discussed, with simple examples, in Chapter 2; it is also mentioned on [202]. Here is a sophisticated example of their use.

```
\def\ifempty#1\\{\ifemp#1\endA\endB}
\def\ifemp#1#2\endB{\ifx#1\endA}
```

Macro `\ifempty` above uses '`\\`' to delimit its single parameter. It expands `\ifemp` with a string equal to its argument followed by the two tokens `\endA\endB`. Macro `\ifemp` accepts two parameters. The first is undelimited (a single token or a group), and the second is delimited by `\endB`. If the argument of `\ifempty` is 'abc', then `\ifemp` will be expanded with the string `abc\endA\endB`. The first argument of `\ifemp` will thus be 'a', and the second, 'bc\endA'. The `\ifx` comparison will fail in this case. However, when `\ifempty` is expanded with an empty argument, it will expand `\ifemp` with the string `\endA\endB`. The first argument of `\ifemp` will thus be `\endA`, and the second one, empty. The comparison will be a success since it will compare `\endA` to itself. Note that `\endA` does not have to be defined.

An interesting feature of `\ifempty` is that it does not terminate the `\ifx`. Thus it must be used in a context such as

```
\ifempty\\ \message{empty}\else\message{not empty}\fi
```

▶ **Exercise 5.5:** What is the result of '`\ifempty\endA\\`'?

5.2 Macro Names

The name of a macro must follow the rules for control sequence names, namely just letters or a single nonletter. However, the serious reader should consider the following:

Once a macro is defined by `\def\abc*{...}`, each expansion must follow the name `\abc` with the delimiter '`*`'. Anything else would trigger the message '`! Use of \abc doesn't match its definition.`' This suggests that macros such as `\mac1`, `\mac2` can be defined, in contradiction of the basic rule (either just letters or a single nonletter).

When this scheme is used, the user should be aware of the following:

1. An expansion such as `\abc *` is considered valid.

2. The case `\expandafter\abc\xyz` would be okay if the definition of `\xyz` starts with an '`*`'.

3. Convenient names such as `\mac1`, `\mac2` still cannot be used since the definition of `\mac2` would be considered a redefinition of `\mac` (with a different parameter delimiter).

4. A definition such as `\def\abc${...}` may interfere with math mode if used carelessly, such as in '`\abc $x=...$`'. (The correct use is, of course, the phrase '`\abc$ $x=...$`', but note that even '`\abc $$x=...$`' is okay.)

5.3 Spurious Spaces

Macros tend to create spurious spaces. Consider the innocent definition

```
\def\a{
abc}
```

When used in H mode, such as in '`A\a B`', it produces 'A abcB'. The reason for the extra space is the end-of-line of the first line of the macro. It gets converted into a space. Changing the definition to

```
\def\a{\dimen0=...
abc}
```

or

```
\def\a{%
abc}
```

or, better still, `\def\a{abc}` eliminates the unwanted space. Macro writers should, therefore, give special consideration to end-of-lines and to spurious spaces in general. The following rules summarize the behavior of TeX regarding spaces.

1. In the vertical modes nothing is typeset, so all spaces are ignored.

In the horizontal modes, the following spaces are ignored:

2. Spaces at the start of a line. (Exception! In the rare case where `\obeyspaces` is in effect **and** the mode is H, spaces at the start of a line are not ignored and are typeset as a single space.)

3. Spaces following a control word. Thus in '`\a b`' the space between the a and the b is ignored.

In the horizontal modes, the following are typeset as spurious spaces:

4. Spaces following a macro parameter token. Thus when '`\def\abc#1{X#1␣Y}`' is expanded, there will be a space between the argument and the 'Y'.

5. Spaces following a control symbol. If `\1` is a defined control symbol, then the sequence '`a\1 b`' creates a spurious space, but '`a\1b`' does not.

(Rules 3–5 can be summarized by saying that a space that's not strictly necessary as a separator becomes spurious.)

6. A space following a brace (left or right) inside a macro. Thus the macro line '`dimen0=0pt␣`' does not create a spurious space, but the line '`{dimen0=0pt}␣`' does. Similarly, the line '`\dimen0=0pt␣\dimen1=1pt␣`' does not create any spurious spaces, the line '`{\dimen0=0pt}␣{\dimen1=1pt}␣`' creates two spaces, and the line '`{␣\dimen0=0pt}␣{\dimen1=1pt}␣`' creates three.

7. An end-of-line following text. A macro line of the form '`...a`' creates a spurious space.

The following experiments should shed more light on this sometimes baffling topic.

```
\def\spurspace{
a
b
c}
```

When tested by **A\spurspace B**, it typesets 'A a b cB'. The first three lines create spurious spaces. Suppressing the spaces and adding an assignment, such as in

```
\def\spurspace{%
\dimen0=0pt
a%
b%
c}
```

shows that the end-of-line of the assignment is not treated as a spurious space. Placing the assignment in braces, however, shows that the right brace following the assignment does create a spurious apace.

```
\def\spurspace{%
{\dimen0=0pt}
a%
b%
c}
```

The same is true for

```
\def\spurspace{%
\def\xxx{x}
}
```

The nested definition of \xxx creates a spurious space because of the right brace. Changing it to '\def\xxx{x}%' suppresses that space.

Note that any spurious spaces created by a macro are appended to the current list. Normally, they are just typeset, but in certain cases they might seriously interfere with the operation of macros. Imagine, for example, a macro using the \lastbox command in a loop, to isolate a sequence of adjacent boxes. A spurious space created by this macro will be appended to the current list, following the boxes being isolated. As a result, the next \lastbox would see the space and return nothing. The spurious space would be more than just a nuisance in this case. It would cause the macro to fail.

It is recommended to expand every new macro in H mode and to look for spurious spaces. An important tool is \tracingcommands. When set to 1, it enters a note in the log file for each space (intentional or spurious) found.

5.4 Space Following a Number

Consider the following: \newcount\ctst \def\digeet{7 days}. It seems that the commands \ctst=1\digeet should set \ctst to 1 and typeset '7 days'. Instead they typeset 'days' and set \ctst to 17. The culprit is the absence of a space between the '1' and the '\'. It causes macro \digeet to be expanded before the assignment is completed. Thus the macro is first used to contribute the digit 7 to the assignment, then to do other things. If this is not what we want, we should write it as \ctst=1␣\digeet. The general rule is that a number should be terminated by a space.

The main idea is that macros may create digits that will be concatenated with other digits to form a number. The two examples below are very similar, differing by one space only.

```
\newcount\temp
\advance\temp by 2\ifnum...
\advance\temp by 2 \ifnum...
```

Yet they may behave very differently! In the second example, the \advance and the \ifnum are separated. In the first, \temp could be advanced by 2 or by something

else, depending on what the \ifnum does. If the \ifnum evaluates to digits, they
will become part of the number that starts with the 2, and \temp will be advanced
by that number.

An example of such an if is \ifnum\a<\b 3 \else 45 \fi. Such an if will
advance \temp by either 23 or 245. On the other hand, a statement such as the
following: \ifnum\a<\b \message{ok}\fi, does not generate any digits and, consequently, nothing will be concatenated to the '2'.

If the replacement text of a macro ends with a number, the number should be
terminated with either a space or a \relax. So \def\x{\advance\temp by 2 } is
okay, as is \def\x{\advance\temp by 2\relax}, but \def\x{\advance\temp by
2} is wrong.

▶ **Exercise 5.6:** What about a macro such as
\def\step{\advance\temp by\ifodd\pageno 1\else2\fi}? It terminates with
a \fi, not with a number. Does it still need a space or a \relax?

To summarize, when TeX encounters digits, it expects the number to be followed by an unambiguous terminator, such as text, a punctuation mark, or a space.
If the digits are followed by a command, it is immediately executed, in an attempt
to find more digits. If the command results in digits, they are appended to the ones
already found; otherwise, the number is considered terminated (but the command
has already been executed).

5.5 Reading the Source File

Imagine a TeX source file containing '...\abc d...'. When TeX recognizes
\abc as a control sequence, it converts it to a token, moves it to the stomach, and
executes it there. The point is that, during this time, TeX does not read ahead in
the source file. It is still positioned in front of the 'd'. Only when \abc has been
fully executed does TeX read the 'd'. An important special case occurs when \abc
requires an argument. In such a case, TeX will read the 'd' (and, possibly, some
more characters) as the argument, **before executing** \abc. A macro writer should
always keep in mind this aspect of TeX's behavior, since it can seriously affect the
results produced by macros. (See also the discussion on page 25.)

Here is an interesting example. Macro \test#1 below tries to change the
catcodes of all dollar signs in its argument. The simple definition
'\def\test#1{\catcode'\$=4..}' is, surprisingly, wrong. An expansion such as
'\test{123$4}' would read the argument '123$4' and assign catcodes to the five
tokens **before** \test is executed. The '$' would thus get its normal catcode of 3.

A better way to define \test is

```
\def\test{\catcode'\$=4 \moretest}
\def\moretest#1{...}
```

Macro \test is defined without any parameters. When \test is expanded,
such as in '\test{123$4}', TeX remains positioned in front of the left brace. The
expansion consists of two steps: (1) change the catcode of '$'; (2) expand \moretest.
Macro \moretest is defined with a parameter, so the argument '123$4' is read
(and catcodes assigned to the five tokens) before step (2), that is, after the catcode

change. The '\$' in the argument is now assigned catcode 4. A similar example is macro \elp, defined in section 5.20.

5.6 Braces

Braces are used for two purposes:
1. to indicate the start and end of a group. Thus '{\bf ...}'.
2. to indicate that a string of tokens should be treated as one unit. Thus in '\def\abc{..}' the braces are used to delimit the replacement text. In '\abc{..}' they are used to delimit the argument, and in assignments such as '\toks0={..}' or '\setbox0=\hbox{..}' they are used to delimit the material that will go into the toks or the box register.

The characters '{', '}' are not hardwired in TEX. Any tokens with catcodes 1 and 2 can be used. The plain format starts [343] by saying \catcode'\{=1 \catcode'\}=2. Tokens with catcodes 1 and 2 are called *explicit braces*. There is also the concept of *implicit braces*. An implicit brace is a control sequence whose replacement text is an explicit brace. Thus the two plain control sequences \bgroup and \egroup are implicit braces. They are defined [351] by: '\let\bgroup={' '\let\egroup=}'. They are used in cases where unbalanced braces are needed.

Simple examples are '\hbox\bgroup...}', which is as good as '\hbox{...}' and '{\it...\egroup', which is identical to '{\it...}'. A better example is a macro to typeset a number of paragraphs in a special way. The macro
\long\def\negIndent#1\endIndent{{\parindent=-10pt #1\par}}
typesets its argument with a negative indentation. This macro requires memory space because the argument (which may be very large) has to be scanned and stored. A better alternative is to define two macros where the first starts a group, and the second one ends it. Here is a naive version:
First macro: '\def\negIndent{{\parindent=-10pt}'
Second macro: '\def\endIndent{\par}}'
This does not work, because the replacement text of a macro has to have balanced braces. A working version is

```
\def\negIndent{\bgroup\parindent=-10pt}
\def\endIndent{\par\egroup}
```

A similar example is a macro that collects material and stores it in \box0. We show two versions: (1) '\def\collect#1\endCollect{\setbox0=\hbox{#1}}'. If we say '\collect\$x=0\$\endCollect', the two dollar signs will go into \box0 with the same catcode. An attempt to change the catcode, such as \collect \catcode'\\$=12 \$x=0\$\endCollect, would not work. When \collect is expanded, the argument is absorbed but no commands are executed, so the two dollar signs go into \box0 with their original catcode. When \box0 is executed, the catcode of a '\$' is changed, but the two dollar signs already inside \box0 retain their original catcode.

Version 2 is more general:

```
\def\collect{\setbox0=\hbox\bgroup}
\def\endCollect{\egroup}
```

Now we can say `\collect$x=0 \catcode'\$=12 $\endCollect`, and the second '$' will go into `\box0` with a catcode of 12.

▶ **Exercise 5.7:** What is the result of executing the above expansion?

Explicit braces have catcodes 1 and 2; implicit braces are control sequences so they don't have any catcode. This is why implicit braces cannot always be used instead of explicit ones. One cannot say, for example, '`\def\abc\bgroup...\egroup`' or '`\toks0=\bgroup...\egroup`'. Careful reading of [App. B] shows that the following commands must have an argument that's ⟨general text⟩: `\message`, `\errmessage`, `\uppercase`, `\lowercase`, `\write`, `\special`, and `\mark`. A ⟨general text⟩ is defined [276] as text with a closing explicit brace. The opening brace can be implicit.

The discussion on [385] shows that macro `\def\leftbrace{{\iffalse}\fi}` has a replacement text of '$\{_1$'. Similarly, `\def\rightbrace{\iffalse{\fi}}` has a replacement text of '$\}_2$'. This is a clever use of `\iffalse` to generate unbalanced braces.

The two primitives `\begingroup` and `\endgroup` can also be used to define a group. However, a group that starts with a `\begingroup` must end with an `\endgroup`. This provides a mechanism for error checking, because TeX can easily tell that the following group

```
\begingroup
...
{
...
\endgroup
```

contains unbalanced braces.

A typical example is a journal accepting articles in TeX. Obviously, errors in one article (such as unbalanced braces) should not affect other articles. Each article should thus be protected from all the other ones, which is achieved by placing a `\begingroup \endgroup` pair around each article.

Note that `\begingroup`, `\endgroup` can only be used to define a group, not to delimit a string. Thus one can say, for example, '`\begingroup\it abc\endgroup`' but not '`\hbox\begingroup\it abc\endgroup`'.

5.7 `\let`

The general form is `\let`⟨control sequence⟩=⟨one optional space⟩⟨token⟩. It defines the control sequence as being identical to the token. Thus defining

```
\def\a{X}
\let\g=\a \def\k{\a}
\g\k
```

produces 'XX'. If we now redefine `\a`, the meaning of `\k` will change (since it was defined by `\def`) but `\g` will not change. Thus `\def\a{*}` `\g\k` produces 'X*'.

Note that the second parameter of `\let` can only be a single token. The code '`\let\^^M=\par\noindent`' is wrong! Instead, the user should write `\def\parnoindent{\par\noindent}`, and follow with `\let\^^M=\parnoindent`.

5.8 \futurelet

The general form [207] is \futurelet\cs⟨token1⟩⟨token2⟩. When TeX encounters it, it executes the sequence \let\cs=⟨token2⟩⟨token1⟩⟨token2⟩. The control sequence \cs is set to ⟨token2⟩ and then TeX executes the sequence ⟨token1⟩⟨token2⟩ in the usual way. This can be useful if ⟨token2⟩ is unknown at the time \futurelet is written, for instance, if \futurelet\cs⟨token1⟩ appears at the end of a macro definition and ⟨token2⟩ is the first token that follows the macro at expansion time. The following macro definition \def\boxt{\futurelet\new\macA} is a good illustration. If the macro is expanded by \boxt[..., then the result will be

- \let\new=[,
- execute the control sequence \macA,
- execute the '[' (unless \macA has already scooped it up).

When a macro is expanded, an argument must be supplied for every parameter. Following are two examples of \futurelet:

1. Imagine a mathematical document that deals with scalar and vector quantities. In order to distinguish between them, the vectors are typeset in boldface. If, for example, something like '$u\mathbf{P}w$' occurs very often in the document, it makes sense to define a macro '\def\bfP{{\bf P}}' so that the author can type the expression above as '$u\bfP w$' instead of '$u{\bf P}w$'. If many such definitions are necessary, it makes sense to generate them with a macro.

We thus develop a macro \multbf, such that the expansion '\multbf aPQ.' will generate the macro definitions '\def\bfa{{\bf a}}', '\def\bfP{{\bf P}}', '\def\bfQ{{\bf Q}}'. Note that the period serves as a delimiter to the string of tokens. We want \multbf to peek at the next token of the string, to create a new definition if that token is not a period, and to stop the operation if it is.

Macro \multbf is thus defined without any parameters. It uses \futurelet to store a copy of the next token of the string in the control sequence \pip, then expands macro \actdef. This macro actually absorbs the token, tests it for a period, generates the required '\bf$x$' macro, and expands \multbf recursively for the next token.

The important point to understand is that TeX absorbs the string token by token, and each token is absorbed only when necessary. Each token is temporarily peeked at by the \futurelet, but is absorbed when \actdef is expanded and an actual argument is needed for its parameter. Each recursive expansion causes TeX to absorb the next token in the string. The result is surprisingly short.

```
\def\multbf{\futurelet\pip\actdef}
\def\actdef#1{\if\pip.\let\next=\relax
  \else \expandafter\def\csname bf#1\endcsname{{\bf #1}}%
       \let\next=\multbf
 \fi\next}
```

The section on \csname explains why the \expandafter is necessary.

2. The second example is a macro with an optional parameter. Macro \boxt is a good candidate. It generates a box of text surrounded by rules on all four sides. The text is a parameter of \boxt and there is another, optional, parameter, indicating the amount of space desired between the text and the rules. If that

parameter is missing, the spacing will be 3pt. To implement \boxt, we first define
two macros: \boxA, which has one parameter, and \boxB, which has two.

```
1 \def\boxA#1{\vbox{\hrule
2 \hbox{\vrule\kern3pt\vbox{\kern3pt#1\kern3pt}\kern3pt\vrule}\hrule}}
3 \def\boxB[#1]#2{\vbox{\hrule
4 \hbox{\vrule\kern#1pt\vbox{\kern#1pt#2\kern#1pt}\kern#1pt\vrule}
5 \hrule}}
```

▶ **Exercise 5.8:** Why not simply call them \box1 and \box2?

   Our macro \boxt checks the first token that follows it when it is expanded. If
that token is a left bracket, \boxt should expand macro \boxB, otherwise it should
expand \boxA. From the definition above we see that \boxt sets \new and then
expands \macA. The only remaining task is to write macro \macA with an \ifx to
examine \new. If \new equals a left bracket, then \macA should cause \boxB to be
expanded, otherwise, \boxA is expanded.

```
\def\macA{\ifx\new[\let\next=\boxB \else\let\next=\boxA \fi \next}
\setbox1=\vbox{\hsize 3in \noindent Here is text that will be boxed
by \boxt twice. The first expansion is \boxt{\copy1}, and the
second, \boxt[9]{\box1}.}
```

> Here is the text that will be boxed by \boxt twice.
> The first expansion is \boxt{\copy1}, and the
> second, \boxt[9]{\box1}.

> Here is the text that will be boxed by \boxt twice.
> The first expansion is \boxt{\copy1}, and the
> second, \boxt[9]{\box1}.

▶ **Exercise 5.9:** Extend the definitions above to allow \boxt to have up to three
parameters. Thus one could expand \boxt[...]{...}[...], where the first and
third arguments are optional.

   Macro \mp below illustrates a simple extension of the ideas of this section. The
macro can have a large number of arguments, which it scans one by one. It distinguishes three types of arguments. Type 1 is a single character, type 2, an argument
in braces, and type 3, an argument enclosed in square brackets. The last argument
must be followed by a ';'. Thus the expansion '\mp A[123]BC{abc}{}[xyz];' scans
and recognizes seven arguments, of which the third (the B) is of type 1, and the
sixth (the empty one) is of type 2.

```
\def\mp{\futurelet\next\macA}
\def\macA{\ifcat\bgroup\next \let\next=\macB
 \else\ifx[\next \let\next=\macC
 \else\ifx;\next \let\next=\macE
 \else\let\next=\macD\fi\fi\fi \next}
\def\macB#1{\message{type 2 '#1'.}\mp}
\def\macC[#1]{\message{type 3 '#1'.}\mp}
```

```
\def\macD#1{\message{type 1 '#1'.}\mp}
\def\macE;{\message{end}}
```

Macro \macA examines the next character in the input stream and expands
\macB or \macC, and so on, depending on what it is. Each of these macros (except
\macE) picks up the next argument, then expands \mp recursively to scan the next
character.

▶ **Exercise 5.10:** Why the ';' in '\def\macE;'?

The conscientious reader may find useful examples of the use of \futurelet
in (Ref. 18, p. 107), and on [357]. A potentially important point is that when
\futurelet peeks at ⟨token2⟩, the catcode of that token is determined.

5.9 \afterassignment

The general format of this command is '\afterassignment <token>'. It is
executed by saving the token and inserting it back into the input after the next
assignment has been performed [279]. Example: \afterassignment\noindent. If
the next assignment is \setbox0=\vbox{...}, then the \noindent will become
the first thing placed in \box0 (it will even precede the tokens of \everyhbox and
\everyvbox). Note that only one token can be saved by \afterassignment.

Simple examples of \afterassignment are the following:
- An assignment to a parameter:

```
\everypar{X}
\afterassignment Z
\baselineskip=24pt
```

following which 'XZ' will be typeset at the start of the next paragraph.
- An assignment that defines a font:

```
\afterassignment Z
\font\dum=cmtt10
\dum Start
```

following which the 'Z' will be typeset in the current font, and the 'Start', in font
\dum.
- An assignment to a box register:

```
\everyhbox{X}\afterassignment Y
\setbox0=\hbox{ab} \showbox0
```

which will show

```
> \box0=
\hbox(6.94444+0.0)x25.55562
.\tenrm Y
.\tenrm X
.\tenrm a
.\tenrm b
```

The most common use of \afterassignment is in a macro whose parameter is ⟨glue⟩ or ⟨dimen⟩. Consider the definition \def\abc#1{\leftskip=#1 \rightskip=#1}. Such a macro can be expanded by \abc{4pt plus5pt minus4pt}, for example, but it's better to be able to say '\abc 4pt plus5pt minus4pt'. To do this, we define

```
\def\abc{\afterassignment\aux \skip0=}
\def\aux{\leftskip=\skip0 \rightskip=\skip0}
```

When \abc is expanded, the first step is to scan the input stream and assign as many tokens as possible to \skip0. Macro \aux is expanded after the assignment. Several plain macros—such as \hglue, \vglue [352], \openup [362], and \magnification [364]—use this method.

Another possible application of \afterassignment is to scan the input stream, token by token, and examine each token individually. This technique is used by macro \ctest on [376] and also in [ex. 11.5]. Macro \scan below scans tokens until a ';' is found, and counts the number of asterisks and of nonasterisk tokens.

```
\newcount\length \newcount\aster
\def\scan{\afterassignment\nextoken\let\next= }
\def\nextoken{\ifx\next;\let\next\relax
 \else\ifx\next*\advance\aster by1
         \else\advance\length by1
  \fi\let\next=\scan
\fi \next}
```

Because of the \afterassignment, macro \scan is executed in two steps: (1) the control sequence \next is \let to the next token in the input stream; (2) macro \nextoken is expanded. If the token was not a ';', \nextoken expands \scan recursively, to read the next token. A typical expansion is '\scan ab c*d;', which yields values of 5 and 1 in \length and \aster, respectively (notice that the space is also counted).

▸ **Exercise 5.11:** What is the result of '\scan\z␣b␣c*d;'?

▸ **Exercise 5.12:** The definition of \scan contains a space after '\let\next= '. What's the purpose of this space?

▸ **Exercise 5.13:** Check \scan for spurious spaces.

▸ **Exercise 5.14:** Modify \scan to count the number of letters, number of spaces, and number of other characters in the input stream, followed by a ';'.

Here is another example of \afterassignment. We define macro \Futurelet as \def\Futurelet{\afterassignment\futurelet\futurelet}. When it is expanded, it should be followed by four tokens: \Futurelet <token1> <token2> <token3> <token4>. The steps of the expansion are

■ The \afterassignment saves the first \futurelet.

■ The second \futurelet is executed as \let <token1>=<token3>. Normally, TeX would execute <token2> and <token3> next, but in our example it inserts the first \futurelet in front.

■ The input now says \futurelet <token2> <token3> <token4>, which is executed as \let <token2>= <token4>, followed by the execution of <token3> and <token4>.

The result is

```
\let <token1>=<token3>
\let <token2>=<token4>
<token3> <token4>
```

5.10 \aftergroup

This is a little-known and greatly underused command. Its general format is \aftergroup⟨token⟩. The token is saved and inserted back after the current group is completed (the user should make sure that there is a current group).

A simple example is a macro that delays reading its argument until it changes some catcodes. It can be done, for example, by

```
\def\abc{\catcode... \xyz}
\def\xyz#1{...}
...
\abc{argument}
```

The argument is actually absorbed by \xyz after the catcodes have been changed. A similar effect can be achieved by \aftergroup:

```
\def\abc{\catcode...\bgroup\aftergroup\xyz\let\dummy=}
...
\abc{...}
```

The \bgroup starts a group. The \aftergroup saves \xyz and will insert it after the group closes. The \let absorbs the left brace in \abc{...}. The effect of \abc is thus to change the catcodes, to open a group, to save \xyz, and to absorb the '{'. Following this, the argument is read (as part of the source file, not as an argument of \abc) and the '}' closes the group. Macro \xyz is then expanded.

Note that the effect of \aftergroup is cumulative. If we say \aftergroup\xyz\aftergroup\lmn, then both tokens will be saved and inserted (first \xyz and then \lmn) after the current group.

See page 134 for another example.

5.11 \expandafter

This control sequence [213] reverses the order of expansion of the two tokens following it. When TEX encounters \expandafter⟨token1⟩⟨token2⟩, it

1. saves ⟨token1⟩,

2. expands ⟨token2⟩. If it is unexpandable, it is left alone.

3. places ⟨token1⟩ in front of the result of step 2 and continues normal processing from ⟨token1⟩.

Rule 2 implies that \expandafter is useful only if ⟨token2⟩ can be expanded, that is, if it is a macro, a primitive, or something similar.

Example: \expandafter\moveright \ifodd\ctst .1in \else -.1in \fi is executed by

- saving the token \moveright,
- expanding \ifodd (using \ctst as its argument). The result is either .1in or -.1in.
- placing the \moveright in front of the result, and continuing as usual.

Developing this example further, we define macros:

```
\def\macA#1{\vbox{\ifodd\ctst \moveright.1in
 \else \moveright-.1in\fi #1}}
\def\macB#1{\vbox{\expandafter\moveright \ifodd\ctst .1in
 \else-.1in\fi #1}}
```

then place something in \box1

```
\setbox1=\vbox{\hsize 2in \noindent A small \vbox with text.
It will be moved left (even into the margin) or right to
demonstrate \expandafter.}
```

Executing either \ctst=0 \macA{\copy1} \ctst=1 \macA{\copy1}
or \ctst=2 \macB{\copy1} \ctst=3 \macB{\box1} results in:

A small \vbox with text. It will
be moved left (even into the margin) or right to demonstrate \ex-
pandafter.

A small \vbox with text. It will
be moved left (even into the margin) or right to demonstrate \ex-
pandafter.

What if ⟨token2⟩ is not expandable? If X, Y are characters, then \expandafter XY has no effect. TEX saves X and recognizes that Y is not expandable. X is then placed in front of the Y, and the string 'XY' is the result of the whole thing. The string may be typeset, be placed in a box for future use, become the parameter of a macro, and so on, depending on the context.

An interesting case is when ⟨token2⟩ is another \expandafter. A typical example appears on [374]:

```
\expandafter\expandafter\expandafter\a\expandafter\b\c
```

To see how TEX executes this, let's write it as \ex1\ex2\ex3\a\ex4\b\c. The steps are as follows:
- When \ex1 is executed, it saves \ex2 and expands \ex3.
- When \ex3 is expanded, it saves \a and expands \ex4. The list of saved items now consists of \ex2\a.
- When \ex4 is expanded, it saves \b and expands \c. Thus \c is the first of \a\b\c to be expanded.
- The list of saved items (\ex2\a\b) is placed in front of the expansion of \c, and TEX continues normal processing. However, the first token that it sees is \ex2. It therefore saves \a, expands \b, and, finally, expands \a.

Our example thus reverses the normal expansion of tokens \a\b\c. Such a thing is not common, but here is an example. The control sequence \lowercase must be followed by an argument in braces. Because of the way \lowercase is executed (see page 21), something such as '\lowercase{\jobname}' results in an expansion of \jobname, but no conversion to lower case. To get the right result we need to: (1) save the \lowercase and the left brace; (2) expand \jobname; (3) place the left brace back; (4) execute the \lowercase. The execution of the three tokens '\lowercase', '{', and '\jobname' should thus be reversed, which is done by:

```
\expandafter\expandafter\expandafter%
 \lowercase\expandafter{\jobname}
```

The same thing applies to \uppercase.

A similar example is \toks0=\expandafter{\expandafter a\the\toks0}. To execute it, TeX

- saves the '{',
- saves the 'a',
- executes '\the\toks0'. The result is a string with a copy of the tokens from \toks0.
- appends the 'a' to the left of that string.
- appends the '{' on the left.
- assigns the whole thing back to \toks0, which has the effect of appending a character to the left of that token register.

Here is an example where ⟨token2⟩ is the primitive \the:

```
\newtoks\t \t={123}
\t=\expandafter{\the\t *}
```

Here, ⟨token1⟩ is a '{', and the steps are as follows:
- The '{' is saved.
- The \the is expanded, using \t as its argument. The result is the value of \t (the string '123').
- The '{' is replaced, generating the string '123*', which is assigned back to \t. A '\showthe\t' can be executed to verify this.

▶ **Exercise 5.15:** What happens without the \expandafter (when the command is \t={\the\t *})?

In the following example, \token2 is a macro that expands to a string of characters. After the definitions

```
\def\strng{pqr}
\def\separateargs#1#2#3{#1,#2,#3;}
```

the test '\separateargs\strng m n' typesets 'pqr,m,n;'. Obviously 'pqr' is the first argument. The test '\expandafter\separateargs\strng m n', however, typesets p,q,r;m n. The \expandafter evaluates \strng by removing the braces, so now p, q, and r are considered separate arguments.

Here is another example of \expandafter. When the value of macro \xyz is to become the argument of macro \abc, we have to say \expandafter\abc\xyz instead of just \abc\xyz. A common example is reading a file line by line by means

of \read (instead of reading the entire file by means of \input). If we have a macro
\abc to examine each input line, then:

```
\immediate\openin\fil=<filename>
\loop
\read\fil to\xyz
\expandafter\abc\xyz
...
\repeat
```

The next example, also from [374], is

```
\newcount\n
\n=1
\uppercase\expandafter{\romannumeral\n}
```

It first converts the value of \n into a roman numeral and then into upper case.

▶ **Exercise 5.16:** What would be the result of the above without the \expandafter?
Note that \expandafter is a one-step expansion. It expands ⟨token2⟩ into
individual tokens but does not expand the tokens further. An example is

```
\def\y{\a}
\expandafter\x\y
```

(assuming that \x and \a are undefined). The \expandafter expands \y into \a
and does not try to expand \a.

▶ **Exercise 5.17:** How can such a claim be proved?
The \expandafter happens to be a very useful command, commonly used in
advanced (and even not-so-advanced) applications. The reader is encouraged to
search the index for more examples of the use of \expandafter, and to practice its
use. The author believes that mastering this command signals the transition of a
user from the state of a struggling TEXhacker to that of a master.

5.12 \csname

A control sequence name is made up of either a string of letters or a single nonletter. The \csname, \endcsname pair of commands makes it possible to change this
basic rule, and to build control sequence names that are different from the above.
This is done by enclosing tokens between a \csname, \endcsname pair. The effect
of such a construct is to add a backslash and make the string of tokens into the
name of a control sequence. This way, new macros can be defined and expanded.

Thus \csname1$=*\endcsname creates \1$=*, which is also a command to
TEX to expand macro \1$=*. Of course, this macro should be defined before it
is expanded, and this should also use \csname, \endcsname. However, we cannot simply write \def\csname1$=*\endcsname{...}. TEX reads such a thing as
\def\csname..., an attempt to define macro \csname. TEX needs to be told to
(1) first execute \csname1$=*\endcsname; (2) then do the \def. This, of course,
is a good example of the use of \expandafter. Macro \1$=* should be defined as
\expandafter\def\csname1$=*\endcsname{...}.

What can we place between \csname and \endcsname? Answer: any characters and any control sequences that expand to characters. Control sequences that expand to commands, rather than to characters, are not allowed. Thus if \def\a{+}, then \csname1$=\a*\endcsname creates the control sequence name \1$=+*.

▸ **Exercise 5.18:** What is the result of

```
\newcount\ctst
\ctst=76
\expandafter\def\csname=\the\ctst\endcsname{abc}
```

Saying \csname1$=\par*\endcsname, however, is wrong. TeX stops at the \par and issues the error message ! Missing \endcsname inserted. The following experiment is worth performing since it shows, in the log file, the definitions of the two macros \a and \1$=+*.

```
\tracingmacros=1
\def\a{+}
\expandafter\def\csname1$=\a*\endcsname{xyz}
\csname1$=\a*\endcsname
\tracingmacros=0
```

Why would one want to use \csname? Why aren't the usual macro names enough? One practical example of the use of \csname is [ex. 7.7]. It describes a macro \ifundefined that determines if any given string is the name of a defined macro. It uses the fact that when a macro name is created by \csname, that macro may be predefined; however, if it is not, TeX sets it equal to \relax. The definition of \ifundefined is short:
\def\ifundefined#1{\expandafter\ifx\csname#1\endcsname\relax} but is interesting because it uses an \ifx without a matching \fi. The test
\ifundefined a \message{yes}\else\message{no}\fi supplies the \fi (and an \else), and displays 'yes' in the log file if \a is undefined.

As a simple exercise, we define a macro \ifdefined#1 that's the opposite of \ifundefined. We want \ifdefined also to supply an if that must be completed outside. We start with the test for an undefined macro:
\expandafter\ifx\csname#1\endcsname\relax
and create either an \iffalse (if the macro is undefined) or an \iftrue (in case it is defined), to be matched outside. The first version is

```
\def\ifdefined#1 {\expandafter\ifx\csname#1\endcsname\relax
\let\next=\iffalse\else\let\next=\iftrue\fi \next}
```

But it fails! The reason is that the \iffalse and \iftrue are considered nested conditionals, and TeX expects each to have its own \fi (see more about this topic at the end of Chapter 6). The next, working, version is

```
\def\maca{\let\next=\iffalse} \def\macb{\let\next=\iftrue}
\def\ifdefined#1 {\expandafter\ifx\csname#1\endcsname\relax
\maca\else\macb\fi \next}
```

following which, we can say

`\ifdefined a \message{yes}\else\message{no}\fi.`

This is an interesting example that illustrates `\expandafter`, `\csname`, `\iftrue`, `\iffalse`, and `\ifx`.

The next example is a macro that checks to see if its parameter is empty (compare to `\ifempty` shown earlier). The definition is

`\def\ifnull#1\\{\expandafter\ifx\csname empty#1\endcsname\empty}`

after which may follow expansions such as:

`\ifnull{}\\ \message{yes}\else\message{no}\fi`
`\ifnull\\ \message{yes}\else\message{no}\fi`

The macro generates the control sequence name `\empty#1` and compares it to `\empty`. If #1 is null, the comparison will be successful. This is much more elegant than the earlier example (`\ifempty`) using delimited parameters. Note that, if #1 is not null (let's say it is 'abc'), the control sequence name generated is `\emptyabc`. It is undefined, but the `\ifx` comparison does not mind.

▶ **Exercise 5.19:** Define macro `\ifundefined` using a different approach.

5.13 Double

The character # is used in the preamble of an alignment, and also to indicate a macro parameter. As a result, when tables and macros are nested (a table inside a macro or a macro inside a macro), the # is sometimes ambiguous, and the notation ## is used instead. The rule [203] is that when a macro is expanded, a pair ## is evaluated to one #.

A simple example is a macro with a table in it. We have to clearly distinguish between the two meanings of the #. The table

1. onceQ	upon	Qa	time
thereQ	lived	Qtwo	birds
inQ	a	Qdistant	nest

was produced by

```
\def\tabl#1#2{
\tabskip=.1in \halign{###1 &## &#2## &## \cr
   1. once&    upon       & a       &time \cr
    there&  lived      &two      &birds\cr
       in&      a        &distant&nest \cr}}
\tabl Q{$\cal Q$}
```

When `\tabl` is expanded, each ## evaluates to a single #, which denotes a preamble item; each #1 denotes a macro parameter. The notation ###1 is thus equivalent to ## #1. The `\matrix` macro of the plain format [361] also uses this feature.

▶ **Exercise 5.20:** Can we nest a macro definition inside a table?

Examples of nested macro definitions are shown next.

5.14 Nested Macro Definitions

The definition of a macro may include the definition of another macro. Thus we may define `def\xyz{...\def\abc{...}...}`. Each time \xyz is expanded, a new definition of \abc is created. This feature is called *nested macro definition* and has been mentioned earlier, in connection with the ambiguity of the #.

The simplest and most common example of a nested macro definition is a test for an empty macro argument. Macro \abc below does it properly:

```
\def\abc#1{\def\aux{#1}%
\ifx\aux\empty
...
\else
...
\fi
...}
```

(See definition of \empty on [351].) Just saying '\ifx#1\empty' does not work because \empty is a control sequence, and the \ifx can only compare a control sequence to another control sequence.

Another simple example is macro \twelve that changes the meaning of \bf and its relatives to 12 point fonts. Assuming that the following fonts have been loaded

```
\font\twerm=cmr12
\font\twebf=cmbx12
\font\tweit=cmit12
\font\twesl=cmsl12
\font\twett=cmtt12
```

Macro \twelve can be defined as

```
\def\twelve{\def\rm{\twerm}\def\bf{\twebf}\def\it{\tweit}%
\def\sl{\twesl}\def\tt{\twett}%
\rm}
```

After saying \twelve, the meaning of \bf will change. Instead of switching to font cmbx10, it will switch to cmbx12.

Another simple example is macro \outdent, which was introduced in connection with paragraphs.

```
\def\outdent#1 {\bgroup\leftskip=-#1pt \hangindent=#1pt \hangafter=1
 \def\par{\endgraf\egroup}\noindent}
```

The primitive \par is redefined inside \outdent so that it closes a group (originally opened by \outdent) when a paragraph is ended. Note that \par is redefined when \outdent is expanded, after the group has been opened, so the redefinition itself is local to the group.

A simple example is macro \deferredSkip that creates a '\kern.1in' each time it is expanded, except the first time. Its definition is simply another, nested, definition of itself. Thus \def\deferredSkip{\def\deferredSkip{\kern.2in}}.

When it is first expanded, it only redefines itself (the new definition is the \kern). Any successive expansions will generate the \kern. This can easily be verified by \tracingmacros=1. This example is easy to adapt to any case where the first expansion of a macro has to be different from all subsequent expansions.

A more complex (albeit less useful) example is macro \nestMD below, which defines a new macro. The name of the new macro is \nestMD's first parameter; its definition is the second parameter.

```
\def\nestMD#1#2{\expandafter\def\csname#1\endcsname{#2}}
```

The \csname, \endcsname pair is needed to make #1 into the name of a control sequence. The \expandafter is needed to delay the \def until the name of the new control sequence is ready. Without the \expandafter, we would be trying to define a new macro whose name is \csname.

The expansion \nestMD{abc}{def} creates \def\abc{def}, so now we can expand \abc and get the string 'def' typeset.

The next step is to generalize \nestMD. It still defines its #1 as the name of a new macro. The new macro typesets #2 and then redefines itself to typeset #2 in quotes.

```
\def\nestMD#1#2{\expandafter\def\csname#1\endcsname{#2
\expandafter\def\csname#1\endcsname{'#2'}
                                            }
           }
```

The expansions \nestMD{abc}{def} \abc \abc now produce: def 'def'.

▷ **Exercise 5.21:** Use nested macro definiton to define a macro \Uppercase that converts a character string to uppercase *including* a conversion of dotless 'ı' to uppercase 'I'. Note that the primitive \uppercase does not do that. A command such as '\uppercase{a\i b}' results in 'AıB'. Such a macro is important when, i.e., a language such as French is used, where accented dotless î's are common, and the designer wants chapter titles in all caps.

Nested macro definitions are more useful when the macros involved have parameters. The definition \def\a#1{..\def\b#1{..#1..}..} is ambiguous. It may mean one of the following:

■ Macro \a has a parameter, and macro \b also has one. In such a case the #1 inside \b is ambiguous. It could mean either the parameter of \a or that of \b. The ## notation should be used to remove the ambiguity.

■ Macro \a has a parameter, and that parameter is used as a delimiter of \b (which itself does not have a parameter). In such a case, each expansion of \b must be followed by the argument of \a.

In the former case (where both macros have parameters), the ambiguity can be removed by saying \def\a#1 {#1 \def\b##1{\dimen0=##1#1}}. Macro \a has one parameter, and when expanded, it (1) typesets its parameter, and (2) defines a macro \b. \b also has a parameter, and it is indicated by the notation ##1. The expansions

```
\tracingmacros=1
\a pt
```

```
\b2
```

generate

```
\a #1 ->#1 \def \b ##1{\dimen 0=##1#1}
#1<-pt
\b #1->\dimen 0=#1pt
#1<-2
```

The tracing shows the definitions of both macros and the actual arguments assigned to each parameter. The definition of \a is straightforward. The definition of \b is \dimen0=#1pt because the parameter of \a was assigned the value pt. When \a is expanded again, the definition of \b changes.

After the expansions

```
\a in
\b2
```

the tracing shows a different definition for \b.

```
\a #1 ->#1 \def \b ##1{\dimen 0=##1#1}
#1<-in
\b #1->\dimen 0=#1in
#1<-2
```

\b is now defined as \dimen0=#1in.

▶ **Exercise 5.22:** Consider the nested definition
\def\a{...\def\b{...\def\x{...}...}...}.
How can we add parameters to the three macros?

5.15 Nested Macro Expansions

The definition of a macro may include any valid commands and quantities (except \end and \outer macros); specifically, it may include expanding a macro. Such a case is called *nested macro expansion*. Macros such as '\def\abc{...\xyz...}' are common. Nested macro expansions are simple, intuitive, and do not present special difficulties (except when a macro expands itself; see next section).

A simple example of such nesting is the plain macro \footnote. It uses one parameter instead of the expected two. Thus '\def\footnote#1{#1\vfootnote}' where #1 is the footnote symbol (the footnote text itself is not a parameter). Macro \footnote just typesets the symbol, and the rest of the work is done by \vfootnote. The idea is that the user can either expand macro \footnote, to get a normal footnote, or expand \vfootnote directly, in which case the footnote is typeset without the symbol.

A more interesting example of nested macro expansion is a macro that expands either macro \a or \b, *depending on a condition*. There are several ways of doing this:

1. If neither \a nor \b needs arguments, we can simply say

```
\def\abc{..\if..\a \else..\b\fi}
```

2. If they need arguments, we can say

```
\def\abc{..\if..\let\next=\a
 \else..\let\next=\b\fi \next}
```

The expansion of \next is the *last step* of \abc, so the argument of \next will simply be the next token in the input stream. Several examples along these lines have already been shown.

3. The effect of method 2 can also be achieved by

```
|\def\abc{..\if..\expandafter\a \else..\expandafter\b \fi}|.
```

The \expandafter guarantees that the nested expansion will occur after the \fi. If either \a or \b is a \relax, this method is simpler than method 2, since it can be written \def\abc{..\if..\else..\expandafter\b\fi} and there is no need to say something like \let\next=\relax.

▸ **Exercise 5.23:** What commands are invalid inside a macro?

5.16 Recursion

A special important case of nested macro expansion is a macro that expands itself. Such a macro is called *recursive* [217]. The following six examples show simple recursive macros. The examples are all bad because the macros expand themselves indefinitely. Some even exceed TeX's capacity. However, the examples should be studied carefully, since the good examples that follow are based on them.

1. Macro \def\Arecur{\Arecur} expands itself indefinitely and thus loops forever. It does not save anything in memory, and thus does not exceed the capacity of any data structure.

2. Macro \def\Brecur{a \Brecur} typesets 'a' and expands itself recursively. We expect it also to loop forever and eject pages full of 'a's. Experiment shows, however, that after a while, it overflows main memory. The reason is that we never finish the paragraph. All the 'a's accumulate in main memory as part of the same paragraph. Changing the definition to \def\Brecur{a\par \Brecur} results in a macro that loops forever and ejects pages full of short paragraphs that each consists of just an 'a'.

3. Macro \def\Crecur{\Crecur a} seems similar to \Brecur but turns out to behave very differently. It expands itself recursively and *then* typesets the 'a'. However, before it expands itself, it has to remember to typeset the 'a' after the recursive expansion is over. It does that by saving the 'a' in a structure called the "input stack" before it expands itself. Since the recursion is infinite, the input stack overflows, and the result is the error message ! TeX capacity exceeded, sorry [input stack size=200].

▉▶ We say that \Brecur does *tail recursion* and \Crecur, *head recursion*. The former is, of course, preferable.

4. Macro \def\Drecur#1{#1 \Drecur{#1}} behaves like \Brecur. Modifying it like \Brecur above, \def\Drecur#1{#1\par \Drecur{#1}}, ejects pages indefinitely.

5. Macro `\def\Erecur#1\endparm{#1 \Erecur#1\endparm}` is identical to
`\Drecur`. The quantity `\endparm` is used as a delimiter, so that long arguments
won't have to be enclosed in braces.

6. Macro `\def\Frecur#1\endparm{\def\endparm{#1\endparm}\endparm}`
produces the same results as `\Erecur` but is tricky. Initially `\endparm` is a delimiter.
Once `\Frecur` has been expanded, however, `\endparm` becomes a recursive macro.
It is both defined and expanded by `\Frecur`. Once expanded, it expands itself
recursively and loops forever. The expansion `\Frecur a\endparm` will result in an
infinite paragraph that will overflow main memory. The (very similar) expansion
`\Frecur a\smallskip\endparm` will eject pages indefinitely.

A little thinking shows that a macro should only expand itself conditionally.
The recursive expansion should always be located inside a conditional. A simple
example is

```
\def\recur{\the\ctst\ \advance\ctst-1
  \ifnum\ctst>0 \recur\fi}
```

When invoked by `\ctst=5 \recur`, it typesets 5 4 3 2 1. However, when
invoked by `\ctst=220 \recur`, it typesets the error message ! TeX capacity exceeded, sorry [input stack size=200]. The reason is that each time `\recur`
expands itself, its next token, the `\fi`, remains waiting for the recursion to end.
All those tokens are saved in a stack (the input stack) whose size is normally 200.
Any recursion that goes deeper than that would overflow the stack. To solve the
problem, the `\fi` should be executed, by means of `\expandafter`, before going into
the recursion. (A better solution is to recompile TeX with a larger stack.) The next
version is

```
\def\recur{\the\ctst\ \advance\ctst-1
  \ifnum\ctst>0 \expandafter\recur\fi}
```

And this does not leave anything in the stack when `\recur` expands itself. This
idea appears in (Ref. 17). Another possibility is to define a macro `\next`, inside
`\recur`, such that `\next` would be identical to `\recur` except at the end, when
`\next` is made equal to `\relax`.

```
\def\recur{\advance\ctst-1 \ifnum\ctst>0
  \let\next=\recur \else\let\next=\relax\fi
\next}
```

This way, the `\ifnum` is fully executed before `\next` starts, so there is nothing
to leave on the stack.

We now generalize `\recur` by adding a parameter:

```
\def\recur#1{#1\ \the\ctst \advance\ctst-1
  \ifnum\ctst>0 \recur{#1} \else \relax\fi}
\ctst=4 \recur{\ Now hear this}
```

which yields

Now hear this 4 Now hear this 3 Now hear this 2 Now hear this 1
`\recur` is now modified in the spirit of `\next` above, by adding a `\let`.

```
\def\recur#1{#1\ \the\ctst \advance\ctst -1
  \ifnum\ctst>0 \let\new=\recur \else \let\new=\relax\fi \new{#1}}
\ctst=4 \recur{\ Now hear this}
```

The result:

Now hear this 4 Now hear this 3 Now hear this 2 Now hear this 1 Now hear this
is different! The difference is caused by the **last** assignment \let\new=\relax.
When we expand \new for that last time, it does nothing (since it is now =\relax),
but its parameter #1 is typeset (since \relax requires no parameter). In fact, this
parameter is now considered the parameter of \recur.

Recursive macros can get really confusing.

A practical example of recursion involves a compound macro argument. Macro
\compndArg accepts a compound argument and breaks it down into its components.
The argument should be of the form xxx,xxx,...,xxx; (the ',' separates the individual components and the ';' delimits the entire argument). The macro accepts
the argument (without the ';', of course), appends ',;,' to the argument, and makes
the whole thing the argument of \pickup, which is then expanded.

```
\def\compndArg#1;{\pickup#1,;,}
\def\pickup#1,{% Note that #1 may be \null
\if;#1\let\next=\relax
\else\let\next=\pickup
  \message{'#1'}% use #1 in any way
\fi\next}
```

Macro \pickup expects its argument to be delimited by a ',' so it ends up
getting the first component of the original argument. It uses it in any desired way
and then expands itself recursively. The process ends when the current argument
becomes the ';'. Note the following:

■ This is also an example of a macro with a variable number of parameters. The
compound argument may have any number of components (even zero).

▷ **Exercise 5.24:** Does this method work for an empty argument?

▷ **Exercise 5.25:** Define a macro \separg#1 whose argument should be a string of
characters. The macro should use recursion to typeset those characters separately.
Thus the expansion '\separg{abc}' should produce 'a' 'b' 'c'.

5.17 \edef

There is an important distinction between defining a macro and using (expanding) it. When a macro is defined, by means of \def, its definition (the replacement
text) is saved as it is. TEX does not try to execute it, to expand it, or even to look
for errors (there are some exceptions to that; see \outer in section 5.18). A bad
definition such as \def\abc{\hs*ip1in} will not produce any error message. Only
when we attempt to *expand* this macro (by saying \abc) will we get

```
! Undefined control sequence.
\abc ->\hs
        *ip1in
```

Similarly, the definition \def\xyz{\abc456}, where \abc has not been defined, is bad, but will cause no error message. Only when we try to expand \xyz, will a message be generated.

Now consider the definitions: \def\abc{123} \def\xyz{\abc456}. When \xyz is defined, its definition is saved as the string \abc456. Macro \abc, which is part of the definition, is not expanded. If we later expand \xyz, then the first step would be to expand \abc.

The command \edef (expanded definition) can be used, like \def, to define macros [215]. It behaves differently, though. Writing \def\abc{123} and defining \edef\xyz{\abc456} expands \abc inside \xyz immediately, and the definition of \xyz is saved as the string 123456. This is easy to verify by using \tracingmacros. However, not everything in the definition is immediately expanded. In the case \def\abc{123} \edef\xyz{\abc\hsize=1in} macro \xyz is saved as the string 123\hsize=1in. The assignment is not executed immediately but only when \xyz is expanded.

The reader should get slightly confused at this point. What things are expanded immediately in an \edef and what are not? Pages [212–215] contain the answer in the form of a long list of control sequences that are expanded whenever expansion has not been inhibited (macros, conditionals, \number, etc.). Anything not on this list will not be expanded immediately in an \edef.

▶ **Exercise 5.26:** What is the replacement text of
\edef\abc{\romannumeral8 \meaning A\string\abc \TeX}?

Another way to explain \edef is to say that, with \def, the results produced by a macro depend on conditions at the time the macro is *expanded*. With \edef, the results depend on conditions at the time the macro is *defined*. Example: \edef\look#1#2{\ifodd\count0#1 \else#2 \fi}. \look is a macro that expands to either #1 or #2 depending on the page number at *the time of definition*. If we define \look at the start of a chapter, we can find out *later* if the chapter has started on an odd- or even-numbered page.

Yet another way to look at \def and \edef is through anatomical considerations. \def is executed in the stomach, whereas \edef is executed earlier, in the gullet. Thus when a macro is defined by means of an \edef (Edefined), the replacement text is executed immediately (in the gullet), so it may contain only quantities that are known to TeX at that point.

This also means that an \edef macro cannot be recursive. It cannot expand itself since its value is not yet known at the time it is being defined. If \n is a \count register with a value of 5, then the simple recursive macro \def\asts{\ifnum\n>0 *\advance\n-1 \asts\fi} typesets five asterisks each time it is expanded, but when the same macro is Edefined

\edef\asts{\ifnum\n>0 *\advance\n-1 \asts\fi}

the result is an error message (undefined control sequence), since the definition contains the recursive expansion of \asts, which is still unknown.

The similar version \edef\asts{\loop\ifnum\n>0 *\advance\n-1 \repeat} also fails, even though \asts does not expand itself recursively. The reason is that macro \loop defines a macro \body, and the \edef requires only known quantities.

Before we see a solution, the advantage of \edef over \def should be explained. The \def version of \asts typesets five asterisks each time it is expanded, but a lot of work has to be done during each expansion. This is a slow way to produce five asterisks. An \edef version, assuming we could create one, would be evaluated to five asterisks in the mouth and would be saved as five asterisks (its replacement text would be the asterisks). Each expansion would only have to typeset the asterisks. This would be a fast way to do the same job.

A solution (see [373–374] for more solutions) is to use \aftergroup to delay the execution of the \edef while the asterisks are being created by the \loop..\repeat.

```
\begingroup
\aftergroup\edef\aftergroup\asts\aftergroup{
\loop\ifnum\n>0 \aftergroup*\advance\n-1 \repeat\aftergroup}
\endgroup
```

The first three aftergroups save the three tokens '\edef\asts{' in memory. The \loop..\repeat is then executed, creating five '\aftergroup*', so five asterisks are also saved. Finally, the last \aftergroup saves the '}'. The saved tokens are thus '\edef\asts{*****}'. They are executed after the \endgroup, thereby Edefining a simple macro \asts whose replacement text is just five asterisks (the reader should verify this by using \tracingmacros).

The number of asterisks typeset by \asts depends on the value of \n at the time of definition. Saying '\n=6 \asts' will still typeset five asterisks. This is in contrast to defining \asts with \def, where it will typeset \n asterisks, depending on the value of \n at the time of expansion. In general we can say that \def is executed late, which makes it more "flexible" than \edef.

5.18 Other Ways to Define Macros

\gdef is the same as \global\def. \xdef is the same as \global\edef. See the summary on [275].

\long\def is used when the macro may be expanded with a long (several paragraphs) argument.

\outer\def or \outer\edef is used to indicate that the macro being defined should not be expanded in certain situations. Such a macro cannot be part of the argument of another macro or part of the replacement text of another macro, and it cannot appear in the preamble of an alignment and in other places (see [206]). Examples:

```
\outer\def\abc{123}
\def\xyz{\abc456}
```

Since \abc is defined as \outer, it cannot appear inside the definition of \xyz. This is one case where TeX checks the validity of a definition, and it issues the message:

```
Runaway definition?
->
! Forbidden control sequence found while scanning definition of \xyz.
```

\outer is a safety feature. The plain macro \bye is defined as outer. This way, if we forget the closing brace in a macro definition \def\abc{..., TeX will read the

remainder of the document as part of the definition and will complain when it finds
\bye, since \abc is not supposed to have an outer macro as part of its definition.

A similar example is the plain macros \newxx. They are also defined as outer,
so something like '\def\abc{\newcount\xx...}' would cause an error message.
This is a safety feature since otherwise \abc would cause a new count register to
be allocated each time it is expanded..

Things such as \outer\gdef, \outer\xdef, and \outer\long\def are okay.

What if a useful macro has been defined as \outer, but we want to use it in
a forbidden place? Important examples of such a case are the various \new macros
mentioned above. If we need to declare new count registers inside our macros, we
can copy the definition of \newcount, delete the \outer, and give it a new name,
such as \noutercount. The original definition is
\outer\def\newcount{\alloc@0\count\countdef\insc@unt}, and we can change
it to

```
{\catcode'\@=11
\gdef\noutercount{\alloc@0\count\countdef\insc@unt}
\catcode'\@=12
}
```

5.19 Debugging Aids

Complex macros tend to behave in unexpected ways, so TeX provides debugging aids in the form of \tracingmacros [212], \show [10], \message [216–218],
\showthe [215], and \meaning [213].

The \tracingmacros command behaves like a parameter that can be assigned
different values. Normally it is set to 0, disabling all tracing. To trace a macro,
\tracingmacros should be set to either 1 or 2 before the macro is expanded. It
can be set back to zero after that.

When \tracingmacros=1, TeX will show, in the log file, the definitions of all
macros expanded, and the values assigned to all parameters. Setting parameter
\tracingmacros to 2 (or any value > 2) will do the same and will also show all
expansions of token lists, such as \output or \everypar. For example,

```
\def\test#1#2{ABC#2 \ifnum#1>0 \test 0{#2}\fi}
\tracingmacros=1
\test 0{XYZ}
\test1{ZZZ}
\tracingmacros=0
```

will place in the log file:

```
\test #1#2->ABC#2 \ifnum #1>0 \test 0{#2}\fi
#1<-0
#2<-XYZ

\test #1#2->ABC#2 \ifnum #1>0 \test 0{#2}\fi
#1<-1
#2<-ZZZ
```

```
\test #1#2->ABC#2 \ifnum #1>0 \test 0{#2}\fi
#1<-0
#2<-ZZZ
```

The \show command can be used to write the definition of a macro in the log file. Thus \show\abc shows the definition of macro \abc once (\tracingmacros shows it every time the macro is expanded). Moreover, \show can be used to show the values of the arguments of a macro. Thus

```
\def\abc#1#2{\def\one{#1} \def\two{#2}
 \show\one \show\two...
```

will show the values of the two arguments each time \abc is expanded.

Another important use of the \show command is to easily get a copy of any plain macro, in case the user wants to modify it. \show\obeylines, for example, lists the definition of \obeylines in the log file, from which it can be cut, pasted into the source file, and modified in any desired way.

The \message command displays character strings and values of internal parameters in the log file. An example is \message{\the\count0; \xxx; done}, which displays the value of \count0, expands macro \xxx, and displays done (the semicolons are also displayed). The next message continues on the same line, but it is easy to add carriage returns by changing the definition of \newlinechar [228]. Example:

```
\newlinechar='@
\message{\string\hsize=,\the\hsize;@}
\message{\string\vsize=,\the\vsize;@}
```

This defines '@' to be the new \newlinechar; following that, each '@' in a \message will start a new line. Each of the two messages above thus ends with a carriage return.

The \showthe command displays values of tokens in the log file. It differs from \message in that it stops after each display, waiting for a user's response.

▶ **Exercise 5.27:** What are valid responses?

The control sequence \meaning produces the meaning of its parameter. In the absence of any other command, the meaning gets typeset. To understand its use, try an experiment such as:

```
{\tt(\meaning\par) (\meaning\bye)
 (\meaning\sevenrm) (\meaning A)}
```

which typesets

```
    (\par) (\outer macro:->\par \vfill \supereject \end )
    (select font cmr7) (the letter A)
```

It is also possible to use \meaning in order to get the definition of any macro into the log file. The following:

```
\toks0=\expandafter{\meaning\ooalign}
\showthe\toks0
```

places 'macro:->\lineskiplimit-\maxdimen \oalign' in the log file. This can later be edited, printed, or whatever. It is also possible to write the \meaning of the macro on any other file by

```
\newwrite\tmpfile
\openout\tmpfile=\jobname.mean
\write\tmpfile{\meaning\ooalign}
\closeout\tmpfile
```

5.20 \catcode

This is a TEX primitive used to assign category codes to characters [39]. Its general format is: \catcode⟨internal code of a character⟩=⟨category code⟩. The internal code of a character (its ASCII code) can easily be obtained in TEX (see [44]) by a left quote followed by the character (or followed by a control sequence made up of one character). Thus 'b is 98, '\b is also 98, and the definition \chardef\%='\% defines the single-character control sequence \% as the ASCII code of '%'.

The concept of a category code is central to TEX and illustrates its design philosophy. Normally, a '\' is reserved as the escape character. However, as far as TEX is concerned, the escape character is any character that has a category code of 0. In a document where the '\' is a common character (see last paragraph on [48]), another, less frequently used character, can be assigned catcode 0, and the '\' can be assigned any other code (perhaps 11).

The category code of a character can be typeset by \the\catcode'A. It can be displayed in the log file by \showthe\catcode'A.

Examples (see [343]): \catcode'\{=1, \catcode'\$=3 are executed by the plain format to assign the usual category codes to '{', '$' and the other special characters; \catcode'\ =9 is executed on [390] to assign code 9 (ignored character) to space. This causes all spaces to be (temporarily) ignored.

When TEX reads the source file, it converts (in its "mouth") the input string into tokens. A token is either a single character or a control sequence. Any character token gets a category code immediately attached to it, and this attachment is permanent. The code tells TEX later (in the "stomach") what to do with the character. A control sequence token does not get a category code and, in order to execute it, TEX has to look up its current definition in a table. This implies that changing the category code of a character should be done carefully. Suppose the code of '{$_1$' is changed to 12. All the '{$_1$' tokens inside TEX that haven't been processed at that moment will retain their original code of 1.

A practical example of a catcode change is a macro \elp, whose single parameter is delimited by the end of the line. It is used to pick up an entire line at a time. Such a macro could be the basis for a set of macros to format a letter (similar to [App. E]). We cannot simply write \def\elp#1^^M{...} since the ^^M would terminate the current line and send TEX looking for the definition {...} on the next line.

So we try to change the catcode of the end-of-line character. Initially we change it to 13 (active). The first attempt is '\def\elp#1^^M{\catcode'\^^M=13...}', but this does not work since, when TEX finds the catcode change, it has already

scanned and determined what the argument is. We have to change the catcode
before \elp is expanded. The definitions \catcode'\^^M=13 \def\elp#1^^M{...}
work, but this means that \elp can only be used when the catcode change is in
effect (i.e., inside a group).

A similar solution is to define \elp in \obeylines mode [352] (in which ^^M is
active). Thus '{\obeylines\def\elp#1^^M{...}...}'. It has the same disadvantage as before.

A better solution is to define \elp *without a parameter*, change the catcode
inside \elp (by means of \obeylines), and then expand another macro, \getpar,
that actually picks up the argument. The result is

```
\def\elp{\begingroup\obeylines\getpar}
{\obeylines
\gdef\getpar#1
{'#1'\endgroup}}
```

Macro \elp performs the catcode change, and expands \getpar. \getpar is
thus always expanded when TeX is in \obeylines, but \getpar is also defined
inside an \obeylines. The fact that its definition is on a separate line means
that its parameter, #1, is delimited by an end-of-line. The \endgroup in \getpar
terminates the effect of the catcode change.

The next step is to realize that the catcode of ^^M can be changed simply to
12 (other), and there is no need to bother with active characters. Perhaps the best
solution is

```
\def\elp{\begingroup\catcode'\^^M=12 \elpAux}
{\catcode'\^^M=12\gdef\elpAux#1^^M{\message{'#1'}\endgroup}}
```

The catcode change is localized by means of \begingroup and \endgroup. An
auxiliary macro, \elpAux, is used to actually pick up the argument. The macro can
be used anywhere.

5.21 \jobname

The \jobname parameter [213] contains the name of the source file (without the
extension). Thus if the input is read from file 'book.tex', the contents of \jobname
will be the four tokens 'book'. Here is a simple example of using \jobname. (See
the discussion of tokens for an extension of this example. See also the discussion of
file I/O in Chapter 8 for more examples of the use of \jobname.)

When writing a book, each chapter is normally saved in a separate .tex file.
When printing draft versions, the author may want the chapter number and name
to appear on every page, for quick identification. The name and number should
also be typeset on the first page of the chapter. A good idea is to have the name
of the source (.tex) file composed of the chapter number and name. When a job
is typeset, the file name is available in the \jobname parameter, so a macro can be
developed to scan \jobname and separate the chapter number and name.

A possible approach to such a macro is to have file names like 3-Departure.tex
or 24-Back-home.tex. The '-' can be used as a parameter delimiter, to help the
macro separate the chapter number and name. The trouble with this approach is

that a chapter may have no number, forcing the user to use an unnatural name such as '-Introduction.tex'.

A more general approach is to have the chapter number as the first one or two characters in the file name. Thus source files may be named Introduction.tex, 3Departure.tex, or 24Back-home.tex. The macro should test the first two characters in \jobname and, if they are digits, separate them from the rest of the name. If none is a digit, no separation should be done.

Macro \title below appends '\\' to the string in \jobname, then expands \jobtitle with the new string as argument. Macro \jobtitle is straightforward. It separates the first two characters of its argument and uses catcodes to test for digits. The two control sequences \hnum and \hnam are defined as the chapter number and name, respectively.

```
\def\title{\expandafter\jobtitle \jobname\\}
\def\jobtitle#1#2#3\\{%
\ifnum\the\catcode'0=\the\catcode'#1
  \ifnum\the\catcode'0=\the\catcode'#2
    \def\hnum{#1#2}\def\hnam{#3}
  \else
    \def\hnum{#1}\def\hnam{#2#3}
  \fi
\else \def\hnum{}\def\hnam{#1#2#3}
\fi}
```

Every job should start with an expansion of \title. If a format file is used, it should contain the code '\everyjob{\title}', to automatically expand \title.

▶ **Exercise 5.28:** What if the source file name is one-character long?

▶ **Exercise 5.29:** Extend the definition of \jobtitle above to check for up to three-digit chapter numbers.

5.22 \chardef

To print the character '¶', we can say \font\q=cmsy10 {\q\char'173}. If '¶' is used a lot, we can define a control sequence \def\Q{\char'173} and use \Q instead of \char'173. However, it is also possible to say \chardef\Q='173. The control sequence \chardef [44] is used to define another control sequence whose value is a number (in the range 0...255).

Another example is \chardef\active=13 [343], after which you can write \catcode'?=\active to change the category code of '?' to 13 (active).

See the Index of this book and of *The TeXbook* for other examples of \chardef. Chapter 13 mentions the \mathchardef control sequence, the mathematical analog of \chardef.

5.23 Unexpandable Control Sequences

When a macro is defined, by means of \def, its replacement text is saved in a table, for future use. To execute a macro, it is *expanded*: Its replacement text is brought from the table, the arguments are substituted for the parameters, and the result (which is a string of tokens) is executed in the usual way. This is why a macro is an *expandable* control sequence.

In general, a token is either expandable or unexpandable. The former category consists of macros, active characters, and a few primitives, such as conditionals, \number, and \jobname (see list on [213–214]).

The latter category includes everything else, such as character tokens, anything defined by \chardef, and most primitives.

5.24 Active Characters

"Active characters behave like control sequences, but they are not prefixed by a '\' " [40]. In the plain format, only '~' is an active character and is equivalent to the control sequence "*typeset a space but don't break a line at that space*" [25, 91–92]. Note that the control sequence \~ is different. It typesets a tilde accent, such that '\~n' is typeset as 'ñ' [38]. How do you typeset a '~'? Answer, with \char'176 [427].

An active character does not behave exactly as a control sequence (see answer to [ex. 7.3]) but can nevertheless be useful. Any character can be turned into an active character, but this must be done very carefully. One doesn't want to turn characters such as '\', '%', '&' into active characters, since they are needed all the time for other tasks. Also, once a letter is turned into an active character, it is impossible to typeset later. Typically, new active characters are defined locally and are used temporarily, within a group. Creating an active character is done by first changing its category code to 13 [37] and then defining its meaning. Examples:

- Hanging punctuations [394–5]. The commands

```
\setbox0=\hbox{,} \catcode`,=\active
\def,{,\kern-\wd0\kern\wd0}
```

change the catcode of the comma from a letter (catcode 11) to an active character whose definition is a comma followed by the two \kerns. From this point on, whenever a comma is used in the text, the two \kerns will cancel each other, and a regular comma will be typeset. If a comma is used at the end of a line, however, the last \kern is discarded and the comma is left hanging out on the right.

- The \obeyspaces control sequence [352] works by changing the catcode of space from 10 to 13. \def\obeyspaces{\catcode`\ =\active}. Next, \obeyspaces is expanded to make the character 'space' an active character. The last step is to define '\ ' (backslash space) as the printable character 'space'. This is done by \let\ ={\ } or, rather [351], \global\let\ =\space \def\space{\ }.

- Chapter 20 shows three examples of useful active characters: (1) the circumflex '~' is declared active (outside of math mode) and is used to generate an index item; (2) the vertical bar '|' is declared active, and is used for verbatim listing; (3) the left quote is also declared active in connection with verbatim listing.

5.25 Summary

Macros are a very powerful feature of TeX. It takes time and effort to master this topic with all the commands and tools associated with it. However, the results are rewarding, since an extensive use of macros is the only way to exercise the capabilities of TeX. Any reader who wants to master TeX and get the maximum out of this book is encouraged to read the next two chapters carefully, and try to work out all the exercises.

... Then there are young men who dance around and get paid by the women, they're called 'macros' and aren't much use to anyone ...

— *Noël Coward, To Step Aside (1939)*

6. Conditionals

As a general rule, people with little or no experience in computer programming find it hard to use TeX to its full potential. They prefer to use it on a higher level, with the aid of LaTeX or a similar macro package. The reason for this is that TeX is, in a sense, a programming language, and writing a TeX source file is like writing a computer program. Conditionals are an important part of any programming language, since they are the key to making decisions, and to performing loops. This is why conditionals have been included in TeX, and they are an important part of every nontrivial TeX job.

There are 17 types of conditionals (not including the \newif; section 6.2) to be used in different contexts to test different situations, but they all have the same format, namely, '\if⟨test⟩ *then* part \else *else* part \fi', where ⟨test⟩ is different for each type of \if, and either the *then* part or the '\else *else* part' is optional. Thus it is okay to write conditionals such as '\if⟨test⟩ *then* part \fi' or '\if⟨test⟩ \else *else* part \fi', but the \fi must always appear. Both the *then* part and the *else* part may contain text and commands. Specifically, they may contain other conditionals. This means that conditionals can be nested.

Lines in the source file may be indented for easier reading, but TeX ignores the indentations. Note that every inner condition must be completely enclosed in an outer one. Each \fi encountered is assumed to match the most recent unmatched \if.

```
\if<test>                          \if<test>
...                                ...
   \if<test>...\else...\fi            \if<test>
...                                ...
\else                                 \if<test>
...                                ...
   \if<test>...\else...\fi              \fi
...                                  \fi
\fi                                \fi
```

6.1 Types of Conditionals

All conditionals are described on [209–210]. Most are easy to use; they are briefly explained in this section and the next one. Only three types, namely, \if, \ifx, and \ifcat, require detailed discussion. They are treated in later sections. The reader should use the index to locate the many examples of conditionals in this book.

- \ifnum⟨num1⟩⟨relation⟩⟨num2⟩ is used to compare two integers. The relation must be '<', '=' or '>'. Examples: '\ifnum\count0<0', '\ifnum`#1=\uccode`#1'.
- \ifdim⟨dimen1⟩⟨relation⟩⟨dimen2⟩ compares two dimensions. The ⟨relation⟩ is the same as above. Examples: '\ifdim\dimen0=1.2in', '\ifdim\dp255=1sp'.
- \ifodd⟨number⟩ evaluates to true if the ⟨number⟩ is odd. A common example is the test '\ifodd\pageno...' for the parity of the page number, done in the OTR. Note that there is no \ifeven. To test for an even number, just say \ifodd⟨number⟩\else...\fi.
- \ifvmode, \ifhmode, \ifmmode can be used to determine if TeX is currently in V or IV mode or H or RH mode or one of the math modes, respectively. Examples:

```
\ifvmode
  \vbox to0pt{...}
  \nointerlineskip
\else
  \raise\step\hbox{\Dot}%
\fi
```

'\ifhmode\lower\dp0\fi \hbox{...}', '\ifmmode\sqcup \else$\sqcup$\fi'.
- \ifinner is used to distinguish between the inner modes (IV, RH, and math inline) and the outer ones (V, H, and math display).
- \ifvoid, \ifhbox, \ifvbox test box registers for the specified conditions. Each must be followed by the number of a box register.
- \ifeof⟨number⟩ tests for end-of-file. This should be used when a file is opened, to make sure that the file exists, and also when \read is used to read a file record by record, to detect the end of the file.
- \iftrue, \iffalse always have the same values but are nevertheless useful in certain situations (see the Index for examples).
- \ifcase is similar to the case statement found in many programming languages. It is divided into several parts, only one of which is selected and executed, depending on the value of an integer. The general format of this conditional is

\ifcase⟨number⟩\or⟨text for case 0⟩\or⟨text for case 1⟩\or...
        \or⟨text for case n⟩\else⟨text for all other cases⟩ \fi

There are $n+2$ cases, of which one (and only one) is executed. The first $n+1$ cases correspond to values of ⟨number⟩ from 0 to n. The last one (the \else) is executed if ⟨number⟩ has a value outside the range 0–n.

Examples:

1. The following (see [406]) converts the integer value of parameter \month to the name of the current month:

```
\ifcase\month\or
  January\or February\or March\or April\or May\or June\or
  July\or August\or September\or October\or November\or December\fi
```

2. The different magsteps are defined, on [349], by

```
\def\magstep#1{\ifcase#1 1000\or
  1200\or 1440\or 1728\or 2074\or 2488\fi\relax}
```

3. From [373] comes
\edef\asts{\ifcase\n\or*\or**\or***\or****\or*****\else\bad\fi} (this is a limited macro for typesetting up to five asterisks).

6.2 Boolean Variables

A different type of conditional is the \newif, which is not a primitive. It is defined, on [348], as part of the plain format. It is used to create a quantity similar to a boolean variable used in programming languages. The user can say, for instance, \newif\ifabc, and this creates the three control sequences \ifabc, \abctrue, and \abcfalse. Following the \newif, it is possible to say \abctrue (to set variable abc to true), \abcfalse (to set it to false), and '\ifabc...\fi' (to use it as a boolean variable).

A simple example of the use of this if is

```
                              \newif\ifcont \conttrue
                              \newread\rec
\newread\rec                  \openin\rec=myfile
\openin\rec=myfile            \loop
\loop                         \ifeof\rec\contfalse\fi
\ifeof\rec\else               \ifcont
\read\rec to\xyz              \read\rec to\xyz
...                           ...
\repeat                       \repeat


            a                              b
```

We want to set up a loop, such as in 'a' above, to read records from file \rec. The loop should repeat as long as the \ifeof\rec condition is false. However, because of the way the \loop-\repeat construct is defined, it is impossible to say \if..\else..\repeat. A simple, readable solution is to write the loop as in 'b' above.

6.3 \ifx

Of the three "difficult" comparisons, \ifx is the simplest and most useful one. The \ifx is followed by two comparands, which are compared without looking too deeply into their values or meaning. The test

```
\def\qwe{trip} \def\rty{trip}
\ifx\qwe\rty
 \message{yes}\else\message{no}
\fi
```

will compare the two macros and, since their definitions are the same, the *then* part will be executed, displaying 'yes' in the log file and on the terminal. Similarly, the test \ifx\qwe\rty True\else False\fi results in the tokens 'True'.

A comparison always results in the expansion of either the *then* or the *else* parts. As mentioned before, each part may contain any tokens. Thus we may have, for example,

```
\baselineskip=\ifx\a\b 24pt\else 36pt\fi
\pageno=\count\ifx\a\b 0\else 1\fi
\message{\ifx\a\b success\else failure\fi}
\ifx\a\b true\else false\fi
\def\M{\ifx\a\b yes\else no\fi}
```

or even something more sophisticated, such as:

```
\newif\ifSome
\csname
 Some\ifx\a\b true\else false\fi
\endcsname
\ifSome ...
```

The \csname, \endcsname pair creates one of the control sequences \Sometrue, \Somefalse, following which, the test \ifSome is meaningful. References 20 and 21 discuss \csname.

The following example is interesting. It shows the meaning of the words "... the *then* or *else* part is *expanded*."

```
\def\x#1{'#1'} \def\y#1{[#1]}
\ifx\a\b \x<argument> \else \y<argument> \fi
```

Depending on how \a, \b are defined, either \x or \y is expanded and its argument used in the expansion. However, if the argument is left outside the \ifx, it is not used in expanding either macro.

```
\def\x#1{'#1'} \def\y#1{[#1]}
\ifx\a\b \x \else \y \fi <argument>
```

If \x is expanded, its argument will be the \else; if \y is expanded, its argument will be the \fi. This is easy to verify with \tracingmacros=1.

The macros compared may have parameters. Thus

```
\def\qwe#1{samething}
\def\rty#1{samething}
\ifx\qwe\rty
```

evaluates to 'yes'. This suggests one use for \ifx, namely, comparison of strings.
To compare two strings, place them in macros and compare the macros. \ifx is,
in fact, heavily used in (Ref. 19) for this purpose. However,

```
\def\qwe#1{samething}
\def\rty{samething}
\ifx\qwe\rty
```

will result in 'no'. Macros must have the same number of parameters to be considered equal by \ifx.

Can \ifx compare a macro and a character? After defining \def\a{*}, both
tests \ifx\a* and \ifx*\a are, surprisingly, a failure. However, after defining
\def\aster{*}, both \ifx\a\aster and \ifx\aster\a are successful.

A similar example is a test for a null macro parameter. A straight comparison
\ifx#1\empty... does not work. We first have to define a macro \inner whose
value is #1, and then compare \ifx\inner\empty.

```
\def\testnull#1\\{\def\inner{#1}%
 \ifx\inner\empty
  \message{yes}\else\message{no}%
 \fi}
```

▸ **Exercise 6.1:** What is \empty?
The test '\testnull \\' displays 'yes' on the terminal, while '\testnull *\\'
displays 'no'.

Interestingly, the auxiliary \inner macro is not necessary. It is possible to
say '\ifx\empty#1\empty..\fi'. If '#1' is empty, the \ifx would compare the
first \empty to the second one; otherwise, the second \empty would be expanded,
causing no harm.

To understand these results, we obviously need to know the rules for evaluating
\ifx. They are (1) if both quantities being compared are macros, they should have
the same number of parameters, the same top-level definition, and the same status
with respect to \long and \outer; (2) in any other case, the quantities compared
should have the same category code and the same character code, for the \ifx to
result in a match. A macro does not normally have either a character or a category
code. However, for the purpose of rule 2, a macro is considered to have character
code 256 and category code 16. So when a macro is compared to a character, they
will not match.

▸ **Exercise 6.2:** What is the result of \def\abc{Something} \edef\xyz{Something}
\ifx\abc\xyz...?
Rule 1 implies that \ifx can be used to compare macros, but the \ifx does not
expand them and does not look too deeply into their meanings. They are considered
equal if they look the same on the surface (see examples later). Rule 2 implies that

\ifx can compare two characters, but not a character and a string, or two strings. Thus

- \ifx AA is a match by rule 2;
- \ifx A#1 can be used, inside a macro, to see whether the first parameter is the letter 'A';
- \ifx Aa is a failure since the comparands have different character codes;
- \ifx {abc}{abc} fails since it compares a '{' to an 'a';
- \ifx A{B} fails since it compares the 'A' to the '{';
- the test \ifx*\a above fails because of rule 1.

With these rules in mind, the following discussion and examples are easy to understand. To compare a macro \a to a string {abc}, we first define \def\b{abc}, and then compare \ifx\a\b. If the string is a parameter of a macro, we can say \def\mac#1{\def\inner{#1} \ifx\a\inner...}. However, to compare #1 to a single character, we can simply say \def\mac#1{\ifx*#1...}, after which the expansion \mac* will be successful. There are some complex examples using this construct on [375–377].

To understand the meaning of "top-level definition," consider the definitions \def\a{*} \def\b{*}. The test \ifx\a\b is a success, but the test

```
\def\a{\b} \def\c{\d}
\def\b{tests} \def\d{tests}
\ifx\a\c...
```

is a failure, since \ifx compares only the top-level definitions and does not bother to expand \b and \d to find out that they are identical. Both tests '\def\qwe{\par} \def\rty{\par} \ifx\qwe\rty' and '\let\qwe=\par \let\rty=\par \ifx\qwe\rty' are successful, but

```
\let\xxx=\par \def\qwe{\par}
\def\rty{\xxx} \ifx\qwe\rty
```

fails, since \ifx does not expand its comparands to find their deep meaning.

This is an important feature of \ifx and has several consequences. One consequence is that \ifx is not bothered by undefined control sequences. It simply considers them all to be equal, so an \ifx comparison of two undefined control sequences always results in a match. Another consequence of the same feature is that two defined control sequences—such as, for example, \if and \ifcat—can be compared by \ifx without worrying about side effects resulting from their expansions. The comparison \ifx\if\ifcat is a failure, whereas \ifx\if\if is a success. It is also possible to compare the control sequence \ifx to itself, by means of \ifx. Thus \ifx\ifx\ifx yes\else no\fi results in 'yes'. Note that the \fi matches the first \ifx, and the other ifs shouldn't have any matching \fis since they are being compared, not executed. Consequently, the test \ifx\fi\fi yes\else no\fi also produces 'yes', as does \ifx\message\message \message{yes} \else \message{no}\fi.

An interesting effect occurs when we try (perhaps as a serendipitous error) '\ifx\message \message{yes}\else\message{no}\fi'. The \ifx compares the two tokens following, which are \message and \message. They are equal, so the

word 'yes' is typeset. It is not displayed in the log file because the control sequence
\message that would have normally displayed it has been used in the comparison. Continuing along the same lines, the test \ifx \message{yes} \else \mes-
sage{no}\fi compares the control sequence \message to the '{'. They are not
equal, because of rule 2 and, as a result, the *else* part is expanded, displaying 'no'
in the log file. Note that the string 'yes}' becomes part of the *then* part and is
skipped.

The reader should be able, at this point, to easily figure out the results of the
following tests:

```
    \def\qwe#1{trip}  \def\rty#1{trip}
1.  \ifx\qwe\message{yes}\else\message{no}\fi
2.  \ifx\qw\message{yes}\else\message{no}\fi
3.  \ifx\q\message{yes}\else\message{no}\fi
4.  \ifx\\message{yes}\else\message{no}\fi
5.  \ifx\message{yes}\else\message{no}\fi
6.  \ifx message{yes}\else\message{no}\fi
7.  \ifx mmessage{yes}\else\message{no}\fi;
```

Tests 1–3 are similar; the control sequences \qwe, \qw, \q are compared to
\message, which results, of course, in a 'no'. The undefined \qw and \q do not
produce any errors. In 4, the control sequence \\ (which is normally undefined) is
compared to the first 'm' of 'message', and in 5, \message is compared to the '{'.
Test 6 compares the first two letters 'me' of 'message' to each other. The six tests
result in a 'no' being displayed in the log file.

Test 7 compares the first two letters 'mm' of 'mmessage' to each other. They
are equal, so everything up to the \else is expanded. This results in the tokens
'essageyes'.

▶ **Exercise 6.3:** What are the results of

```
    \def\\{message}
(a) \ifx\\message{yes}\else\message{no}\fi
(b) \ifx\\\\ yes\else no\fi
```

and why?

More about undefined control sequences. The test \ifx\a\b, where \a, \b
are undefined, results in a match. This means that all undefined control sequences
have the same meaning. On the other hand, the test \ifx\a\relax, where \a is
undefined, is a failure. This means that an undefined control sequence is not equal
to \relax (at least not its upper-level meaning).

However, when the name of an undefined control sequence is synthesized by a
\csname–\endcsname pair, that control sequence is made equal to \relax. Thus if
\a is undefined, the construct \csname a\endcsname (that creates the name \a) is
set equal to \relax, and the test \expandafter\ifx\csname a\endcsname\relax
is a success. This is the basis of [ex. 7.7]. It describes a macro \ifundefined that
determines if any given string is the name of a defined macro.

```
\def\ifundefined#1{\expandafter\ifx\csname#1\endcsname\relax}
```

The test `\ifundefined a \message{yes}\else\message{no}\fi` writes 'yes' in the log file if `\a` is undefined.

▸ **Exercise 6.4:** Define a macro `\ifdefined#1` that will be the opposite of `\ifundefined` and will be used in the same way.

What if `\a` has been defined as `\relax`? Predictably, the test '`\let\a=\relax \ifundefined a`' is successful. It is (somewhat) more surprising that the test '`\def\a{\relax} \ifundefined a`' is a failure. This difference is a direct consequence of the difference between `\let` and `\def`. Following is a short discussion of that difference, which is important in advanced applications, where macros are defined and compared.

The general form of `\let` is '`\let⟨control sequence⟩=⟨token⟩`'. It defines the control sequence as being identical to the token. This is similar, but not identical, to `\def<control sequence>{<token>}`, and the following illustrates that difference. After `\def\a{X} \let\g=\a \def\h{\a}`, the sequence `\g\h` produces 'XX'. If we now redefine `\a`, the meaning of `\h` will change (since it was defined by `\def`) but `\g` will not change. Thus `\def\a{*} \g\h` produces 'X*'.

As a result, we can say that `\let\a=\b` assigns `\a` that value of `\b` which is current at the time the `\let` is executed, and this assignment is permanent. In contrast, `\def\a{\b}` assigns `\a` the name `\b`. When `\a` is expanded, its expansion causes an expansion of `\b`, so the result is the value of `\b`. Each expansion of `\a` may, therefore, be different since `\b` may be redefined.

A more formal way of saying the same thing is as follows: A `\let` makes a copy of the definition of `\b`, and that copy becomes the definition of `\a`; in contrast, `\def` sets a pointer to point to the definition of `\b`, and that pointer becomes the definition of `\a`.

Back to `\ifx`. The comparands of an `\ifx` are not limited to just macros, primitives, or characters. They can also be

- font names. `\font\abc=cmr10 \font\xyz=cmr10 \relax \ifx\xyz\abc` produces 'yes'.

▸ **Exercise 6.5:** Why the `\relax`?

- active characters (see [ex. 7.3]). The result of `\let\a=~ \ifx\a~` is a match.

▸ **Exercise 6.6:** Why does `\def\a{~} \ifx\a~` fail?

- names of the same TeX register. A test such as '`\countdef\me=3 \countdef\you=3 \ifx\you\me`' is a success.

- macros defined at run time, such as in:

```
\def\tone{\count0=9 A }%
\message{Enter a definition}%
\read16 to\note
\ifx\tone\note
 \message{yes}\else\message{no}
\fi
```

If the user enters '`\count0=9 A`' from the keyboard, in response to the message, there will be a match. Entering anything else, such as '`\count0=9 a`', will result in a failure. In either case the value of `\count0` will not be changed (by the way, what

is it?), nor will the letters 'A' or 'a' be typeset. Notice that a message entered from
the keyboard must terminate with a carriage return, which, in turn, is converted
by TeX into a space. This is why the definition of \tone must end with a space (to
avoid that, change the value of \endlinechar as explained on [48]).

6.4 \if

The second "difficult" comparison, \if, is executed in a completely different
way. TeX expands the token following the \if (if it is expandable), then expands
its expansion (if possible), and so on until only unexpandable tokens (characters
or unexpandable control sequences) are left. If less than two unexpandable tokens
are left, the process is repeated with the next input token. The process ends when
there are two or more unexpandable tokens to be compared, or when an \else or
an \fi are encountered. The final result is a string of unexpandable tokens, the
first two of which are compared by *character code* but not by category code. The
rest of the tokens, if any, are added to the *then* part.

If a comparand is an unexpandable control sequence, rather than a character, it is assigned a character code of 256 and a catcode of 16. Thus the tests
\if\hbox\vbox, \if\hskip\vskip, and \if\hbox\kern succeed. (See [209] for exceptions regarding the use of \let.) This also implies that comparing a primitive
to a character always fails.

There is also the case where evaluating the comparands results in just one
unexpandable token. Such a comparison should not be used since its result is
undefined. Unfortunately, no error message is given by TeX. The advanced reader
is referred to [§495] for the details of such a case.

The first example starts with \def\a{*}. Both tests \if\a*, \if*\a are successful (compare with the similar \ifx test above).

After \def\a{\b}, \def\c{\d}, \def\b{*}, \def\d{*}, the test \if\a\c is a
'yes', but \def\a{\b}, \def\c{\d}, \def\b{testing}, \def\d{testing}, \if\a\c
will fail, since the two tokens compared are the first two characters resulting from
the expansion of \a, which are 'te'. As mentioned above, the rest of \a (the string
'sting') and the whole of \c (the string 'testing') do not participate in the comparison, are added to the *then* part, and are therefore skipped. More insight into the
working of \if is provided by the test

```
\def\a{\b} \def\c{\d}
\def\b{ttsting} \def\d{ttsting}
\if\a\c \message{yes}\else \message{no}\fi
```

It compares the first two t's of \a. They are equal, so TeX expands everything up
to the \else. It displays 'yes' and also typesets the rest of \a ('sting') and the
whole of \c ('ttsting'). Note that, again, \c is not used in the test.

Similar results are obtained in the experiment

```
\def\tone{*}
\message{Enter a}\read16 to\note%
\if\tone\note
```

Assuming that the user enters '*\count90=89', the result will be a match,
and \count90 will also be set to 89. However, if the user enters '?\count90=89',

the comparison will fail, and \count90 will not be affected. Similarly, if the user enters '*abc', the comparison will be successful, and the string 'abc' will be typeset. Entering, '?abc' however, will result in 'no', and the string 'abc' will be skipped.

The test \if\s, where \s is undefined, results in the message '! Undefined control sequence', since \if always tries to expand its comparands.

Defining \def\w{xyz}, the test '\if x\w yes\else no\fi' is a success, since the first token of \w is an 'x'. However, the other two tokens are added to the *then* part, and the result of the test is the string 'yzyes'. Sometimes it is desirable to discard that part of \w that does not participate in the comparison. This is a special case of the general problem of how to extract the first token of a macro \w and discard the rest.

One way of doing it is

```
\def\tmp#1#2\\{#1}
\ifx\w\empty
 \def\W{}
\else
 \def\W{\expandafter\tmp\w\\}
\fi
```

When \W is expanded, the first step is to expand \w, and the second, to expand \tmp. The first argument of \tmp is thus the first token of \w, and the second argument, the rest of \w. The result of expanding \tmp is thus the single token 'x', and that token becomes the definition of \W. The test \if x\W yes\else no\fi now results in the string 'yes'. This method works even if \w is \empty.

▸ **Exercise 6.7:** Perform the test

```
\if\the\count90 \the\count90
 \message{yes}\else\message{no}\fi
```

for \count90 set to 1, 11, and 12.

The next example is the two tests \let\a=~ \if\a~, \def\b{~} \if\b~. In the first test, the \let makes \a equivalent to the active character '~'. In the second one, the \def makes \b a macro whose definition is the same active character '~'. The \if expands its comparands, so it ends up comparing '~' to '~'. Both tests thus result in a match.

Having mentioned active characters, let's use them to further illustrate the behavior of \if. The following

```
\def\a{*~}
\hbox{Mr. Drofnats}
\hbox{Mr.\if*\a\fi Drofnats}
\hbox{Mr.\if+\a\fi Drofnats}
```

results in

Mr. Drofnats
Mr. Drofnats
Mr.Drofnats

which is easy to explain. The test \if*\a\fi expands \a and only uses its first character (the '*'). The second character (the tilde) remains and affects the space between 'Mr.' and 'Drofnats' (it has the effect of \frenchspacing). In contrast, the test \if+\a\fi expands \a and, since there is no match, *skips* the second character. As a result, there is no space between 'Mr.' and 'Drofnats'.

The \noexpand command can be used to suppress expansion during an \if. Assuming the definitions \def\q{A}, \def\p{9}, the test \if\p\q fails since it compares the characters 'A', '9'; however, the test \if\noexpand\p\noexpand\q is a success (even if the macros involved are undefined).

6.5 \ifcat

The third comparison, \ifcat, is less useful. It works like \if, expanding its comparands, and resulting in a string of characters, of which the first two are compared by category codes [37], but not by character codes. For example, the catcode of '&' is 4 (alignment tab) and the catcode of '8' is 12 (other). If we change the catcode of '8' to 4 and compare '\catcode'8=4 \ifcat 8&', we get a 'yes'. It is hard, however, to find simple, practical examples for \ifcat (the examples on the notorious [377] are hardly simple or practical).

Similar to an \if, there is also the case where expanding the comparands results in an unexpandable control sequence, rather than a character. In such a case, TeX assigns it a character code 256 and a catcode of 16. Thus all the following comparisons, \ifcat\hbox\vbox, \ifcat\hskip\vskip, and \ifcat\hbox\kern, succeed. (Again, see [209] for exceptions concerning the use of \let.)

The category code of a character can be typeset by \the\catcode'A. It can be displayed in the log file by \showthe\catcode'A. This does not work for control sequences since they have no catcode. When comparing control sequences with an \ifcat, they are first expanded, and the first two tokens are compared. For example, after defining \def\a{&} \def\b{+=} \def\c{true}, the comparison \ifcat\a\b fails, since the catcodes of '&' and '+' are different. However, the comparison \ifcat\b\c is a 'yes' since the comparands are '+' and '='. The string 'true' is typeset.

It is possible to compare macros without expanding them. Assuming the definitions of \a, \b above, the test \ifcat\noexpand\a\noexpand\b results in a match since it does not expand the macros, and they are treated as undefined (category code 16).

▶ **Exercise 6.8:** With \c defined as above, what is the result of \ifcat\c?

Perhaps the simplest practical example of \ifcat is a test for a letter. Assuming that the parameter of macro \suppose is supposed to be a letter or a string starting with a letter, the macro can be defined as
\def\suppose#1{\ifcat A#1...\fi...}.

▶ **Exercise 6.9:** If the parameter of \suppose is a string, only the first character will be used by the \ifcat, and the rest will be added to the *then* part, perhaps interfering with the rest of the macro. Generalize the definition of \suppose to suppress the rest of the parameter during the \ifcat.

The examples

```
\def\a{~} \ifcat\a~
\let\b=~ \ifcat\b~
```

are identical to the ones shown earlier, in connection with \if. They behave the same as in that case, both resulting in a 'yes'.

Another example of \ifcat is to find out if a given token is a '{'. An \ifcat can be used to compare the token's catcode to that of a '{'. Since the user may change the catcode of '{', it is better to compare to the catcode of \bgroup. Thus '\ifcat{\x...' or '\ifcat\bgroup\x...' or, better yet, '\ifcat\bgroup\noexpand\x...'.

Active characters may also be compared with an \ifcat, since they all have the same catcode (13). After defining \catcode'?=13 \def?{:}, \catcode'!=13 \def!{;}, the test \ifcat?! is a success, seemingly confirming the above statement. A deeper look, however, shows that the test expands the two active characters and compares the catcodes of their values! The values just happen to have the same catcode.

To actually compare the catcodes of the active characters, a \noexpand should be used to prevent their expansions. Thus the test \ifcat\noexpand?\noexpand! compares the catcodes of the active characters without expanding them (and is also a success).

6.6 Nested ifs

In principle, it is possible to nest ifs one inside another. An if may be a comparand of another if, or it may be nested in either the *then* or the *else* part of another if. However, because our three ifs work in different ways, not every combination of nested ifs is valid. In general, a nested if is written as '\if..\if<inner>\fi..\else..\if<inner>\fi..\fi', where any of the inner ifs may have an *else* part and may itself be nested by other ifs. However, as the examples below show, such an if should be carefully analyzed before it is used, since it tends to produce unexpected results.

Since \if evaluates its comparands, they can be other ifs. Defining \def\a{}, \def\b{**}, the test '\if\ifx\a\b1\else\if\a\b23\fi\fi\else4\fi' is an \if with an \ifx as a comparand (see [ex. 20.13g]). The \ifx, in turn, has another \if nested in its *else* part.

The process starts when the outer \if evaluates its comparands in order to come up with two tokens for comparison. It activates the \ifx, which, in turn, compares \a and \b. They are not equal, so the '1' is skipped, and TeX starts executing the *else* part of the \ifx. This part contains the inner \if, which evaluates \a and \b, compares the two asterisks, and results in the '23'. The outer \if is now equivalent to \if23\else4\fi, which typesets the '4'.

The test '\if\ifx\a\b1\else\if\a\b22+\fi\fi\else3\fi' is similar, it has the '+' left over after the comparison, so it gets typeset.

▶ **Exercise 6.10:** What is typeset by the following?
\if\ifx\b\b1\else\if\a\b2\fi\fi+\else3\fi

The \ifcat comparison is similar to \if in that it first evaluates all its comparands. As a result, other comparisons may be used as comparands and may also be nested inside an \ifcat. Compare the following tests to the ones above:

\ifcat\ifx\a\b1\else\if\a\b234\fi\fi\else5\fi
\ifcat\ifx\a\b1\else\if\a\b22+\fi\fi\else3\fi
\ifcat\ifx\b\b1\else\if\a\b2\fi\fi\else3\fi

Since \ifx does not evaluate its comparands, they cannot be other \ifs. In the test \ifx\if\a\b..., the \ifx would simply compare the \if to the \a. We cannot even use braces to separate the inner and outer ifs \ifx{\if\a\b...}... since the \ifx will compare the '{' with the \if, and TeX will eventually complain of an 'Extra }'.

We can, however, nest an if (of any type) in either part of an \ifx. The test '\ifx\a\b\else\if\a\b ok\fi\fi' (with \a, \b defined as above) typesets 'ok'. The \ifx compares \a and \b and finds them different. It skips to the *else* part and expands it. The inner \if is thus executed in the usual way; it finds two identical tokens (the two asterisks of \b), and typesets 'ok'.

A note about user-defined ifs. The \ifdefined macro discussed earlier is used almost like any of the other 17 conditionals; each \ifdefined must have a matching \fi. However, because of the way it is defined, \ifdefined cannot be nested inside another if. Trying to nest it in the *then* part of an if, such as in

```
\iftrue
\ifdefined x Yes\else No\fi
\fi
```

works, but nesting it in the *else* part

```
1 \iftrue
2 \else
3 \ifdefined x Yes\else No\fi
4 \fi
```

produces the error message '! Extra \fi'. TeX has to skip all the material between the \else and the \fi. As a result, it assumes that the \fi on line 3 matches the \iftrue. The \fi on line 4 is thus considered extra.

One last point: When evaluating a condition, the first step is to evaluate the comparands. In the nested conditional '\ifnum1=2\ifnum3>4 8\else9\fi\fi' the second comparand, '2', is not properly terminated, so the inner \ifnum will be expanded as part of evaluating the second comparand of the outer \ifnum. The outer if will then be executed as '\ifnum1=28\fi' or '\ifnum1=29\fi'.

6.7 Examples

1. A practical example is macro \flexins below. It lets the user decide, *at run time*, whether any floating insertion should be a \midinsert or a \topinsert. The user is prompted to enter either 'mid' or 'top' from the keyboard. In response, the macro uses a nested \ifx to create either a \midinsert or a \topinsert.

```
\def\flexins{%
\def\b{mid } \def\d{top }
\message{mid or top? }\read-1 to\a%
\csname
  \ifx\a\b mid%
  \else
    \ifx\a\d top\fi
  \fi insert%
```

```
\endcsname
            }
\flexins
Insertion material
\endinsert
```

What if the user enters none of these inputs? Clearly \flexins should be extended so it can recover from a bad input. It is a good idea to expand \flexins recursively, in such a case, to give the user another chance to enter a valid input. The first try is

```
\def\flexins{%
\def\b{mid } \def\d{top }
\message{mid or top? }\read-1 to\a%
\csname
\ifx\a\b mid\else
   \ifx\a\d top\else \flexins\fi
\fi insert\endcsname
}
```

It does not work! When TEX expands \flexins recursively, it is still inside the \csname. During the recursive expansion it finds \def\b, but \def is not expandable and thus is not supposed to be inside a \csname [40]. The result is an error message (which one?).

We now realize that we have to delay the recursive expansion of \flexins until we get out of the \csname–\endcsname pair. The final version is

```
\def\flexins{%
\def\b{mid } \def\d{top }
\def\badinsert{\flexins}
\message{mid or top? }\read-1 to\a%
\csname
\ifx\a\b mid\else
   \ifx\a\d top\else bad\fi
\fi insert\endcsname
}
```

In the case of bad input, the \csname–\endcsname pair creates the control sequence name \badinsert. We predefine it to simply expand \flexins, which then asks the user for another input.

Proposed by R. Whitney.

2. This is a generalization of the previous example. Macro \yesno below prompts the user to respond with a 'Y' or an 'N', but also accepts the responses 'y', 'n'. It does the following:

- Prompts the user with a question where the response can be 'Y', 'N', 'y', or 'n'.
- Reads the response into \ans.
- Uses \ifx to compare \ans to macros containing one of the valid responses.
- If a match is found, uses \csname to create the name of, and expand, one of the macros \yesresult, \noresult. These macros should be predefined to do anything desirable.

■ If no match is found, expands \badresult, which, in turn, should expand \yesno recursively.

```
\def\y{y } \def\n{n } \def\Y{Y }
\def\N{N } \def\badresult{\yesno}
\def\yesresult{<whatever>}
\def\noresult{<whatever>}
\def\yesno{%
\message{Respond with a Y or N! }
\read-1 to\ans
\csname
\ifx\y\ans yes\else
  \ifx\Y\ans yes\else
    \ifx\n\ans no\else
        \ifx\N\ans no\else bad%
        \fi
      \fi
  \fi
\fi result\endcsname
}
\yesno
```

A different version of \yesno uses \if instead of \ifx. We start with

```
\def\badresult{\yesno}
\def\yesresult{<whatever>}
\def\noresult{<whatever>}
\def\yesno{%
\message{Respond with a Y or N! }
\read-1 to\ans
\csname
\if y\ans yes\else
  \if Y\ans yes\else
    \if n\ans no\else
      \if N\ans no\else bad%
      \fi
    \fi
  \fi
\fi result\endcsname
}
```

It compares \ans to the token 'y' instead of the macro \y, but it does not work! Macro \ans contains a 'y' (or 'Y' or whatever), followed by a space. The space gets added to the *then* part, which then becomes '␣yes', creating the control sequence \␣yesresult. To get this to work, the first token of \ans has to be extracted, and all the other ones discarded. This can be done (see also page 167) by

```
\def\tmp#1#2\\{#1}
\def\sna{\expandafter\tmp\ans\\}
```

Macro \sna now contains just one character, and the next version is

```
\def\badresult{\yesno}
\def\yesresult{<whatever>}
\def\noresult{<whatever>}
\def\yesno{%
\message{Respond with a Y or N! }
\read-1 to\ans
\def\tmp##1##2\\{##1}
\def\sna{\expandafter\tmp\ans\\}%
\csname
\if y\sna yes\else
  \if Y\sna yes\else
    \if n\sna no\else
      \if N\sna no\else bad%
      \fi
    \fi
  \fi
\fi result\endcsname
}
```

Note that it works for any response that's a string starting with one of the four valid characters.

▶ **Exercise 6.11:** Extend this example. Define a macro \triresponse that accepts the responses 'left', 'right', 'center', or any strings that start with 'l', 'r', or 'c'. The macro then expands one of the (predefined) macros \doleft, \doright, \docenter, or \dobad.

3. A practical example of the use of \ifcat arises when style files or format files are used. If such a file has internal macros, they can be made private by declaring \catcode`@=11 and giving the macros names that include the '@'. At the end of the file, a matching \catcode`@=12 should be placed. The problem occurs when such a style file, say b.sty, is \input by another file, a.sty, that also contains the pair \catcode`@=11, \catcode`@=12. A simple experiment should reveal the problem to readers who still don't see it. The solution is to place the test

```
\ifcat @A\else
 \chardef\catcount=\catcode`@
 \catcode`@=11
\fi
```

at the beginning of b.sty, and reset at the end to \catcode`@=\catcount. (This example actually shows how hard it is to find good examples of \ifcat. The test \ifcat @A could produce unexpected results if the '@' happens to be active. A better solution is '\ifnum\catcode`\@=\catcode`A'.)

▶ **Exercise 6.12:** Use \ifcat to solve the following: Given \def\foo#1{...}, devise a test to see if, in the expansion \foo..., the argument is delimited by a space. Normally, such a space is automatically absorbed by TeX and cannot be recognized.

4. A compound macro argument.

Macro `\compndArg` accepts a compound argument and breaks it down into its components. The argument should be of the form xxx,xxx,...,xxx; (the ',' separates the individual components and the ';' delimits the entire argument). The macro accepts the argument (without the ';', of course), appends ',;,' to the argument, and makes the whole thing the argument of `\pickup`, which is then expanded.

```
\def\compndArg#1;{\pickup#1,;,}
\def\pickup#1,{% Note that #1 may be \null
\if;#1\let\next=\relax
\else\let\next=\pickup
  \message{'#1'}% use #1 in any way
\fi\next}
```

Macro `\pickup` expects its arguments to be delimited by a comma, so it ends up getting the first component of the original argument. It uses it in any desired way and then expands itself recursively. The process ends when the current argument becomes the semicolon. Note the following:

- This is also an example of a macro with a variable number of parameters. The compound argument may have any number of components (even zero; see below).
- The method works even for an empty argument. The expansion '\compndArg ;' will cause `\pickup` to be expanded with a null argument.
- The macros do not create spurious spaces. In many macro lines, the end-of-line character is converted to a space, which is eventually typeset if the macro is invoked in horizontal mode. Such lines should be identified, with a test such as C\compndArg g;D, and should be terminated by a '%'. Try the above test with and without the '%' in the first line of `\pickup`. A similar technique is used in macro `\Cboxit` on page 188.

6.8 Evaluation of Conditionals

We now discuss the details of the evaluation of a conditional. The first step is to evaluate the ⟨test⟩ (which itself may involve macro expansions). The next step depends on the result, but it involves a skip over the *then* part or the *else* part. Macros are not expanded during the skip, and braces found don't have to be balanced. However, nested conditionals found during the skip have to be balanced (§500).

The simple example '\if<test>...\else\if....\fi....\fi' illustrates this point. If the test results in **true**, TEX has to skip over the *else* part. Since this part contains a nested conditional, the first `\fi` found is assumed to terminate that conditional, and only the second `\fi` is matched with the outer conditional. The skip, therefore, is done to the **matching** `\fi`.

This point becomes important when a conditional should create another conditional as its final result. Imagine the simple macro '\def\newtest#1{\ifnum1=#1 \iffalse\else\iftrue\fi}'. It compares its argument to the number '1' and should create either an `\iffalse` or an `\iftrue`, depending on the result. An expansion of such a macro should always be used in conjunction with an \fi, so the simple test '\newtest1 ABC\fi' should expand to '\iffalse ABC\fi', and '\newtest2 ABC\fi' should expand to '\iftrue ABC\fi'.

The discussion above shows why this simple definition will not work. When TeX skips over the *then* or the *else* part of the \ifnum, it encounters the \iffalse or the \iftrue and considers it a nested conditional, requiring an \fi. The first \fi (or \else) found (the one matching the \ifnum) is thus matched with that conditional, creating an error.

▶ **Exercise 6.13:** Describe the details of the expansion '\newtest1ABC\fi'.

A correct definition of \newtest should use the fact that macros are not expanded during the skip. The \iffalse and \iftrue should be created (outside the \ifnum) by inner macros. One way of achieving this is

```
\def\createiffalse{\let\next=\iffalse}
\def\createiftrue{\let\next=\iftrue}
\def\newtest#1{\ifnum1=#1 \createiffalse\else\createiftrue\fi\next}
```

There is another, more sophisticated, solution that works by using special macros to properly skip over the \else and the \fi. Once the macros do that, they supply an \fi to match the \ifnum, and create either an \iffalse or an \iftrue.

```
\def\falseskip#1\fi{\fi\iffalse}
\def\trueskip\fi{\fi\iftrue}
\def\newtest#1{\ifnum1=#1 \falseskip\else\trueskip\fi}
```

If the \ifnum is a match, \falseskip skips material up to the \fi (gobbling up the \fi as a delimiter). No inner conditionals are found during the skip. The material skipped over becomes argument '#1', which is ignored. \falseskip then supplies an \fi to match the \ifnum and finally creates an \iffalse. \trueskip works similarly but does not require a parameter.

6.9 Conclusion

The main source of the confusion surrounding the various \if comparisons is the inability to find out exactly what TeX is comparing. In future extensions of TeX it would be useful to have a control sequence \tracingcomparands, such that setting \tracingcomparands>0 would show, on the terminal, the actual quantities compared. This seems to involves changes in §506–508.

Large, small, long, short, high, low, wide, narrow, light, dark, bright, gloomy, and everything of the kind which philosophers term accidental, because they may or may not be present in things,—all these are such as to be known only by comparison.

— *Leon Battista Alberti, On Painting (1435)*

7. Examples of Macros

Computer programming is best studied by practicing it, and writing macros *is* computer programming. The examples presented here have been carefully selected. They are simple to understand; each illustrates an important point, a useful technique, or a common pitfall in macro writing; and most either are useful in practice or can easily be modified to make them useful. They are not arranged in any special order. They range from the simple (such as the slanted lines) to the complex (verbatim listing), and from the useful (dropshadow) to the esoteric (the grid macros).

The conscientious reader should read each example carefully and actually run it. If any code remains unclear, it should be either modified or eliminated, to see how this affects the results. Needless to say, it is important to try to work out all the exercises.

The author welcomes suggestions for other useful examples to be included in future editions of this book.

7.1 Slanted Lines

It is easy to create horizontal and vertical rules in TeX, but slanted lines are sometimes needed. Such a line can be produced by repeatedly printing a dot and moving it in small steps. This idea is due to A. Hendrickson (Ref. 22), the originator of many useful macro ideas. Here we use a small rule instead of a dot, since the size of a rule can be specified and can be adjusted precisely to the resolution of the printer used. The sequence

```
\newdimen\dotsize \dotsize=1pt
\newdimen\step \step=0pt
\def\Dot{\vrule width\dotsize height\dotsize}
\hbox{%
\raise\step\hbox{\Dot}\advance\step by\dotsize
\raise\step\hbox{\Dot}\advance\step by\dotsize
\raise\step\hbox{\Dot}\advance\step by\dotsize
\raise\step\hbox{\Dot}\advance\step by\dotsize}
```

generates ⟋ . This, of course, is possible only for a few dots. All the work is done in the \hbox, which is mode neutral. Without the box, the above code would only work in the horizontal modes, since **\raise** cannot be used in the vertical modes.

▶ **Exercise 7.1:** What are good values for \dotsize?

The next version, macro \slantA, is recursive. Note how it starts a box and expands macro \slantAaux to create the dots and close the box.

```
\newcount\loup
\def\slantA{\hbox\bgroup\slantAaux}
\def\slantAaux{\raise\step\hbox{\Dot}\advance\step by\dotsize
 \advance\loup by1
\ifnum\loup<6 \slantAaux\else\egroup\step=0pt \loup=0 \fi}
```

The sequence '\loup=0 \step=0pt \slantA' generates ⟋.

The next step is to replace the constant 6 with a parameter. This leads us to macro \slantB.

```
\def\slantB#1{\hbox\bgroup\slantBaux{#1}}
\def\slantBaux#1{\raise\step\hbox{\Dot}\advance\step by\dotsize
 \advance\loup by1
\ifnum\loup<#1 \slantBaux{#1}\else\egroup\step=0pt \loup=0 \fi}
```

The tests '\slantB6It $math \slantB8 mode$' generate ⟋It *math*⟋*mode*. An important point about \slantB is that it does not generate any spurious spaces. This should always be verified in a macro that's supposed to work in H mode. The prudent reader should try to insert a space between the \slantB{#1} and the \else, to see how the space is translated into several spaces that are typeset after each slanted line.

Notice also that the macro works in math mode (does it also work in display math mode?) because it does not any features that are prohibited in that mode.

▶ **Exercise 7.2:** Generalize \slantB by adding one more parameter such that #1 is the number of steps and #2, the vertical rise between steps.

7.2 Generating a Grid

We now turn to the generation of a grid, made up of horizontal and vertical rules. The problem is to loop and generate the right number of rules, with the right spacing between them.

Macro `\grida` builds and typesets a `\vbox` with the grid. The `\loop..\repeat` construct is used twice, and the macro works in either V or H mode. In V mode, the generated `\vbox` has zero dimensions, which means that the reference point is not moved. In H mode, the `\vbox` has the width of the grid, thus moving the reference point to the right. The macro has three parameters. #1 is the number of horizontal grid lines, #2 is the number of vertical lines, and #3 is the height and width of each grid box (in points).

```
1  \newcount\kount
2  \def\grida#1 #2 #3.{%
3  \kount=1 \setbox1=\hbox{%
4  \loop                  % build vertical lines in \box1
5  \vrule height#3pt
6  \ifnum\kount<#2 \advance\kount by 1 \kern-0.4pt\kern #3pt
7  \repeat}%
8  %     box 1 is now loaded with short vertical lines |  |  |  |  ... |
9  \kount=1 \setbox2=\vbox{
10 \loop                  % over horizontal lines
11 \hrule width\wd1
12 \ifnum\kount<#1 \advance\kount by 1 \kern-0.4pt\copy1
13 \repeat}%
14 \ifvmode\ht2=0pt\wd2=0pt\fi        %dont move the ref point in v mode
15 \box2\ifvmode\nointerlineskip\fi} %end of grida
```

The macro is straightforward. Lines 3–7 set `\box1` to the pattern| | | | | | | | but with a width of zero (the pattern sticks on the right). The loop on lines 9–13 sets `\box2` to hrules separated by copies of `\box1`.

The tests are

```
\grida 4 6 2.AY\grida 8 4 6.BM \smallskip \grida 5 8 3.Y%
\lower14pt\hbox{\grida 6 7 4.}M \smallskip\grida 4 8 5.
```

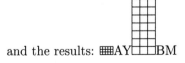

and the results: ▦AY⬚BM

The reader should look carefully at the tests and should try to figure out the modes changes between them (that is, which grids were done in H mode and which ones, in V mode).

▸ **Exercise 7.3:** Modify `\grida` to produce rectangular grid boxes. Parameter `#3` should specify the width of each grid box, and a new parameter, `#4`, should be added to specify the height of a box.

▸ **Exercise 7.4:** Define a macro `\gridb #1 #2 #3.` that builds the grid by placing full-size vertical rules in `\box1`, placing full-size horizontal rules in `\box2`, and overlapping the two boxes. Make sure the macro doesn't create spurious spaces in H mode.

Here is a third method for typesetting a grid. Macro '`\gridC#1x#2(#3)`' uses an `\halign` to typeset a grid of '`#1`' rows by '`#2`' columns. Each grid box is '`#3`' pt on a side. The macro proceeds in three steps:

1. It defines an internal macro '`\tabline`' to create a single row of the alignment. The definition of '`\tabline`' is '`{&&...&\cr}`', where the number of '`&`' is one less than the number of columns. Note the `\edef` and `\aftergroup` trickery used in the definition.

2. It builds an alignment where the preamble uses the '`&&`' notation for an indefinite number of columns. A loop expands `\tabline` once for each row. An `\hrule` is placed below each row (except the last one) by a `\noalign`. The result of this step is an alignment of the form:

```
\halign{#\tabskip=10pt&&\vrule height10pt#\cr
&&&\cr\noalign{\hrule}&&&\cr\noalign{\hrule}&&&\cr}
```

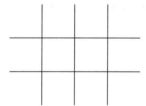

which creates

3. The result is placed in `\box0` and is surrounded by four rules using macro `\boxit`.

```
\newcount\ct

\def\gridC#1x#2(#3){\ct=#2
\begingroup\let\ag=\aftergroup \ag\edef\ag\tabline\ag{%
  \loop\ifnum\ct>1 \ag&\advance\ct-1 \repeat
  \ag\cr\ag}\endgroup
\ct=#1
\setbox0=\vbox{\offinterlineskip
\halign{##\tabskip=#3pt&&\vrule height#3pt##\cr
  \finishgrid}}\boxit{\box0}0}

\def\finishgrid{\tabline\ifnum\ct>1 \noalign{\hrule}
  \global\advance\ct-1 \finishgrid\fi}
```

Note that the macro cannot handle grids of a single column. The reader is encouraged to find out why and to generalize the macro.

7.3 Scanning the Argument

The next few sections present examples of macros that scan their argument, token by token, examining each token, and taking different actions depending on certain characters or specific catcodes found.

The examples include transferring an argument from one macro to another, reversing a string, ignoring the last token of a string, printing a string vertically, writing strings verbatim on a file, listing strings verbatim, and a few more. They are all based on the same principle, explained below.

7.4 Transferring an Argument

Macro \length below (see also [219]) measures the length of its argument, which must be a string (embedded blanks are ignored in the counting).

```
\def\length#1{{\count0=0 \getlength#1\end \the\count0}}
\def\getlength#1{\ifx#1\end \let\next=\relax
  \else\advance\count0 by 1 \let\next=\getlength\fi \next}
```

The key to its operation is the way it transfers its argument, '#1', to macro \getlength. It is done by \getlength #1. Assuming that the argument is ABC, macro \length now reads \count0=0 \getlength ABC\end \the\count0. The point is that the argument ABC is no longer enclosed in braces, so \getlength will only receive the first character, 'A', as its argument.

TeX starts expanding \length by setting \count0=0. It then starts expanding \getlength. The argument of \getlength is, as usual, the next token in the input stream. This, however, happens to be the *single* letter A. So TeX executes \getlength with A as the argument, and with B as the next unread token in the input stream. The \ifx compares the argument A to \end, and they are found to be different. Thus \next becomes \getlength. The last thing in the expansion of \getlength is the expansion of \next.

When TeX starts expanding \next, it finds that \next is equal to \getlength, so it expands \getlength for the second time. Again, the argument is the next source token, which happens to be B. During the second expansion, the next unread source token is C. It is read when \getlength is invoked for the third time. In the fourth expansion, the next token happens to be \end, which results in \next=\relax.

At this point the expansion of \getlength terminates and the execution of \length is resumed. It only contains one more command, \the\count0 (or \number\count0 [40]), which typesets the result. The whole process can be easily traced by setting \tracingmacros=2.

▸ **Exercise 7.5:** Why are there two sets of braces in the definition of \length?

A similar example is macro \dimendemo, which uses the same method to print each (nonblank) character of the argument followed by its dimensions. See also [ex. 11.5]. Look carefully at the results, to see the difference between \the and \number.

```
\def\dimendemo#1{\getlength#1\end \par}
\def\getlength#1{\ifx#1\end \let\next=\relax
```

```
\else\setbox0=\hbox{#1}%
\line{$<$\copy0$>$\ wd=\number\wd0\ ht=\the\ht0\ dp=\the\dp0;\hfil}
\let\next=\getlength\fi \next}
```

The test `\dimendemo{my phi.}` produces
<m> wd=546135 ht=4.30554pt dp=0.0pt;
<y> wd=345886 ht=4.30554pt dp=1.94444pt;
<p> wd=364090 ht=4.30554pt dp=1.94444pt;
<h> wd=364090 ht=6.94444pt dp=0.0pt;
<i> wd=182045 ht=6.67859pt dp=0.0pt;
<.> wd=182045 ht=1.05554pt dp=0.0pt;

Result: The proper way for a macro to transfer its argument to another macro is by enclosing it in braces. Without braces, useful effects (but also bad macros) can be achieved.

7.5 Vertical Printing

Another example along the same lines is macro `\verti`, which typesets its parameter vertically. Such a thing is useful if a font is available, where the characters are rotated 90°. The following macros

```
\newbox\ver
\def\verti#1{\setbox\ver=\vbox{}\vertcl#1\end\endd}
\def\vertcl#1#2\endd{\ifx\end#1\box\ver
                   \else\setbox0=\hbox{#1}%
\setbox\ver=\vbox{\unvbox\ver\kern.3ex\box0}%
                   \vertcl#2\end\endd\fi}
```

will typeset the parameter of `\verti` vertically, such that A\verti{wmit}B will

produce At B. The argument of `\verti` must be a string with no embedded spaces.

▶ **Exercise 7.6:** Use `\verti` to create this pattern: Woodward

▶ **Exercise 7.7:** Change the definition of `\vertcl` such that 'A\verti{wmit}B' will produce A B.

▶ **Exercise 7.8:** Change the definition of `\vertcl` such that 'A\verti{wmit}B' will produce A t B. The idea is to center each character in an `\hbox to1em`, and then place this box vertically in `\box\ver`.

7.6 Reversing the Argument

Macro \revers reverses its argument, which must be a string of characters without spaces or control sequences. The first try is

```
\def\revers#1{\toks0={}\macb#1\\}
\def\macb#1#2\\{\toks0={#1\the\toks0}\def\tmp{#2}
\ifx\tmp\empty\let\next=\macc\else\let\next=\macb\fi\next#2\\}
\def\macc#1\\{\showthe\toks0}
```

Macro \revers simply appends '\\' to its argument and expands \macb with the augmented argument. Macro \macb separates the first character of the argument, places it on the left end of \toks0, and if the rest of the string (#2) is not empty, expands itself recursively. If the rest is empty, \macb expands \macc, which can use the reversed string in \toks0 in any way.

This version does not work because the assignment \toks0={#1\the\toks0} does not expand \the\toks0 (see list of no expansions on [215]). We saw earlier that the assignment \toks0={\the\toks0#1} should be done as \toks0=\expandafter{\the\toks0#1}. The \expandafter reverses the order of expansion of the { and the \the\toks0. To get '\toks0={#1\the\toks0}' we have to reverse the order of expansion of the three quantities {, #1, and \the\toks0. This is done by the code \expandafter{\expandafter#1\the\toks0}, and the next version works.

```
\def\revers#1{\toks0={}\macb#1\\}
\def\macb#1#2\\{%
\toks0=\expandafter{\expandafter#1\the\toks0}
\def\tmp{#2}
\ifx\tmp\empty\let\next=\macc\else\let\next=\macb\fi\next#2\\}
\def\macc\\{\showthe\toks0}
```

▶ **Exercise 7.9:** Rewrite the above macros to build the reversed string in \box0, instead of in \toks0.

Finally, here is a much simpler version that typesets the reverse of its argument, without storing it anywhere. The first version is
\def\reverse#1#2\end{\ifx#2\end\else\reverse#2\end\fi#1}.

The next version uses the same principle but checks for a null #2 by defining an auxiliary macro \aux.

```
\def\reverse#1#2\end{\def\aux{#2}%
 \ifx\empty\aux\else\reverse#2\end\fi#1}
```

7.7 Scanning a String

Recursive macros can be used to perform simple operations on character strings. The idea is to break a string argument into two parts; the first character and the rest. The desired operation is performed on the first character, and the macro is recursively expanded with the rest of the string as the new argument. Macro `\scan` below expands `\onestep` to perform the operation, then expands itself recursively.

```
\def\scan#1#2\end{\def\aux{#1}%
\ifx\aux\empty
\else
 \def\aux{#2}%
 \onestep{#1}%
 \ifx\aux\empty
 \else
  \scan#2\end
\fi\fi}
```

Here are two applications of this technique:

1. Measuring the size of a string. We define

```
\newcount\ccnt
\def\onestep#1{\advance\ccnt by1 }
```

and can write expansions such as 'A\scan abcd\end\the\ccnt B'. If the string contains spaces, `\obeyspaces` should be used, as in:

```
{\obeyspaces
A\scan{}ab cd\end\the\ccnt{}B}
```

Note the empty braces. They replace spaces that otherwise would have been counted. If the string is stored in a `\toks` register, an `\expandafter` is necessary: 'A\expandafter\scan\the\toks0 \end\the\ccnt B'.

▶ **Exercise 7.10:** The space between `\the\toks0` and `\end` is necessary; why?

2. The definition '`\def\onestep#1{\ifx\aux\empty\else#1\fi}`' can be used to delete the last character of the string. (Try it. Also see section 7.13.)

7.8 Adding a Constant

The active character '!' below scans its argument, adds a constant to each character code, and typesets the new character. Such a thing is useful, say, when working with an "exotic" language, where many characters have unusual accents, or double accents. A special font can be developed where the first half (the first 128 positions) is occupied by the normal characters, and the second half, by the same characters with accents. Saying 'A' will typeset an 'A', but '!A' will typeset the character located 128 positions away from 'A'.

In the listing below, the '!' is used with font cmr10, and the constant is 8 instead of 128. Thus the result of 'A!B C' is 'AJC', and that of 'a!bc d' is 'ajkd'.

```
\catcode'\!=13
\newcount\Temp
```

```
\def!#1 {\aux#1\end}
\def\aux#1{\ifx\end#1\let\next=\relax
\else\let\next=\aux
\Temp='#1 \advance\Temp by 8 {\char\Temp}\fi
\next}
```

7.9 Writing Strings on File

When material is written on a file, any expandable tokens are expanded. Thus '\write\aux{\abc}' writes the expansion (replacement text) of \abc. If a token is unexpandable, its name is written. If the token is an undefined control sequence, an error message is issued when TeX tries to expand it.

Sometimes we want to inhibit expansion during file output. This can be done by either '\write\aux{\noexpand\abc}' or '\write\aux{\string\abc}'. The former form writes a space after \abc, while the latter one does not.

Using this idea, we now solve the general problem of writing any token string (containing text and control sequences) on a file without expansion. The solution uses macros \scan and \onestep above. Macro \onestep receives the next token in the string, checks to see if it is a control sequence (by comparing its catcode to that of \relax), and if it is, inserts a \noexpand in front of it. The new string is created, token by token, in \toks0, which is eventually written on the file.

```
\newwrite\out
\immediate\openout\out=tempfile

\def\main#1{#1{\toks0={}\obeyspaces\scan#1\end%
\immediate\write\out{\the\toks0}}}

\def\scan#1#2\end{\def\aux{#1}%
\ifx\aux\empty
\else
 \def\aux{#2}%
 \onestep{#1}%
 \ifx\aux\empty
 \else
  \scan#2\end
\fi\fi}

\def\onestep#1{\ifcat\relax\noexpand#1%
\toks0=\expandafter{\the\toks0\noexpand#1}%
\else\toks0=\expandafter{\the\toks0 #1}\fi}
```

Note the (local) use of \obeyspaces in \main. Without it, spaces are skipped in the scanning process and thus don't end up in \toks0.

A different way of avoiding expansion during file output is to turn an expandable control sequence temporarily into an unexpandable one. The simplest way of achieving this is to \let the control sequence be equal to \relax (which is unexpandable). Thus

```
\let\abc=\relax
\immediate\write\aux{\abc}
```

will write the string '\abc' on the file. Using this method, we illustrate a different solution to the same problem. In this version, \onestep identifies all the control sequence tokens in the string and sets each equal to \relax.

After every control sequence in the string has been changed in this way, the string is written on a file. Readers who have read and understood the preceding examples will find this solution easy to understand.

```
\def\scan#1#2\end{\def\aux{#1}%
\ifx\aux\empty
\else
 \def\aux{#2}%
 \onestep{#1}%
 \ifx\aux\empty
 \else
  \scan#2\end
\fi\fi}


\let\Let=\let
\def\onestep#1{\ifcat\relax\noexpand#1\Let#1=\relax\fi}


\newwrite\out
\immediate\openout\out=tempfile
\def\main#1{{\scan #1\end\immediate\write\out{#1}}}
```

The following points should be mentioned:

■ Macro \main has an extra pair of braces, so everything done in it is local. This way, the setting of control sequences to \relax is only temporary.

■ Imagine the string 'abc\let hjk\x'. The control sequence \let is first identified and is set to \relax. Later the control sequence \x is identified, but saying \let\x=\relax fails because \let is now equal to \relax. This is why the command '\let\Let=\let' has been added. Macro \onestep uses \Let instead of \let. Of course, a string such as '...\Let...\x' would cause the same problem, so our method cannot handle such strings.

▶ **Exercise 7.11:** What about the string '...\Let'?

■ Our macros do not use tail recursion, so each recursive expansion of \scan saves two \fi's in the parameter stack, whose normal size is 60. As a result, a long argument will exceed TeX's capacity.

■ Our method works only for an \immediate\write. A nonimmediate \write is executed in the OTR, where the various control sequences are no longer equal to \relax. For such a case, a different approach is necessary, which changes the catcodes of the special characters. It is described in "Approach 1" on page 202.

7.10 Verbatim Listings

➠   We start with a short review of interword spaces. A space (between words) is glue whose value is determined by the font designer. It is usually flexible, but in a fixed-space font, it should be rigid (its value for font cmtt10, e.g., is 5.25pt). The size of a space is affected by the space factor, so that spaces following certain punctuation marks get more stretch (and sometimes even greater natural size). Naturally, this discussion applies to any character with catcode 10 (space being the only character assigned this catcode by INITEX [343]).

Consecutive spaces are treated as one space. To defeat this, the `plain` format offers macro `\obeyspaces`. The format starts by defining [351] '`\def\space{ }`'. Thus `\space` is a macro whose replacement text is a normal space (affected by the space factor). Next, `\obeyspaces` is defined as a macro that declares the space active '`\def\obeyspaces{\catcode'\ =13}`', and `plain` says (on [352]) '`{\obeyspaces\global\let =\space}`'. This means that when `\obeyspaces` is in effect (when the space is an active character) the space is defined as `\space`.

To get spaces that are not affected by the space factor, one of the following methods can be used:
1. Change the sf codes of the punctuation marks to 1000 by means of the command `\frenchspacing`.
2. Use a control space '`\␣`'. Control space [290] is a primitive that inserts glue equal to the interword space of the current font, regardless of the space factor. Defining the space as a control space is done by saying '`{\obeyspaces\global\let =\ }`'.
3. Assign nonzero values to `\spaceskip` and `\xspaceskip`.

Now we are ready for the verbatim macros. Four macros are discussed here, all extensions of macro `\elp` of section 5.20. The aim is to develop macros that would typeset any given text, verbatim, in font cmtt10. The main problem is that the text may include special characters, such as '`\`' and '`#`', so these have to be turned off temporarily. Another problem is that the text has to be picked up line by line, and each line typeset individually. We shouldn't try to absorb the entire text as a macro argument, since there may be too much of it. Other problems have to do with blank lines and consecutive spaces.

We start with a macro `\sanitize` that is used to change the catcodes of certain characters to 12 (other). It is similar to `\dospecials` [344].

```
\def\makeother#1{\catcode'#1=12\relax}
\def\sanitize{\makeother\#\makeother\%\makeother\~\makeother\\%
\makeother\}\makeother\{\makeother\&\makeother\$\makeother\_%
\makeother\^\makeother\^^M\makeother\ }%
```

Next comes the main macro `\ttverbatim`. We tentatively start with the simple definition '`\def\ttverbatim{\begingroup\sanitize\tt\gobble}`', but the final definition below also contains '`\def\par{\leavevmode\endgraf}`' (and a few other things). This is necessary because of blank lines. A blank line becomes a `\par` in the mouth, and `\par` has no effect in V mode. We thus have to switch to H mode and do an `\endgraf`, which is the same as `\par`.

Macro \gobble gobbles up the end-of-line following the \ttverbatim and expands \getline to get the first line of verbatim text. Without gobbling, \getline would read the end-of-line and translate it into an empty line in the verbatim listing.

Macro \getline gets one line of text (à la \elp), typesets it, executes a \par, and expands itself recursively. When it senses the end of the verbatim text, it should simply say '\endgroup' to revert to the original catcodes. The end of the text is a line containing just '\endverbatim' (without any preceding blanks), and the main problem facing \getline is to identify this line. The identification is done by means of \ifx, which compares two strings, stored in macros, character by character. The point is that an \ifx comparison is done by character code *and* category code. When the '\endverbatim' is read, sanitizing is in force, and the '\' has catcode 12 (the eleven letters have their normal catcode, 11).

We thus cannot say '\def\endverb{\endverbatim}' and '\ifx\endverb\aux', because the string in macro \endverb starts with '\$_0$' instead of '\$_{12}$'. The solution is to define \endverb in a group where the '\' has catcode 12, thus '{\catcode'*=0 \makeother\\ *gdef*endverb{\endverbatim}}'. Now \getline can say \ifx\endverb\aux.

One of the verbatim methods below uses the vertical bar '|' to delimit small amounts of verbatim text. This is done by declaring '|' active. Since we want to be able to include the '|' in verbatim listings, we sanitize it in \ttverbatim by saying '\makeother\|'.

After using the macros for several years, I was surprised one day to see a ? listed as ¿. On a closer look I discovered that it was the pair ?' that was listed as ¿. It took a while to figure out that, in the cmtt fonts, the combinations ?' and !' are considered ligatures and are replaced by ¿ and ¡, respectively (Ref. 4, p. 36).

The solution is to declare the left quote active and to define it as the macro \lq. This is why \ttverbatim and its relatives say '\catcode'\'=13' and why the code '{\catcode'\'=13\relax\gdef'{\relax\lq}}' is also necessary (see also [381]).

The definitions of the two macros should now be easy to understand.

```
\def\ttverbatim{\medskip\begingroup\tt\def\par{\leavevmode\endgraf}%
 \catcode'\'=13\makeother\|\sanitize\gobble}
{\makeother\^^M\gdef\gobble^^M{\getline}%
\gdef\getline#1^^M{\def\aux{#1}%
\ifx\endverb\aux\let\next=\endgroup\medskip%
\else#1\par\let\next=\getline\fi\next}%
\obeyspaces\global\let =\ \catcode'\'=13\relax\gdef'{\relax\lq}}
{\catcode'\*=0 \makeother\\ *gdef*endverb{\endverbatim}}%
```

▶ **Exercise 7.12:** The verbatim text above contains \endverbatim, but this string terminates verbatim listings. How was the text produced?

Note the two \medskip commands. They create some spacing around the entire listing, and the first one also makes sure that the listing is done in v mode. They can be replaced, of course, by any vertical skip (flexible or rigid), depending on specific needs and personal taste.

Readers trying these macros will very quickly discover that they typeset '␣' instead of spaces. This is because the space (whose character code is '40) has been sanitized (it is now a regular character, of catcode 12) and font cmtt10 has '␣' in position '40. This feature is sometimes desirable, but it is easy to modify \ttverbatim to get blank spaces in the verbatim listing.

The new macro is called \verbatim, and the main change is to say \obeyspaces instead of sanitizing the space. In verbatim listings, of course, we don't want the space to be affected by the space factor, so '{\obeyspaces\global\let␣=\␣}'.

```
\def\makeother#1{\catcode'#1=12\relax}
\def\sanitize{\makeother\#\makeother\%\makeother\~\makeother\\%
\makeother\}\makeother\{\makeother\&\makeother\$\makeother\_%
\makeother\^\makeother\^^M}%
```

Macros \verbatim and \getline are defined by

```
\def\verbatim{\medskip\begingroup\tt\def\par{\leavevmode\endgraf}%
 \catcode'\'=13\sanitize\makeother\^^M\makeother\|\obeyspaces\gobble}
{\makeother\^^M\gdef\gobble^^M{\getline}%
\gdef\getline#1^^M{\def\aux{#1}%
\ifx\endverb\aux\let\next=\endgroup\medskip%
\else#1\par\let\next=\getline\fi\next}%
\obeyspaces\global\let =\ \catcode'\'=13\relax\gdef'{\relax\lq}}
{\catcode'\*=0 \makeother\\ *gdef*endverb{\endverbatim}}%
```

▸ **Exercise 7.13:** Why do we have to place '{\obeyspaces\global\let =\ }' outside the macros? It seems more elegant to have it included in the definition of \verbatim.

Preventing line breaks. Each line of a verbatim listing is typeset by saying (in \getline) '#1\par'. The line becomes a paragraph and may be broken if it is too wide. If this is not desirable, then the code above may be changed to \hbox{#1}. Macro \verbatim changes the mode to V, which means (1) the boxes will be stacked vertically; (2) a wide box will not cause an 'overfull box' error.

Numbering lines. The definition of \verbatim is now generalized also to typeset line numbers with the verbatim text. Macro \numverbatim, which follows, uses the same \sanitize as \verbatim, and a new count register is declared, to hold the current line number. The line numbers are typeset on the left margin, by means of an \llap, but this is easy to modify.

```
\newcount\verbline
```

```
\def\numverbatim{\medskip\begingroup\tt\def\par{\leavevmode\endgraf}%
 \catcode'\'=13\sanitize\makeother\|\obeyspaces\verbline=0
\everypar{\advance\verbline1
 \llap{\sevenrm\the\verbline\ \ }}\gobble}
{\makeother\^^M\gdef\gobble^^M{\getline}%
\gdef\getline#1^^M{\def\aux{#1}%
\ifx\endverb\aux\let\next=\endgroup\medskip%
```

```
\else#1\par\let\next=\getline\fi\next}%
\obeyspaces\global\let =\ \catcode'\'=13\relax\gdef'{\relax\lq}}
{\catcode'\*=0 \makeother\\ *gdef*endverb{\endverbatim}}%
```

Verbatim in H **mode.** The macros developed above are suitable for "large" verbatim listings, involving several, or even many, lines of text. Such listings are normally done in V mode, between paragraphs. The next approach declares the vertical bar '|' active and uses it to delimit small amounts of text (normally up to a line) that should be listed verbatim within a paragraph. This is convenient notation, commonly used, whose only disadvantage is that the '|' itself cannot appear in the text to be listed.

The first step is not to sanitize the space and the end-of-line:

```
\def\makeother#1{\catcode'#1=12\relax}
\def\sanitize{\makeother\#\makeother\%\makeother\~\makeother\\%
\makeother\}\makeother\{\makeother\&\makeother\$\makeother\_%
\makeother\^}%
```

Next, the '|' is declared active and is defined similarly to \verbatim above. The main differences are as follows:
1. Macro \moreverb can pick up the entire text as its argument, since there is not much text.
2. Instead of defining the space as a control space, we preempt the space by assigning nonzero values to \spaceskip and \xspaceskip.

```
\catcode'\|=13
\def|{\begingroup\obeyspaces\catcode'\'=13\sanitize\moreverb}
\def\moreverb#1|{\tt\spaceskip=5.25pt%
  \xspaceskip=5.25pt\relax#1\endgroup}
```

(The value 5.25pt is the interword space of font cmtt10. If a different font is used, this value should be replaced by its interword space. An alternative is to use .51em, which gives good results in most sizes of cmtt.)

▷ **Exercise 7.14:** Why is the \relax necessary after the 5.25pt?

The reader should note that '|' cannot be used in the argument of a macro. If \abc is a macro, we cannot say, for example, '\abc{...|\xyz|...}'. The reason is that all tokens in the argument get their catcodes assigned when the argument is absorbed, so '|' cannot change them later. Using '|' in a box, however, is okay.

▷ **Exercise 7.15:** Change the definition of '|' to typeset '␣' instead of blank spaces.

A different approach. Here is a completely different approach to verbatim listing, using the primitive \meaning. This primitive typesets the meaning of its argument. If the argument is a macro (e.g., '\def\abc#1;{A\B$#1^&\C_}'), then '{\tt\meaning\abc}' results in 'macro:#1;->A\B $#1^&\C _'. The unnecessary tokens at the beginning can easily be stripped off by using > as a delimiter. We end up with the simple macros (compare with [382])

```
\long\def\verbatim#1{{\tt\expandafter\strip\meaning#1}}
```

```
\def\strip#1>{}
```

(See another application of \strip on page 215.) Since a macro is a token list, we can get verbatim listing of tokens this way, but with the following limitations: (1) extra spaces are automatically inserted by \meaning at certain points; (2) end-of-lines become spaces in the verbatim listing; (3) a single '#' cannot be included in the verbatim text.

7.11 Fancy Verbatim

Sometimes it is necessary to typeset parts of a verbatim listing in a different font, or to mix verbatim and nonverbatim text. The following extensions of the verbatim macros can read and execute commands before starting on their main job. The commands are typically catcode changes but, in principle, can be anything. The commands are specified in two ways. Commands that should apply to all verbatim listings of a document are placed in the toks register \everyverbatim. Commands that should apply just to certain listings are placed between square brackets right following \verbatim, thus '\verbatim[...]'.

Macro \verbatim uses \futurelet to sneak a look at the token following the 'm'. If this is a left bracket, the commands up to the right bracket are executed. Sanitization is done before the commands are executed, so the user can further modify the catcodes of sanitized characters. However, since the commands start with a '\', sanitization of this token should be deferred. The code below shows how \verbatim places the next token into \nextc, how \options expands \readoptions if this token is a '[', and how \readoptions scoops up all the commands and executes them. Macro \preverbatim sanitizes the '\', and performs the other last-minute tasks, before expanding \gobble.

```
\def\verbatim{\medskip\begingroup\sanitize\the\everyverbatim%
 \makeother\^^M\futurelet\nextc\options}
\def\options{\ifx[\nextc\let\next=\readoptions
 \else\let\next=\preverbatim\fi\next}
\def\readoptions[#1]{#1 \preverbatim}
\def\preverbatim{\def\par{\leavevmode\endgraf}%
 \makeother\\\makeother\^^M\tt\obeyspaces\gobble}
```

Note that the left quote is made active very late (together with the sanitization of the '\'). This means that the optional commands can use it in its original meaning, but they cannot change its catcode. It is possible to say '\verbatim[\catcode`\*=11]', but something like '\verbatim[\makebgroup\`]' won't work because the left quote will be made active at a later point.

Advanced readers may easily change the macros such that the left quote would be made active early (perhaps by \sanitize). In such a case, the effect of

\verbatim[\catcode`*=11] can be achieved by defining
\def\makeletter#1{\catcode`#1=11 }, then saying
\verbatim[\makeletter*]

Similar remarks apply to the curly braces. Saying '\verbatim[\everypar={...}]' is wrong because the braces are sanitized early. The solution is to define '\def\temp{\everypar={...}}' and then say '\verbatim[\temp]'.

The simplest example is '`\verbatim[\parindent=0pt]`', which prevents indentation in a specific listing. A more sophisticated example introduces the concept of meta code. The idea is that certain pieces of text in a verbatim listing may have to be typeset in a different font (we use cmr10). Such text is identified by enclosing it in a pair of angle brackets, `<>`. The following simple code implements this idea:

```
\def\enablemetacode{\makeactive\<}
{\enablemetacode \gdef<#1>{{\tenrm#1}}}
```

And the test:

```
\verbatim[\enablemetacode]
\halign{<...preamble...>\cr \beginCont
<...1st line...>\cr
<...>
<...last line...>\endCont\cr}
\endverbatim
```
produces

\halign{...preamble...\cr \beginCont
...1st line...\cr
...
...last line...\endCont\cr}

(It's easy to modify `\enablemetacode` to typeset the brackets also.) Next we introduce fancy comments. Suppose we want to typeset comments in a verbatim listing in italics. A comment is anything between a '%' and the end-of-line. Again, the following simple code is all that's needed to achieve this. It declares the '%' active (preempting the action of `\sanitize`) and defines it as a macro that typesets its argument in `\it` (the '%' itself can also be in `\it`, if desired).

```
\def\itcomments{\makeactive\%}
{\itcomments \gdef%#1\par{\char"25{\it#1}\par}}
```

The test

```
\verbatim[\itcomments]
line 1 %comment 1
line 2  % Comment #2
line 3 % Note \this
\endverbatim
```
results in

line 1 %*comment 1*
line 2 % *Comment #2*
line 3 % *Note "this*

The next example typesets selected parts of a verbatim listing in `\bf`. The '!' is declared a temporary escape character, and the two parentheses, as temporary braces. The result of

```
\verbatim[\makeescape\!\makebgroup\(\makeegroup\)]
```

```
  (!bf while) \lineno>\totalines
    \lineshiped:=\totalines
    (!bf extract) \temp (!bf from) \Bsav
    \totalines:=\totalines+\temp
  (!bf end while);
\endverbatim
```
is

    while \lineno>\totalines
      \lineshiped:=\totalines
      extract \temp **from** \Bsav
      \totalines:=\totalines+\temp
    end while;

Sometimes a mixture of visible and blank spaces is required in the same verbatim listing. Here are two simple ways of doing this:

```
{\tt\makeactive\!\gdef!{\char32 }}
\verbatim[\makeactive\!]
a 1!!2   3
x!4  5!!!6
\endverbatim
```
results in

```
a 1␣␣2   3
x␣4  5␣␣␣6
```

```
{\makeactive\!\gdef!{\ }}
\verbatim[\vispacetrue\makeactive\!]
a 1!!2   3
x!4  5!!!6
\endverbatim
```
results in

```
a␣1  2␣␣␣3
x 4␣␣5   6
```

One more example, to convince the skeptics, shows how math expressions can be placed inside a verbatim listing. We simply say

```
\verbatim[\makeescape\!\catcode`\$=3 \catcode`\^=7]
prolog $!sum x^2$ epilog
\endverbatim
```

And the result is

```
prolog ∑ x² epilog
```

The concept of optional commands is powerful and can be extended to create verbatim listings that are numbered or that show visible spaces. This way, macros \ttverbatim and \numverbatim are no longer necessary and are replaced by \verbatim[\vispacetrue] and \verbatim[\numbered], respectively.

The difference between \verbatim and \ttverbatim is that the former says '\obeyspaces', whereas the latter says '\makeother\␣'. We add a boolean variable \ifvispace that selects one of the choices above.

Macro \numverbatim says
'\everypar{\advance\verbline1 \llap{\sevenrm\the\verbline\ \ }}'
We therefore define the two macros

```
\def\numbered{\everypar{\advance\verbline1
```

```
 \llap{\sevenrm\the\verbline\ \ }}}
\def\notnumbered{\everypar{}}
```

which can turn the numbering on and off. The final version of \verbatim is shown below.

```
\def\verbatim{\medskip\begingroup\tt\sanitize%
 \makeother\^^M\makeother\|\futurelet\nextc\options}
\def\options{\ifx[\nextc\let\next=\readoptions
 \else\let\next=\preverbatim\fi\next}
\def\readoptions[#1]{#1 \preverbatim}
\def\preverbatim{\def\par{\leavevmode\endgraf}%
 \lastasks\ifvispace\makeother\ \else\obeyspaces\fi\gobble}
```

The following tests are especially interesting (the left column is the source; the right one, the results).

```
\everyverbatim{\numbered\vispacetrue}
\verbatim
abc 123 \x %^?'!'
@#$% ^& *(   )_
\endverbatim
```

```
\verbatim[\vispacefalse]          1  abc␣123␣\x␣%^?'!'
abc 123 \x %^?'!'                 2  @#$%␣^&␣␣*(␣␣␣)_
@#$% ^& *(   )_
\endverbatim
                                  1  abc 123 \x %^?'!'
\verbatim[\notnumbered]           2  @#$% ^& *(   )_
abc 123 \x %^?'!'
@#$% ^& *(   )_                      abc␣123␣\x␣%^?'!'
\endverbatim                         @#$%␣^&␣␣*(␣␣␣)_
```

The vertical bar can also take optional arguments. Below we show how to generalize the definition of '|', so things like

```
|[\enablemetacode]...<text>..|
|[\vispacetrue]..A B..|
```

will work.

The method is similar to the one used with \verbatim, with one difference: The backslash must be sanitized before \futurelet peeks at the next token. Consider the simple example '|\abc|'. The \futurelet will scoop up \abc as one (control sequence) token. Later, when \moreverb typesets its argument (when it says '#1'), there will be an error since \abc is undefined.

If the \futurelet reads a '[', the backslash has to be restored (by means of a '\makeescape\\'), so that macro \readOptions can read and execute the optional commands. Following that, macro \preVerb expands \lastasks to resanitize the backslash before the rest of the verbatim argument is read.

```
\makeactive\|
\def|{\begingroup\tt\obeyspaces\sanitize\makeother\\%
```

```
\futurelet\nextc\Voptions}
\def\Voptions{\ifx[\nextc\makeescape\\\let\next=\readOptions
 \else\let\next=\preVerb\fi\next}
\def\readOptions[#1]{#1\relax\preVerb}
\def\preVerb{\lastasks\ifvispace\makeother\ \else\obeyspaces\fi%
 \moreverb}
\def\moreverb#1|{\spaceskip=5.25pt%
 \xspaceskip=5.25pt\relax#1\endgroup}
```

▷ **Exercise 7.16:** Why is the `\relax` necessary in macro `\readOptions` but not necessary in the similar macro `\readoptions` (expanded by `\verbatim`)?

▷ **Exercise 7.17:** Now that the '[' is special, how can we typeset it verbatim?

▷ **Exercise 7.18:** What is the effect, if any, of `|[\numbered]...|`?

7.12 Complete Verbatim Macros

Following is the complete code for all the verbatim macros necessary to implement the concepts discussed here. Note the new macro `\verbfile`. It can be used to list the contents of a given file verbatim. The argument '#1' is the name of the file. This is a simple modification of `\verbatim`, without optional commands (but see exercise 7.20).

```
\newtoks\everyverbatim \newcount\verbline
\newif\ifvispace \vispacefalse
\def\makeescape#1{\catcode`#1=0 }
\def\makebgroup#1{\catcode`#1=1 }
\def\makeegroup#1{\catcode`#1=2 }
% can have similar \make.. macros for catcodes 3--10
\def\makeletter#1{\catcode`#1=11 }
\def\makeother#1{\catcode`#1=12 }
\def\makeactive#1{\catcode`#1=13 }
\def\makecomment#1{\catcode`#1=14 }

\def\sanitize{\makeother\#\makeother\&\makeother\%\makeother\~%
 \makeother\}\makeother\{\makeother\$\makeother\_\makeother\^%
 \the\everyverbatim}%
\def\lastasks{\makeactive\`\makeother\\}

\def\verbatim{\medskip\begingroup\tt\sanitize%
 \makeother\^^M\makeother\|\futurelet\nextc\options}
\def\options{\ifx[\nextc\let\next=\readoptions
 \else\let\next=\preverbatim\fi\next}
\def\readoptions[#1]{#1 \preverbatim}
\def\preverbatim{\def\par{\leavevmode\endgraf}%
 \lastasks\ifvispace\makeother\ \else\obeyspaces\fi\gobble}

{\makeother\^^M\gdef\gobble^^M{\getline}%
```

```
\gdef\getline#1^^M{\def\aux{#1}%
\ifx\endverb\aux\let\next=\endgroup\medskip%
\else#1\par\let\next=\getline\fi\next}%
\obeyspaces\global\let =\ \makeactive\'\relax\gdef'{\relax\lq}}
{\makeescape\*\makeother\\ *gdef*endverb{\endverbatim}}%


\makeactive\|
\def|{\begingroup\tt\obeyspaces\sanitize\makeother\\%
  \futurelet\nextc\Voptions}
\def\Voptions{\ifx[\nextc\makeescape\\\let\next=\readOptions
  \else\let\next=\preVerb\fi\next}
\def\readOptions[#1]{#1\relax\preVerb}
\def\preVerb{\lastasks\ifvispace\makeother\ \else\obeyspaces\fi%
  \moreverb}
\def\moreverb#1|{\spaceskip=5.25pt
  \xspaceskip=5.25pt\relax#1\endgroup}


\def\verbfile#1 {\medskip\begingroup\tt\def\par{\leavevmode\endgraf}%
  \sanitize\lastasks\makeother\|%
  \obeylines\obeyspaces\input#1\endgroup\medskip}


\def\enablemetacode{\makeactive\<}
{\enablemetacode \gdef<#1>{{\tenrm#1}}}
\def\itcomments{\makeactive\%}
{\itcomments \gdef%#1\par{\char"25{\it#1}\par}}
\def\numbered{\everypar{\advance\verbline1
  \llap{\sevenrm\the\verbline\ \ }}}
\def\notnumbered{\everypar{}}
```

▸ **Exercise 7.19:** Why is the \relax necessary in '\makeactive\'\relax\gdef...'
(normally a space is enough to terminate a number)?

▸ **Exercise 7.20:** Extend the definition of \verbfile#1 {...} to detect and execute
optional commands.

■▶   A last word. The verbatim macros described here are very useful when writing
a book or an article on TEX. It is the belief of the author, however, that a complete understanding of these macros elevates the reader one step toward the title of
TEXmaster, regardless of the document being worked on. This is why these macros
have been described in such detail. When these macros are included in a large
format, the '@' may be added to the names of the macros, to make them "private".

7.13 Ignoring the Last Character

Here is a macro that ignores the last character in its argument. The argument must be a string of characters.

```
\def\ignoreLast#1{\toks0={}\macb#1\\}
\def\macb#1#2\\{\toks0=\expandafter{\the\toks0#1}\length{#2}%
\ifnum\number\count20<2
  \let\next=\macc\else\let\next=\macb\fi\next#2\\}
\def\length#1{\count20=0 \glength#1\end }
\def\glength#1{\ifx#1\end \let\nxt=\relax \else
  \advance\count20by1 \let\nxt=\glength\fi \nxt}
\def\macc#1\\{\showthe\toks0}
```

Macro \ignoreLast, similarly to \revers, appends '\\' to its argument and expands \macb. Macro \macb breaks up the argument into the first character (#1) and the rest (#2). It appends the first character to \toks0, which will eventually become the final result. It then measures the length of the rest and if ≥ 2 expands itself recursively; otherwise, it expands \macc, and the result should be in \toks0. Note that the ignored character becomes the argument of \macc.

▶ **Exercise 7.21:** Macro \ignoreLast does not work properly when its argument is a string of size 1 or 0. Correct this problem.

7.14 Spreading a String

Macro \Spread#1 spreads its argument (a string) in an \hbox to2in such that each character is followed by an \hfil, and each space is replaced by an \hskip0pt plus 0.5fil (an extra half skip). The result of \Spread{one two three} should be 'o n e   t w o   t h r e e'.

The macro starts by changing the catcode of space to 12. Macro \Aux then absorbs the argument and presents it to macro \aux, which scans it character by character. The characters (plus the appropriate skips) are appended to \box0. The last character is followed by an unwanted \hfil, which is removed by the \unskip. Note the careful placement of the '%' signs, to prevent spurious spaces.

```
\def\Spread{\catcode'\ =12 \Aux}
\def\Aux#1{\setbox0=\null \aux#1\end}
\def\aux#1{%
\ifx\end#1%
  \hbox to2in{\unhbox0 \unskip}%
  \catcode'\ =10
\else
  \if\space#1%
  \setbox0=\hbox{\unhbox0 \hskip0pt plus0.5fil}%
  \else
  \setbox0=\hbox{\unhbox0 #1\hfil}%
  \fi
  \expandafter\aux
\fi}
```

Because of the catcode change, there should be no intervening spaces between \Spread and its argument. If present, such spaces would become the argument.

▶ **Exercise 7.22:** Change the definitions of the macros above such that the width of the final result is given as argument '#2' of \Spread. Thus the expansion \Spread{one two three}{2in} should create a result identical to the above.

7.15 Placing Commas between Digits

Macro \insertcommas accepts one argument, a string of digits, and inserts commas preceding each group of three digits. Thus \insertcommas{1234} results in '1,234'. The steps are

1. Macro \getlength is expanded. It expands itself recursively, looking at each digit and appending it, packed in a small \hbox, to \box0.

2. Macro \insertcommas then opens \box0 and starts a loop where, in each iteration, it uses \lastbox to remove the rightmost digit and append it on the left of \box2. Each fourth digit is appended with a comma. At the end of the loop, \box2 is typeset.

```
\newcount\ct \newif\ifmore
\def\insertcommas#1{\setbox0=\null \getlength#1\end
\hbox{\unhbox0 \setbox2=\null \ct=0 \moretrue
\loop
\advance\ct 1
\setbox1=\lastbox
\ifvoid1
 \morefalse
\else
 \setbox2=\hbox{\unhbox1 \ifnum\ct=4,\ct=1\fi \unhbox2}%
\fi
\ifmore\repeat
\box2}}
```

```
\def\getlength#1{\ifx#1\end \let\next=\relax
\else\let\next=\getlength \setbox0=\hbox{\unhbox0\hbox{#1}}\fi\next}
```

Note the following:

1. After saying \unhbox0, the macro should be careful not to generate any spurious spaces. Such spaces would be appended to the current list (which consists of the components of \box0) and would interfere with the operation of \lastbox.

2. The command \lastbox is horizontal. Saying \lastbox when the macro is expanded in v mode causes an error. To make \insertcommas robust, all its operations are carried out inside an \hbox. This \hbox also helps isolate the loop from any boxes that happen to be present in the current list just before the \unhbox0 is executed. Imagine an expansion such as \par\insertcommas{123}. The \unhbox0 would place the contents of \box0 following the paragraph indentation. Instead of repeating three times, the loop would repeat four times, picking up the '3', the '2', the '1', and the paragraph indentation (which is also a box).

3. The macro typesets its result in this isolation box. Therefore, if expanded in V mode, it would not start a paragraph.

7.16 A Table with No Markup

The data for an \halign requires markup information in the form of '&' and '\cr' at the right places. If the data for the table is simple, we sometimes wish we could insert it as shown below, where each source line corresponds to a table row and spaces are used to separate table items. A typical input may look like:

```
\Table{%
a bc de
f g hi
jkl m n}
```

The macros below use \catcode, \ifx, and recursion to achieve this result. They are fairly easy to read, understand, and modify. Macro \Table has no parameters. It changes the catcodes of carriage return '^^M' and of a space to active (this is done inside a \begingroup that acts as a quarantine) and expands \getpar. Macro \getpar has one parameter, the entire data for the table. When that parameter is scanned, all carriage returns are changed to \par (since TeX is in \obeylines mode), and all spaces are considered active (i.e., they are not ignored). The macro has four parts:

1. It starts the preamble of the \halign (note that this changes the mode to V).

2. It expands \insertItems#1\relax to insert the data, with all the necessary '&' and '\cr' into the \halign.

3. It closes the \halign with a '}'.

4. It executes \endgroup, which returns the catcodes to their normal values.

Macro \insertItems receives the data for the table, token by token, in its #1 parameter. It uses \ifx to distinguish four cases:

1. The #1 parameter is \relax. This is the end of the table, and the last \cr is inserted. \insertItems terminates at this point.

2. The parameter is \par. This is the end of a row. A '\cr' is inserted, and \insertItems expands itself recursively.

3. The parameter is a space. An '&' is inserted, followed by a recursive expansion.

4. The parameter is none of the above. It is assumed to be a piece of data for the table; it is inserted in the table, and \insertItems expands itself recursively.

```
\def\Table{\begingroup\obeylines\catcode` =\active\getpar}
\def\endtable{\cr}
\def\endrow{\cr\insertItems}
\def\column{&\insertItems}
\def\getpar#1{\halign{##&##&##\cr
 \insertItems#1\relax}\endgroup}
\def\insertItems#1{%
 \ifx\relax#1\let\next=\endtable
\else\ifx\par#1\let\next=\endrow
```

```
\else\ifx\space#1\let\next=\column
\else#1\let\next=\insertItems
\fi\fi\fi\next}
```

The only thing that makes the use of the macros unnatural is the fact that the preamble is built into macro `\getpar`. It is easy to generalize the above macros such that the preamble becomes a parameter of `\Table`. Macro `\Table` saves that parameter in `\toks0`, and `\getpar` simply says `\the\toks0`. There are a few more differences, and the details are shown below.

```
\def\Table#1\cr{\begingroup\toks0={#1\cr}%
 \obeylines\catcode` =\active\getpar}
\def\endtable{\cr}
\def\endrow{\cr\insertItems}
\def\column{&\insertItems}
\def\getpar#1{\the\toks0
 \insertItems#1\relax\egroup\endgroup}
\def\insertItems#1{%
 \ifx\relax#1\let\next=\endtable
\else\ifx\par#1\let\next=\endrow
\else\ifx\space#1\let\next=\column
\else#1\let\next=\insertItems
\fi\fi\fi\next}
```

Macro `\Table` now has two parameters, and a typical expansion is

```
\Table\halign\bgroup#&#&#\cr{%
a bc de
f g hi
jkl m n}
```

General notes:

■ These macros were inspired by ideas due to C. G. van der Laan, who used this technique to typeset a crossword puzzle (Ref. 25).

■ Normally, spaces at the beginning of a line are ignored. However, once the space is changed into an active character, it is no longer ignored, and source lines cannot be indented for easy reading.

■ An `\halign` changes the mode to v, so macro `\Table` terminates the current paragraph.

■ The data may include control sequences such as `\omit`.

■ Catcode changes play an important role in this example. Interested readers should at the Index for more applications of catcode changes.

7.17 Looking at the Next Word

A macro `readword` **is automatically** invoked at the beginning of every paragraph. It collects (about) 2 inches' worth of text and typesets it in boldface.

It has one parameter, delimited by a space. The parameter is thus the next word of text. The boldface version of that word is appended to `\linebox` and `\readword` expands itself. When the width of `\linebox` exceeds 2 inches, it is printed, together with the next word, and `\readword` expands `\relax`.

```
\newbox\linebox \newdimen\linesize
\def\readword#1 {\ifdim\wd\linebox>2in%
 {\box\linebox} #1 \let\next=\relax
\else
 \setbox\linebox=\hbox{\unhbox\linebox\bf\ #1}\let\next=\readword \fi
\next}
\everypar{\readword}
```

7.18 Bracketing Text

Macro `\bracket` accepts text (bracketed by a `\bracket`–`\endbracket` pair) and typesets it in brackets. The principle is to place the text in a `\vbox` and make that box the (only) element of a `\matrix`. The matrix is then delimited by square brackets. The macro enters math display mode and, therefore, cannot be invoked from inside math mode.

```
\newtoks\brack \newbox\brkt \newdimen\nwidth
\nwidth=\hsize \advance\nwidth by-.21in
\def\bracket#1\endbracket{\brack={#1}
\setbox\brkt=\vbox{\hsize=\nwidth \the\brack}
\ifdim\ht\brkt<\baselineskip \setbox\brkt=\hbox{\the\brack} \fi
$$\left[\matrix{\box\brkt\cr}\right]$$}
```

The text to be bracketed is first placed in a `\toks` register and then in a `\vbox` of width `\hsize-.21in`. The height of the `\vbox` is determined by the amount of text. This does not work when there is very little text, since the `\vbox` always has the same width. With very little text (less than one line), the `\vbox` would be too wide for the text, and the text would be separated from the right bracket.

$$\left[\begin{matrix} \text{This case is handled by testing the height of the \textsf{vbox} and, if it has only one line} \\ \text{of text in it, placing the text in an \textsf{hbox}. The width of an \textsf{hbox} is determined} \\ \text{by the amount of text in it, resulting in a text that is always nicely bracketed.} \end{matrix} \right]$$

[Here is an example of a single bracketed line.]

▸ **Exercise 7.23:** Generalize the definition of `\bracket` to make it possible to have any large operators as brackets.

7.19 A DropShadow

Here is a useful macro that takes a box with text, frames it with the \boxit macro [ex. 21.3], and creates a dropshadow to the right and below. The first argument of \dropshadow is the box with the text; the second one is the thickness of the shadow (such as 2pt).

```
\def\dropshadow#1#2{\setbox0=\boxit{#1}3%
\vbox{\offinterlineskip
  \hbox{\copy0\smash{\lower#2\hbox{\vrule width#2height\ht0}}}
  \hbox{\kern#2\vrule width\wd0height#2}}}
```

The test \dropshadow{\hbox{This is a test of ... macro.}}{3pt} produces

| This is a test of the dropshadow macro. |

If more than one line of text is needed, the first argument should, of course, be a \vbox with the text.

▶ **Exercise 7.24:** Why the \smash?

▶ **Exercise 7.25:** Write a macro \leftdropshadow such that the expansion '\leftdropshadow{\hbox{This is a test of a left shadow}}{3pt}' will create

| This is a test of a left shadow |

7.20 Combining Text and Diagrams

Hanging indentation is a feature that can be used to combine text with drawings. Here is a simple macro that uses hanging indentation to reserve room for a drawing on the left side of a paragraph of text.

Figure 7.1 shows how this is done. If the paragraph is short, it is typeset in a narrow margin and placed to the right of the drawing (part a). If the paragraph is longer, hanging indentation should be used to create space at the top left corner of the paragraph for the drawing (part b).

a   b

Figure 7.1

Hanging indentation has already been discussed, but here is a recap. Two parameters control hanging indentation. The first one, '\hangafter= n', specifies how many lines are to be indented (for negative n, indentation is applied to the first n lines. For positive n, it is applied to lines $n + 1$, $n + 2, \ldots$). The second parameter, '\hangindent x', specifies a $\langle$dimen$\rangle$ x for the amount of indentation. A positive x produces a left indent and a negative x produces a right one.

Macro \leftdraw creates a paragraph with a hanging indentation on the top left, to fit around a diagram. The height and width of the diagram are given by the user as arguments of the macro.

1.   When the macro is expanded, it sets the paragraph in a box of width \hsize$-\langle$width of diagram$\rangle$. If the height of the paragraph $\leq$ the height of the diagram, the paragraph is simply typeset to the right of, and aligned with the bottom of, the diagram.

2.   Otherwise, \leftdraw sets \hangindent to the width of the drawing and computes a (negative) value for \hangafter that will create a hanging indentation with a height $\geq$ that of the diagram.

```
\newdimen\temp
\def\leftdraw#1 #2 #3\endpar{%#1=height, #2=width, #3=text
\temp=\hsize \advance\temp by-#2
\setbox0=\vbox{\parindent=0pt\hsize=\temp #3}
\ifdim\ht0<#1
 \line{\vbox to#1{\hrule width#2 \vfil\hrule}\box0}
\else
  \temp=#1 \divide\temp\baselineskip
  \setbox0=\vtop{\hangafter=-\temp \advance\hangafter-1
    \hangindent=#2\parindent=0pt#3}
  \setbox1=\vtop to#1{\hrule width#2 \vfil\hrule width#2}
  \wd1=0pt \ht1=0pt
  \line{\box1 \box0}
\fi}
```

The user, of course, still has to paste in the diagram. This can be done either by hand or with a \special. Possible improvements to this simple macro are

1. It only works if the text consists of one paragraph. It is possible to extend it to work on more than one paragraph (see Ref. 27), but this is messy.

2. It may look better if there is some space separating the text and the diagram. The size of the space either may be specified by the user or may be a percentage of the size of the diagram. This is easy to do.

3. It is also easy to add another parameter to specify whether the diagram should be placed to the left or to the right of the text.

4. It is possible to place the diagram either on the left or on the right, depending on the page number. This is an OTR problem, since the page number is known only in the OTR. Chapter 17 discusses techniques for solving such a problem.

7.21 A Variation on \boxit

We illustrate a macro \Cboxit that generates [figure] in response to '\Cboxit\hbox{1},\hbox{2}\hbox{34},\hbox{5};' . It takes any number of arguments, separated by commas and terminated by a ';'.

The macro illustrates variable macro arguments, recursion, and the use of \lastbox. The main problem is to find the maximum height and depth of all the arguments. The task is done in three steps:

1. Macro \Cboxit appends ',;,' to its argument and expands \PickupNext with the resulting string as argument.

2. Macro \PickupNext uses recursion to break up the argument into individual components that are appended, each as a separate box, to \box0. The macro also counts the components in \numArg. When done, it expands \insertrules.

3. Macro \insertrules is also recursive. It measures the vertical size of \box0 (the maximum size of all the individual components), breaks each component from \box0, expands \boxit to surround it with rules, and appends the result to \box1, which is eventually typeset.

In H mode, the user may say \lower6pt\hbox{\Cboxit...;} to drop the whole thing, if desired.

Note the percent signs on three lines. They guarantee that no spurious spaces be inserted when the macro is used in H mode.

```
\newcount\numArg
\def\Cboxit#1;{\setbox0=\null\PickupNext#1,;,}
\def\PickupNext#1,{% Note that #1 may be \null
\if;#1\let\next=\insertrules
\else\let\next=\PickupNext
  \setbox0=\hbox{\unhbox0\vbox{#1}}%
  \advance\numArg1
\fi\next}
\def\insertrules{\setbox1=\null
\dimen0=\ht0 \advance\dimen0 \dp0
\unhbox0
\loop
 \setbox2=\lastbox
 \advance\numArg-1
 \setbox1=\hbox{\boxit{\vbox to\dimen0{\vss\box2\vss}}6%
  \kern-0.4pt\unhbox1}%
 \ifnum\numArg>0
\repeat
\box1}
```

7.22 Fitting Text in a Given Box

Given a piece of text and the width and height of a box the text should fit in, how can we fit the text to the box? The only variable is the font size. Macro \fontSize below illustrates a simple solution to this problem. It will not work in all cases and is only intended as an illustrative example of a loop.

The three parameters are the width and height of the box (dimensions), and the text. A typical expansion is '\fontSize 3in 3.5in {Here is ...}', following which we can say 'Best font size is \the\size pt'.

A count register \siz is declared and initialized to 7 (we start at 7pt, but it is possible, of course, to start at a smaller font size). The text is saved in \toks0. During each iteration the text is "poured" from \toks0 to \box0, whose width is set to #1. If the box is not high enough, the font size is incremented (by 1pt or even a smaller step), and the process continues.

The loop terminates when \ht0 is greater than #2, so the final font size is too big and is decremented by one step. The code is surprisingly readable.

```
\newcount\size
\long\def\fontsize#1 #2 #3{{\toks0={#3}%
\size=5 \baselineskip=\size pt
\loop
  \font\fnt=cmr10 at\size pt
  \setbox0=\vbox{\tolerance=10000 \hsize=#1 \fnt
  \the\toks0 \vss}%
\ifdim\ht0<#2 \advance\size by 1
\advance\baselineskip by1pt
\repeat
\global\advance\size by -1}}
```

 Another example along the same lines is the binary search (discussed on [387]). Two pieces of text are stored in boxes that should be typeset side by side. For better appearance, the two boxes should have the same height, and the problem is to calculate widths for the two boxes that will give them identical heights.

7.23 The Sierpinski Triangle

The Sierpinski triangle is defined recursively. Start with any triangle, find the midpoint of each edge, and connect the three midpoints to obtain a new triangle, fully contained in the original one. Cut the new triangle out. The newly created hole now divides the original triangle into three smaller ones. Repeat the process on each of the smaller triangles. At the limit, there is no area left in the triangle. It resembles Swiss cheese without any cheese, just holes.

The Sierpinski triangle is used here to illustrate a recursive problem solved with a nonrecursive macro. We start with

```
\setbox0=
\hbox{\vrule height2pt width0pt\vrule height1pt width1pt\kern2pt}
```

which sets \box0 to □. We then define macro \tri to create a triangle made of three copies of \box0, and assign it as the new value of \box0. This way, each expansion of \tri creates a bigger triangle in \box0. The following code

```
\setbox0=
 \hbox{\vrule height2pt width0pt\vrule height1pt width1pt\kern2pt}
\def\tri{%
 \setbox0=\vbox{\hbox{\kern.5\wd0\copy0}\hbox{\copy0\copy0}}%
 \vbox{\copy0\vskip12pt}}

{\offinterlineskip
\hbox{\tri\tri\tri\tri\tri\tri}}
```

creates

7.24 Macros for Looping

Looping is an important feature of any programming language. In TEX, it is done by means of recursion. The \loop \repeat construct [219] is one way of doing it, and the following examples illustrate a similar approach, using nested macro definitions. We start with macro \maxt#1, which defines another macro, \next, and expands it.

```
\def\maxt#1{%
 \def\next#1=##1 or ##2 do ##3 \endd #1{%
  \ifnum##1>##2 ##1\else##2\fi##3}
 \next #1}
```

Macro \next has three parameters delimited in a special way; its definition is simple. It compares its first two arguments (which should be numbers), and typesets the larger of the two, followed by the third one.

The Expansion	Creates
\maxt\x=2 or 3 do Here is the text \endd\x	3Here is the text
\maxt\y=4 or 3 do Here is more text \endd\y	4Here is more text

In the above examples, the argument of \maxt is \x. The rest (e.g., 2 or 3 do Here is the text \endd\x) is only read when \next is expanded. The confused reader should use tracing to find out the arguments assigned to each macro.

Note that the third argument of \next is normally just text, but can also include commands. One can say '\long\def\next' if the third argument should be more than one paragraph long.

▶ **Exercise 7.26:** There is a simple way to have a multiparagraph third argument without a \long definition of \next. What is it?

Next comes macro \aloop, which does simple looping. It must be expanded with arguments of the form
$$\text{\cs}=num1 \text{ step } num2 \text{ until } num3 \text{ do } text \text{ \endloop\cs,}$$
where \cs should be a predeclared \count register. The macro sets \cs to num1 and loops—incrementing \cs by num2 (which must be positive)—until it reaches num3 (or a value very close to it). During each iteration, the count register is typeset, together with a copy of text. This is a good introduction to a general loop macro.

```
\def\aloop#1=#2 {
  \long\def\next step ##1 until ##2 do ##3 \endloop#1{%
    ##3
    \advance#1 by ##1
    \ifnum#1>##2 \else \next step ##1 until ##2 do ##3 \endloop#1 \fi
    }                % end of \next
  #1=#2 \next}       % end of \aloop
```

Macro \aloop can only have a positive step since it stops when the loop counter becomes greater than the until parameter.

Macro \bloop can have a negative step. If the step is negative, the macro stops when the loop counter becomes smaller than the until parameter.

```
\def\bloop#1=#2 {
  \long\def\next step ##1 until ##2 do ##3 \endloop#1{%
    ##3
    \advance#1 by ##1
    \ifnum##1>0
      \ifnum#1>##2
      \else \next step ##1 until ##2 do ##3 \endloop#1 \fi
    \else
      \ifnum#1<##2
      \else \next step ##1 until ##2 do ##3 \endloop#1 \fi
    \fi}             % end of \next
  #1=#2 \next}       % end of \bloop
```

It is a good idea to expand such macros with \tracingmacros=2 and also \tracingcommands=2. Try the following experiments:

```
\newcount\t
\bloop\t=7 step -2 until 2 do text=\the\t: \endloop\t
```

```
\newcount\c
\aloop\c=1 step 1 until 4 do text=\the\c; \endloop\c
```

Our simple approach to loops does not allow double loops. If we try

```
\bloop\t=7 step -2 until 2 do row=\the\t:
  \bloop\c=1 step 1 until 4 do col=\the\c; \endloop\c x
\endloop\t
```

we run into an error situation because of the way \next is used in the two loops. However, it is possible to implement nested loops in TEX. Reference 24 shows an elegant way of doing this, and also shows how to implement a \while loop.

7.25 Exercise

▸ **Exercise 7.27:** Write a macro \inwords that, given a number, produces the translation of the number into English, suitable, say, for writing checks. Thus the expansion, \inwords{163} dollars and \inwords{84} cents should print out the phrase you would see on a check from DEK if you find a bug in TEX.

Few things are harder to put up with
than the annoyance of a good example.
— *Mark Twain*

8. Tokens and File I/O

It has been mentioned elsewhere (page v) that TeX is programmable. It has features that are found in most programming languages, and three such features are discussed in this chapter.

- Data structures are a central concept in programming languages. Token strings are the only data structure supported by TeX. Such strings can be stored in token registers and manipulated in a limited way.
- File input/output is also an important feature. Data can be read from a file by a TeX job. Files can also be created, for later use by TeX or any other program. Interaction with the user is an extension of file I/O.
- Data can be input from a standard input device (normally the keyboard) and output to a standard output device (normally a screen).

8.1 \toks

When reading the source file, the first step performed by TeX is to convert input characters to tokens. A token is either a single character or a control sequence (but there is a third type, a macro parameter, such as '#1'). A character token also gets a category code attached to it, and that attachment is permanent. The tokens are generated in an early stage (in the mouth). Control sequence tokens don't have any catcodes.

Operations in the gullet and in the stomach depend on the catcodes. When the stomach receives a token, it first looks for a catcode. If there is one, it tells the stomach what to do with the token. If there is none, the stomach has to look up the meaning of the token. This means that once a catcode is assigned to a character token, the token has a meaning that cannot be changed later. The meaning of a control sequence, however, can always be changed because of the lack of a catcode.

▸ **Exercise 8.1:** When a blank space is input, is it also converted into a token?

Tokens can be stored in a \toks register and then used as many times as necessary. Token registers are named \toks0 through \toks255 and can be declared and assigned names by a \newtoks command. There is also a \toksdef command that can be used to assign a name to a specific \toks register. Thus \newtoks\xyz assigns the first available \toks register and names it \xyz, whereas \toksdef\abc=1 assigns the name \abc to \toks1.

▸ **Exercise 8.2:** What if the control sequence \abc is already defined (it is a macro or a primitive)?

Tokens can be saved in a \toks register by an assignment \toks2={abc \xyz}. They can later be executed by \the\toks2.

▸ **Exercise 8.3:** How many tokens does \toks0 contain after '\toks0={abc \xyz}'?

Saying \the\toks2 will expand its contents, so it will typeset 'abc' and expand \xyz. Note that \xyz must be defined at that point, but it does not have to be defined during the original assignment. This is an aspect of a general rule [215] that says that tokens are not expanded when they are assigned to a token register.

To illustrate this rule, consider the assignment \toks3={\the\toks2}. This assignment will simply store the string '\the\toks2' in \toks3 and will not expand the contents of \toks2. This is easy to verify with a \showthe\toks3. To copy \toks2 to \toks3, just say \toks3=\toks2.

▸ **Exercise 8.4:** How many tokens does the string '\the\toks2' contain?

A similar example is '\toks0={\x} \showthe\toks0'. This will show \toks0 containing '\x'. A later attempt to expand '\the\toks0' will result in the error message '! Undefined control sequence'.

The most common operations on strings are appending a string to an existing one, and concatenating strings.

How can the string 'abc' be appended to the *right end* of \toks0? Just saying '\toks0={\the\toks0 abc}' won't work since '\the\toks0' is not expanded. To force the expansion of \the\toks0, an \expandafter should be used. Thus we end up with \toks0=\expandafter{\the\toks0 abc}. The \expandafter expands \the\toks0 before it sees the '{'.

How can a single token 'a' be appended to the *left end* of \toks0? Answer: with the command: \toks0=\expandafter{\expandafter a\the\toks0}. This will do the job, since it expands the three tokens '{', 'a' and '\the\toks0' in reverse order.

To append the string 'abc' to the left end of \toks0, first assign '\toks1={abc}' then concatenate \toks1 and \toks0. So let's discuss string concatenation.

The operation of string concatenation is a basic one. It should be allowed in every language that supports strings. Assuming that \toks2 and \toks3 contain strings, how can we concatenate them into \toks4? The following simple tests are very illuminating.

```
\toks2={abc}\toks3={xyz}
\toks4=\expandafter{\the\toks2 \the\toks3}                                    a
\showthe\toks4
```

```
\toks2={abc}\toks3={xyz}
\toks4=\expandafter{\the\toks2\the\toks3}                            b
\showthe\toks4
```

```
\toks2={abc}\toks3={123}
\toks4=\expandafter{\the\toks2\the\toks3}            c
\showthe\toks4
```

Test **a** results in 'abc \the\toks3' because the effect of the \expandafter stops at the space following '\toks2. In test **b**, that space is eliminated, so after TeX reads the '2', it reads ahead, expanding '\the\toks3', looking for more digits. It does not find any, but the result of expanding '\the\toks3' becomes part of \toks4. The end result is 'abcxyz', a true concatenation. Test **c** is similar, except that the expansion of '\the\toks3' results in more digits. This causes TeX to end up with '\the\toks2123' and the message '! Bad register code (2123)'.

So how can two or more strings be concatenated? Answer: with the use of a macro as an intermediate quantity.

```
\toks2={abc}\toks3={123}
\edef\tmp{\the\toks2 \the\toks3}
\toks4=\expandafter{\tmp}
```

▸ **Exercise 8.5:** Why \edef and not simply \def?

When using the contents of a \toks register, the original data remains in the register and can be reused. This is especially useful when a piece of text has to be typeset repeatedly (perhaps with different hsizes; see the example of column balancing by binary search in [App. D]).

Tokens can also be stored in token lists, such as \everypar [105], \everycr [362], \everyhbox, \everyvbox [275], and others.

Tokens can also be stored in box registers, but there are important differences between \toks and \box registers.

1. The assignment \toks0={abc} stores just the three tokens in the register. In contrast, the assignment \setbox0=\hbox{abc} also places a kerning in the box, between the 'b' and the 'c'. The idea is that material stored in a box should be ready for typesetting.

2. The assignment \toks0={ab\c} places the three tokens in the register without expanding \c. In contrast, the assignment \setbox0=\hbox{ab\c} expands \c, and results in the error message

! Argument of \c has an extra }

(Recall that \c is the **plain** macro for çedilla and requires an argument.) The reason for the different behavior is the same as above.

3. A \toks register has no structure or dimensions. A box has, of course, structure (lines of text with glues and penalties) and dimensions.

This is why box registers are not always sufficient, and \toks registers are necessary.

8.2 Example

This example presents two macros \title and \jobtitle, useful in formatting a book. Imagine a book designer deciding to use roman numerals for chapter numbers. Each chapter is saved in a file, with file names such as Introduction.tex, iiiDeparture.tex, xxivBack-home.tex. A macro is needed that scans \jobname character by character until the first uppercase letter is found. Everything preceding that point is considered the chapter number.

The macro below follows the last example on [219]. Macro \jobtitle gets the next character of the \jobname string each time it is expanded. It tests for an uppercase letter by comparing the character code '#1 to the \uccode of the same thing. Depending on the result, the next character from \jobname is appended to either \hnum or \hnam.

```
\newtoks\hnum \newtoks\hnam
\newif\ifupper
\def\title#1{\upperfalse\hnum={}\hnam={}\jobtitle#1\end}
\def\jobtitle#1{%
\ifx#1\end
 \let\next=\relax
\else
 \ifnum'#1=\uccode'#1 \uppertrue\fi
  \ifupper\hnam=\expandafter{\the\hnam#1}%
 \else
   \hnum=\expandafter{\the\hnum#1}%
 \fi
 \let\next=\jobtitle
\fi \next}
```

8.3 Toks and Macros

The main use of toks registers is for saving strings of tokens, and operating on them. For tokens with catcodes 11 and 12 (which we denote *regular tokens*), this can also be done with macros. A simple example is the two commands \def\tst{xyz123} \toks0={xyz123}. Following these, the commands \tst and \the\toks0 produce identical results. Even simple string processing can be done successfully with both toks and macros, as long as we limit ourselves to regular tokens. The two sections

```
\toks0={abc}                              \def\a{abc}
\toks0=\expandafter{\the\toks0 123}       \edef\a{\a 123}
\showthe\toks0                            \show\a
```

show the same string, namely, 'abc123'.

Another application of toks (and macros) is the creation of a *list* in a toks register. Each component of the list is preceded by an undefind control sequence, such as '\\'. An example is '\toks0={\\{a 19}\\{c 12}\\{md 1}\\{x 7}}'. Certain list operations are now easy to perform. Examples:

a. Count the number of list elements. Just say

```
\newcount\length
\def\\#1{\advance\length by1}
\the\toks0
```

b. Find the first occurrence of a list element starting with 'md' and return the number following it.

```
\def\\#1{\aux#1.}
\def\aux#1 #2.{\def\md{md}\def\tp{#1}\ifx\md\tp '#2'\fi}
\the\toks0
```

(Even this application can be implemented with macros instead of toks.)

The reader should modify the above technique to store the string in a macro instead of a toks.

In general, however, macros and toks registers behave differently when control sequences and tokens of other catcodes are involved.

1. The first obvious difference is the behavior of the \the command. It cannot be used with macros but is easy to use with toks registers, where it produces tokens of catcode 12 (except spaces). Here is an example of its use.

Saying '\immediate\write\out{\the\toks0}' writes the contents of \toks0 on the file as a string of ASCII codes. The file becomes a regular text file. In contrast, the command '\immediate\write\out{\aux}' (where \aux is a macro) completely expands \aux (including expansions of all nested control sequences) and writes the results on the file.

2. Another difference between toks and macros has to do with the behavior of tokens in an \edef. Defining \def\w{\x} \def\x{\y} \def\y{\z} \def\z{*} \edef\a{\w} does a complete expansion of \w in \a, resulting in '*' as the replacement text of \a. Saying, on the other hand, \toks0={\w} \edef\b{\the\toks0} \show\b shows the replacement text of \b to be \w. The expansion of \toks0 inside \b is done at top level only. (Even this top-level expansion can be suppressed by saying \edef\b{\noexpand...}.)

With this in mind, we present our next macro, \append. This macro appends a given string to a given toks register.

```
\newtoks\str
\str={\a}
\def\append#1to#2{\toks0={#1}
 \edef\aux{\noexpand#2={\the#2 \the\toks0}}\aux}
\append\b \ {$&^_ \c}& to\str
\showthe\str
```

Defining '\edef\aux{\noexpand#2={\the#2 #1}}' would try to expand '#1' and would result in an error if '#1' contains undefined control sequences or special characters. The example above results in '\a \b \ {$&^_ \c }&'. Note that the

strings involved may contain undefined control sequences and the special characters
'$^&_{}' (except that braces must be balanced).

This operation is easy to do with macros, but only if the strings are limited to
regular tokens.

8.4 File I/O

It is easy to create and read text files from inside a TEX document. The
command \input⟨filename⟩ is well known and is commonly used. In addition, a
\read command makes it possible to read a file record by record. New files can
be created by TEX, and data written on them, record by record. There can be
a maximum of 16 input and 16 output files open at any given time. Each file
is identified internally by means of a file number. The \newread and \newwrite
commands generate the next available file number. The \openin and \openout
commands associate a file number with an actual file name, following which the file
can be used.

All file I/O commands are listed on [217, 226]. Output is done by
\write⟨number⟩{⟨material⟩} or \immediate\write⟨number⟩{⟨material⟩};
input is done either by \read⟨number⟩ to⟨control sequence⟩, or \input⟨filename⟩.
Each \write creates a record on the file, whose maximum size is usually limited by
the operating system of the computer (see Ref. 26 for a way around this limitation).

An important feature of file output is that expandable tokens are expanded
during a \write. If the name of a control sequence, rather than its expansion,
should be written on a file, either \noexpand or \string should be used to inhibit
the expansion. If a control sequence is unexpandable, its name is written on the
file. If it is undefined, an error message is issued when TEX tries to expand it during
the \write. It is also possible to avoid expansion during a \write by changing the
catcode of '\'. This way, anything that starts with a '\' is no longer considered a
control sequence (see page 202 for an example of this approach).

To complicate matters a bit more, the actual write is deferred until the current
page is shipped out in the OTR. The reason for that is that the user may want
to write the page number on the file (this is common when a table-of-contents
file or an index file is generated) and this number is only known inside the OTR
(see section 16.2 on the asynchronous nature of the OTR). If no page numbers are
involved, the user can force the record to be written on the file immediately by
\immediate\write.

Interaction with the user. Valid file numbers are between 0 and 15. File numbers
outside this range refer to the standard I/O devices. Thus \read-1 to\note reads
from the keyboard (without a prompt) into \note. The quantity \note doesn't have
to be predefined or declared. Once input is read into it, \note can be expanded
like a macro. \write16{...} displays information on the screen.

▸ **Exercise 8.6:** What is the difference between \write16 and \message?

A few simple examples follow. More extensive examples of file I/O appear in
Chapter 9, in connection with multipass techniques.

8.5 Example: Keyboard Input

Here is a simple example that asks the user to enter a number n from the keyboard. It then typesets n asterisks using a recursive macro \aste.

```
\newcount\ctst
\def\aste#1{\ifnum#1>0 *\advance\ctst-1 \aste\ctst\fi}
\message{How many asterisks? }\read-1to\mtst \ctst=\mtst
\aste\ctst
```

▸ **Exercise 8.7:** Why is it necessary to assign \ctst=\mtst?

▸ **Exercise 8.8:** Does macro \aste work for zero asterisks?

There are similar examples on [373, 374].

8.6 Example: File Input/Output

We now turn to an example of file I/O that also illustrates \noexpand. It shows that when a file is \input, its contents is treated by TEX as if it came from the source file. Any commands read are executed, and any text is typeset. We define a macro \test#1, which starts by opening a file with an internal name \abc and external name 'myfile'. It writes a record on the file and expands itself. This is repeated five times, and the result is five similar records on 'myfile', each with these six fields:

1. the value of its parameter (a string), followed by ':␣'
2. the characters \ctst=
3. the value of \ntst, followed by ':' (produced by \the)
4. the value of \ctst—produced by \number—followed by '"'
5. the word '\number', produced by the \noexpand
6. the word '\ctst'

The macro is defined by:

```
\newwrite\abc
\immediate\openout\abc=myfile        % see p. [227] for \immediate
\def\test#1{
\immediate\write\abc
{#1:\space\ctst=\the\ntst:\number\ctst"\noexpand\number\ctst}
\advance\ntst-1
\advance\ctst 1 \ifnum\ctst<6 \test{#1}\fi}
```

▸ **Exercise 8.9:** To get the word \number on the file, we had to use \noexpand. Why didn't we have to use \noexpand to get the word \ctst on the same file?

It is expanded by

```
\newcount\ctst \newcount\ntst
\ctst=1 \ntst=9 \test{Now hear this}
```

which generates the file

```
Now hear this: \ctst =9:1"\number \ctst
Now hear this: \ctst =8:2"\number \ctst
```

```
Now hear this: \ctst =7:3"\number \ctst
Now hear this: \ctst =6:4"\number \ctst
Now hear this: \ctst =5:5"\number \ctst
```

The file is then \input by

```
\immediate\closeout\abc
{\obeylines
\input myfile
\smallskip}
```

which generates

> Now hear this: :1"9
> Now hear this: :2"8
> Now hear this: :3"7
> Now hear this: :4"6
> Now hear this: :5"5

When the file is \input, the second field—which is the control sequence \ctst—is executed. Fortunately, it is followed by a number (the third field), so no error message is issued. The result of this execution is to assign \ctst the values $9, 8, \ldots, 5$. Next ':1"' is typeset and, finally, \number is executed, causing the value of \ctst to be typeset. As a result, what gets typeset is different from the contents of the file.

An alternative to \input is to read the file record by record. The \ifeof test should be used to detect the end of the file. Here is an example using file myfile:

```
\newcount\ctst
\newif\ifcont \conttrue
\newread\rec
\openin\rec=myfile
\loop
\ifeof\rec\contfalse\fi
\ifcont
\read\rec to\xyz
'\xyz'
\repeat
```

Note the following:

■ It would be simpler to say \ifeof\rec\else\repeat, but this is wrong because of the way the \loop-\repeat construct is defined. This is why the auxiliary boolean variable cont has to be introduced.

■ A novice would tend to place the \ifeof test at the end of the loop, just before the \repeat. Placing it at the beginning of the loop, however, is better since it does not try to read anything in the case of an empty file.

Here is a similar version using a macro \ifnoeof, that makes the loop easier to read:

```
\def\falseskip#1\fi{\fi\iffalse}
\def\trueskip\fi{\fi\iftrue}
```

```
\def\ifnoeof#1{\ifeof#1 \falseskip\else\trueskip\fi}
```

```
\newcount\ctst
\newread\rec
\immediate\openin\rec=myfile
\ifeof\rec\message{no such file}\fi
\loop
\read\rec to\xyz
\ifnoeof\rec
process the record
\repeat
```

8.7 Writing Arbitrary Strings on a File

We start with a short review of \edef. In '\edef\abc{\xyz \kern1em}', the control sequence \xyz is expanded immediately (when \abc is defined), but the \kern command is only executed later (when \abc is expanded).

The same thing happens when \abc is written on a file. '\write\aux{\abc}' writes the replacement text that would have been created by \edef\abc{...}.

Sometimes it is desirable to write the name of a control sequence on a file, rather than its expansion. This can be done by either '\write\aux{\noexpand\abc}' or '\write\aux{\string\abc}'. The former form writes a space following \abc, while the latter one does not.

With this in mind we now consider the following problem: Given an arbitrary string containing text, control sequences, active characters, and special characters (such as #%{}), first write it on a file without expansion (verbatim), then expand it.

Before delving into the details, here are some examples that show that this problem is practical:

1. When writing a textbook with exercises and answers, the author would like to be able to say

```
\exercise...
\answer...
...
\endanswer
```

and have the answer written verbatim on a file. Later the file can be input, to typeset all the answers in an appendix. However, while the book is being written, the author may also want, for proofreading purposes, to typeset the answer right after the exercise. Note that an answer may contain many control sequences and may be long.

2. When writing a book on TEX, the author would like to have an active character (say '^'), such that '^{\baselineskip}=24pt' would write \baselineskip on a file (perhaps with the page number, for later preparation of an index) *and also* execute '\baselineskip=24pt'.

We develop two approaches to this problem. The first one uses catcode changes to suppress the special meanings of certain characters before the string is read by

TEX. It is then easy to read the string and write it verbatim on a file. However, in order also to expand the string, all characters should have their normal catcodes. This is done by writing the string on another file and reading it back immediately. This way, the string is parsed into tokens that get their normal catcodes and can later be expanded.

In the second approach no catcodes are changed; the string is input as usual and tokens are created. The string is then scanned, token by token, to identify the control sequence tokens, and they are temporarily redefined as \relax (which is nonexpandable). The string can then be written on a file and no control sequences will be expanded. After that, all control sequences get back their original meanings, and the string can be expanded in the usual way.

Before we continue, here is an important disclaimer: The macros presented here are simple. Each has its limitations and can be used only for certain applications. The macros should therefore not be copied and used verbatim. They should be carefully studied and fully understood, so that they could be modified for specific applications.

Approach 1: We present a number of macros, all based on catcode changes. The first two change the catcodes of all the special characters. The other three change the catcodes of just a few characters. In between the two groups, we illustrate how the macros can be modified to handle a specific problem, namely, writing index items, with page numbers, on a file.

To avoid expansion we change the catcodes of the special characters, such as '$', '#' and '\', to 12 (other). This way, the '\' is no longer the escape character, so TEX does not recognize any control sequences, and there is nothing to expand. The catcode changes should, of course, be done locally, in a group. Macro \VwriteA below starts a group, does the catcode changes, and expands \aux. Macro \aux absorbs the argument, does the \write, and closes the group.

```
\newwrite\out
\immediate\openout\out=filename

\def\VwriteA{\begingroup\sanitize\aux}
\def\aux#1{\write\out{#1{\folio}}\endgroup}

\def\makeother#1{\catcode`#1=12\relax}
\def\sanitize{\makeother\ \makeother\\\makeother\$\makeother\&%
\makeother\#\makeother\_\makeother\%\makeother\~\makeother\|}
```

A typical expansion may be '\VwriteA{te xt#$%&_~\a\x fin}'. The argument, which seems to belong to \VwriteA, is actually absorbed by \aux. Also, since the actual writing is done in the OTR, it is possible to write the page number on the file, even in braces, as above.

Note that the vertical bar '|' was also sanitized since it is often declared active and is used for verbatim listings. The braces, on the other hand, were not sanitized, which makes it possible to enclose the argument in braces (but then the argument cannot contain arbitrary braces, only balanced ones).

Macro \VwriteB below is a slightly different version that does sanitize the braces. The argument can now contain arbitrary braces, but it must be delimited by something else (the string 'endP' in our case).

```
\newwrite\out
\immediate\openout\out=filename

\def\VwriteB{\begingroup\sanitize\aux}
\def\aux#1endP{\write\out{#1{\folio}}\endgroup}

\def\sanitize{\makeother\ \makeother\\\makeother\$\makeother\&%
\makeother\#\makeother\_\makeother\%\makeother\~\makeother\|%
\makeother\}\makeother\{}
```

As a practical example, we use \VwriteA to illustrate the creation of a raw index file. The following macros declare the '^' an active character to write index items (with page numbers) on a file. They also use \futurelet to allow silent index items.

Expansions such as ^{kerning} or ^[B.~L.]{User} or ^[\kern]{} are allowed. Up to two arguments can be specified and are written on the index file as one string (with the page number). However, only the main argument (in braces) is typeset. The optional argument, in brackets, is silent.

```
\def\Caret{\ifmmode\def\next{^}\else\let\next=\indexT\fi\next}
\catcode'\^=\active \let^=\Caret

\def\indexT{\futurelet\new\macX}
\def\macX{\ifx\new[\let\Next=\inxB\else\let\Next=\inxA\fi
\begingroup\sanitize\Next}
\def\inxA#1{#1\writeinx{#1}}
\def\inxB[#1]#2{#2\writeinx{#1#2}}
\def\writeinx#1{\write\inx{\string\indexentry{#1}{\folio}}\endgroup}
```

The only problem is that an item such as '^\TeX' is expanded when the \futurelet sees it (before sanitizing). Therefore, its expansion, rather than its name, is written on the file. When the item is enclosed in braces '^{\TeX}', the \futurelet only sees the '{', so the item is not immediately expanded. After sanitizing, its name is written on the file. However, because of the sanitizing, the name of the item, rather than its expansion, is also typeset in the document. In the case of index items, the user can write the control sequence twice, once outside the '^' to expand it, and once inside the '^' to write its name on the file. Thus '\TeX^[\TeX]{}'.

In general, a way is needed to write the contents of any string, with no expansion, on a file, then expand it. Unfortunately, sanitizing is done by changing catcodes, and once a catcode is assigned to a token, this assignment is permanent [39] and cannot be changed. A solution exists, however, and is developed below.

The development proceeds in three steps. In step 1, a simple macro is developed that can write strings on a file without expanding control sequences. Its limitations are that (a) the macro sanitizes certain characters so, after writing the string on a file, it (the string) cannot always be expanded; (b) a multiline string is written on the file as one line.

In step 2, limitation (a) above is overcome. The string is written on the file as before and is then written on another file, which is immediately \input. When the string is read back, all characters get their normal catcodes, and the string can be expanded as usual.

In step 3, the macro is modified, à la \elp (section 5.20), to scoop up one input line at a time. This way, a multiline string is written on a file line by line. The string can also be long, since only one line need be saved at a time.

Step 1: To write a string on a file without expansion, it is placed in a toks register and then written from the register.

```
\toks0={...string...}
\immediate\write\out{\the\toks0}
```

This works since the control sequence '\the' creates a string of tokens, all of catcode 12, except spaces. Fortunately, '\the' can be applied to a \toks register. Note that this illustrates another difference between macros and \toks registers, since '\the' cannot be applied to a macro. Trying to say

```
\def\aux{...string...}
\immediate\write\out{\aux}
```

would expand \aux and all the commands in it. Things such as '\noexpand\aux' or '\string\aux' would simply write the name of the macro, not its contents, on the file.

In practice, the string to be written on a file is the argument of a macro, so the actual code is

```
\def\aux#1\endP{\toks0=\expandafter{#1}
  \immediate\write\out{\the\toks0}
  ...}
```

This works since \expandafter is a one-step expansion. It expands '#1' into individual tokens but does not expand the tokens further. Our first version is thus

```
\newtoks\str \newwrite\out
\immediate\openout\out=filename
\def\VwriteC{\begingroup\sanitize \aux}
\def\aux#1\endP{\str=\expandafter{#1}\immediate\write\out{\the\str}%
  \endgroup}
\def\sanitize{\catcode'\ =12 \catcode'\%=12 \catcode'\#=12
  \catcode'\{=12 \catcode'\}=12 }
```

After this, the macro can be expanded by, for instance,
\VwriteC a$x\le0$\c C\vrule 1\endP or \VwriteC X\hskip10pt$\sum$\endP

Step 2: The string written on a file cannot, in general, be expanded by our macros, since the catcodes of the characters '#%{}' (and of the space) were changed. Trying to say, for example,

```
\def\aux#1\endP{\str=\expandafter{#1}\immediate\write\out{\the\str}%
```

```
\endgroup #1}
```

might produce wrong results, or an error message, if the string contains a '#', a '&', or any braces.

Macro \VwriteD below solves this problem by writing the string on a second file and reading it back immediately. Upon reading, the string is parsed into tokens that get their normal catcodes. The string can now be expanded. This is a general (albeit slow) solution to the problem. TeX is coerced into scanning the same string twice. The first time with special catcodes, and the second time, under normal conditions.

```
\newtoks\str \newwrite\out \newwrite\Tmp
\immediate\openout\out=\jobname.aux
\def\VwriteD{\begingroup\sanitize \aux}
\def\aux#1\endP{\str=\expandafter{#1}\immediate\write\out{\the\str}%
\immediate\openout\Tmp=\jobname.tmp \immediate\write\Tmp{\the\str}%
 \endgroup \immediate\closeout\Tmp \input\jobname.tmp }
\def\sanitize{\catcode'\ =12 \catcode'\%=12 \catcode'\#=12
 \catcode'\{=12 \catcode'\}=12 }
```

Step 3. The new version, macro \VwriteE, can read a long, multiline, argument and write it on a file line by line. The idea is to have macro \aux scoop up one line of the source string as its argument, write it on the file, then expand \VwriteC recursively until a certain string (\endP in our case) is found that signals the end of the argument.

The main difference between this step and step 1 is the definition of \aux. It is defined in a group where the catcode of ⟨return⟩ is set to 12. A more detailed explanation appears in conjunction with macro \elp.

```
\newtoks\str \newwrite\out
\immediate\openout\out=\jobname.aux
\def\VwriteE{\begingroup\sanitize\catcode'\^^M=12 \aux}
{\catcode'\^^M=12\gdef\aux#1^^M{\def\temp{#1}%
\ifx\temp\enP%
 \gdef\next{\relax}%
\else \str=\expandafter{#1}\immediate\write\out{\the\str}%
 \gdef\next{\VwriteE}%
\fi\endgroup\next}}
\def\enP{\endP}
\def\sanitize{\catcode'\ =12 \catcode'\%=12 \catcode'\#=12
 \catcode'\{=12 \catcode'\}=12 }
```

A typical expansion now looks like

```
Any text...\VwriteE9A{\bf abc} \B
11\halign{#\cr1\cr}\TeX
x$\yy@#%^&_?}\it {\c
\endP
...more text
```

Notes:

1. The `\endP` must be on a line by itself and must start on column 1. The user may, of course, change from '`\endP`' to any other string.

2. Macro `\aux` is defined when the catcode of ⟨return⟩ is 12. Therefore, every line in the definition of `\aux` must be delimited by a '`%`'. Otherwise the end-of-line would be typeset as `\char`'015 in the current font. (The table on [367] shows that '015 is the character code of ⟨return⟩.)

3. The temporary macro `\next` is defined by '`\gdef`' instead of by '`\let`', since it is defined inside `\aux` and `\aux` is defined inside a group.

4. It seems that steps 2 and 3 can be combined. It is suggested that the reader develop a macro `\VwriteF` with the combined features of `\VwriteD` and `\VwriteE`.

5. The three macros above write the value of the toks register `\str` on the file. They therefore cannot use a delayed `\write` and must use `\immediate\write`. Trying to say '`\write\out{\the\str}`' would delay all the write operations to the OTR, where register `\str` may contain the string from the most recent write or may even be undefined.

Approach 2. In this approach there are no catcode changes (except that `\obeyspaces` is used locally, during scanning). The string is input and is parsed into tokens in the normal way. This way, our macros can expand it by simply saying '`#1`'. The first version, macro `\VwriteM`, scans the string of tokens and inserts a `\noexpand` in front of every control sequence token. The second version, `\VwriteN`, performs the same scanning and changes the meaning of every control sequence to `\relax`. The scanning is done with macros `\scan` and `\onestep`, a technique described in Chapter 7.

Version 1. Macro `\onestep` receives the next token in the string, checks to see if it is a control sequence (by comparing its catcode to that of `\relax`), and if it is, inserts a `\noexpand` in front of it. The new string is created, token by token, in the toks register `\str`. The only step that needs detailed explaining is macro `\temp`. It is important to understand why this macro is necessary (why not simply say '`\immediate\write\out{\the\str}`'), and why `\edef` is used?

Imagine the expansion '`\VwriteM{a\TeX}`'. When scanning is complete, register `\str` contains '`a\noexpand`␣`\TeX`␣' (including the spaces). Saying '`\immediate\write\out{\the\str}`' would write that string (including the `\noexpand`) on the file, as in approach 1 above. Defining `\temp` by means of `\def` would make '`\the\str`' the replacement text of `\temp`, so '`\immediate\write\out{\temp}`' would expand `\temp` and would be identical to writing '`\the\str`'. The `\edef`, however, creates '`a\noexpand \TeX `' as the replacement text of `\temp`. During the write operation `\temp` is expanded, which is when the `\noexpand` does its job and prevents the expansion of `\TeX`.

```
\newwrite\out \newtoks\str
\immediate\openout\out=filename

\def\VwriteM#1{'#1'{\str={}\obeyspaces\scan#1\end
\edef\temp{\the\str}\immediate\write\out{\temp}}}

\def\scan#1#2\end{\def\aux{#1}%
```

```
\ifx\aux\empty
\else
 \def\aux{#2}%
 \onestep{#1}%
 \ifx\aux\empty
 \else
  \scan#2\end
\fi\fi}
```

```
\def\onestep#1{\ifcat\relax\noexpand#1%
\str=\expandafter{\the\str\noexpand#1}%
\else\str=\expandafter{\the\str #1}\fi}
```

Note the following:

1. There is a (local) use of \obeyspaces. Without it, spaces are skipped when TEX determines the arguments of \scan.

2. Because no catcodes are changed, the four characters '#%{}' cannot appear in the argument of \VwriteM. A '#' in the argument will become '##' when the argument is absorbed. A '%' will send TEX to look for the rest of the argument on the next line. Unbalanced braces will cause an error message when the argument is absorbed. Balanced braces would be absorbed, would be used to nest groups in the argument, and will not appear on the file. For this reason, the use of \VwriteM is limited to cases where these characters do not appear in the strings to be written on file.

3. The \noexpand command adds an extra space. Thus '\VwriteM{\?M}' writes '\? M' on the file. To suppress the space, use \string instead of \noexpand in '\str=\expandafter{\the\str\noexpand#1}'. This may look better but may give wrong results in some cases. A typical example is the expansion '\VwriteM{\bf M}', which would write '\bfM' on the file.

4. Our macros do not attempt to identify active characters. If the string includes any active characters, their expansions would be written on the file. It is, however, relatively easy to test for tokens of catcode 13 and insert a \noexpand in front of them.

Version 2. A different way of avoiding expansion during file output is to temporarily turn an expandable control sequence into a nonexpandable one. The simplest way of achieving this is to \let the control sequence be equal to \relax. Thus

```
\def\abc{...}
...
\let\abc=\relax
\immediate\write\aux{\abc}
```

will write \abc on the file. Using this method we illustrate a different solution to the same problem. In this version, macro \onestep identifies all tokens in the string that are control sequences, and sets each equal to \relax.

After every control sequence in the string has been changed in this way, the string is written on a file. This version is similar to the previous one, the most important difference being that the final quantity being written on the file is '#1' and not the replacement text of a macro or the contents of a toks register. As a result, any braces in the argument will be written on the file (but see note below for a subtle problem with braces).

```
\newwrite\out
\immediate\openout\out=filename
\def\VwriteN#1{{\scan #1\end\immediate\write\out{#1}}}

\def\scan#1#2\end{\def\aux{#1}%
\ifx\aux\empty
\else
 \def\aux{#2}%
 \onestep{#1}%
 \ifx\aux\empty
 \else
  \scan#2\end
\fi\fi}

\let\Let=\let
\def\onestep#1{\ifcat\relax\noexpand#1\Let#1=\relax\fi}
```

The following points should be mentioned:

1. Macro \VwriteN has an extra pair of braces, so everything done in it is local. This way, the setting of control sequences to \relax is only temporary.

2. Imagine the string 'abc\let hjk\x'. The control sequence \let is first identified and is set to \relax. Later the control sequence \x is identified, but saying \let\x=\relax fails because \let is now equal to \relax. This is why the command '\let\Let=\let' has been added. Macro \onestep uses \Let instead of \let. Of course, a string such as '...\Let...\x' would cause the same problem, so this method cannot handle such strings.

3. The '#' and active characters still cannot be part of the argument of \VwriteN, for the same reason as above.

4. Braces in the argument must still be balanced but will be written on the file as mentioned earlier. There is another, subtle, problem associated with braces. Consider the expansion '\VwriteN{{\bf M}$}'. At a certain step during the scanning, the argument of \onestep becomes the group '\bf M'. The \noexpand#1 thus becomes '\noexpand\bf M', which typesets the 'M'. The '\Let#1=\relax' becomes '\Let\bf M=\relax', which lets \bf to 'M' and typesets the '='. As a result, this version, too, should only be used in limited cases.

5. Macro \scan does not use tail recursion because it has to expand either \onestep or itself with different parameters. As a result, each recursive expansion of \scan saves two \fi's in the parameter stack, whose normal size is 60. A long argument will thus exceed TeX's capacity.

6. This method works only for an \immediate\write. A nonimmediate \write is executed in the OTR, where the various control sequences are no longer equal to \relax.

8.8 Last Words

Tokens and file I/O are two more features that distinguish TEX from most word processors and page layout programs. They contribute to the power of TEX and should be mastered by anyone aspiring to the title of TEXmaster (or even that of a TEXnician).

And it shall be for a token upon thine hand,
and for frontlets between thine eyes:
for by strength of hand the LORD brought us
forth out of Egypt.

— *Exodus 13:16*

9. Multipass Jobs

Certain typesetting problems, mostly those having to do with cross-referencing in a document, can be solved with a multipass job. Normally, two or three passes are necessary. The basic idea is to write a TEX program with macros for all passes and to run it as many times as necessary. In the first pass, certain macros are expanded, and they write the necessary cross-reference data on an auxiliary file. In the second pass, the file is input, and other macros are expanded, based on the information read. Another auxiliary file is written by the second pass, either to be read by the third pass (if one is necessary) or to be ignored.

This chapter contains two examples; a 2-pass job, for cross-referencing equation numbers, and a 3-pass job, to prepare a table-of-contents. These examples can be used for other problems requiring file I/O, such as index preparation, or collecting answers to exercises, to be typeset in an appendix.

9.1 Cross-referencing in a Document

Many times, the author of a document needs to refer to parts of the document from other parts. This is called *cross-referencing* and can be done in TEX by writing the cross-reference material on an auxiliary file. Such a job takes at least two passes, the first generates the auxiliary file and the second reads it. The auxiliary file can have any name allowed by the operating system of the computer. Good choices, however, are names such as \jobname.aux, \jobname.ref, or \jobname.toc, where the control sequence \jobname [213] expands to the name of the source file.

To simplify the entire job, the two passes should be identical. A 2-pass job simply means running the same TeX program twice. Each pass starts by trying to read the cross-reference material from the auxiliary file. Of course, such a file normally does not exist when the first pass starts, so it is important to verify that the file exists before attempting to \input it. If the file does not exist, an appropriate message should be placed in the log file. This is done by

```
\newread\aux
\immediate\openin\aux=\jobname.aux
\ifeof\aux \message{! No file \jobname.aux;}
\else \input \jobname.aux \immediate\closein\aux \fi
\newwrite\aux
\immediate\openout\aux=\jobname.aux
```

Notice that a new output file of the same name is immediately opened to write the new cross-reference material. Normally, the first pass does not find the auxiliary file, so nothing is \input. During this pass, the file is created, record by record, with cross-reference material. The second pass finds the file and inputs it, so it has all the cross-reference material necessary. It generates the typeset output and also creates another version of the auxiliary file, which normally is not used and is later deleted by the user.

We illustrate the principle of cross-referencing by applying it to equation numbers. Many equations, like Eq. (1) below, are numbered, so they can be referred to.

Definition: A polynomial of degree n in t is the function

$$(1) \qquad P(t) = \sum_{i=1}^{n} B_i t^{i-1} = B_1 + B_2 t + \cdots + B_n t^{n-1}.$$

The equation above was written

```
$$P(t)=\sum_{i=1}^nB_it^{i-1}=B_1+B_2t+\cdots+B_nt^{n-1}.
\leqno(1)\label1{Poly}$$
```

and the reference was done by '....Many equations, like Eq.~(\ref{Poly}) below, are numbered...'.

It should be clear from this example that an item to be cross-referenced is written by macro \label on the auxiliary file. After the file is input, in the second pass, macro \ref can be used to refer to any labeled items. Since the two passes are identical, macro \ref is also expanded in the first pass, where the cross-reference material is not yet available. Each expansion of \ref should, therefore, make sure that the required reference exists. If it does not, \ref should place a message in the log file.

For example, the first pass of this job has placed the following in the log file:

```
\aux=\read0
! No file cross-ref.aux;
\aux=\write0
! No ref. to Poly;
```

The main question is: What exactly should be written on the auxiliary file for each cross-reference item? In a conventional programming language, each cross-reference item is typically written on the file as a two-part record with the name and the value of the item, thus "Poly:1". The file is then read into a table, and \ref searches the table for a certain name. When finding it, \ref returns the corresponding value. In TeX, however, there are no tables. We are limited to registers of type \count, \dimen, and so on. Also, there are only 256 registers of each type, and a document may have more than that number of cross-references.

The solution is to declare each cross-reference item as a macro. The number of macros is limited only by the available memory [300]. Macro \label thus writes records such as "\def\Poly{1}" on the auxiliary file. When that file is \input, each record read creates a new macro definition. To reference such a macro, \ref simply expands it (after checking that it exists).

```
\def\label#1#2{\immediate\write\aux
{\noexpand\def\expandafter\noexpand\csname#2\endcsname{#1}}}
\def\ref#1{%
\ifundefined{#1}\message{! No ref. to #1;}%
 \else\csname #1\endcsname\fi}
```

If the reference material (parameter #1 of \ref) is undefined, a message is placed in the log file. Otherwise, \ref simply expands the macro whose name is #1.

To check for an undefined #1, the \csname-\endcsname pair is used to attach a '\' to it, thereby turning it into a control sequence, which is then compared to \relax [40] (see also [ex. 7.7] for the definition of \undefined):

```
\def\ifundefined#1{\expandafter\ifx\csname#1\endcsname\relax}
```

At the end of the document it is important to close the auxiliary file by saying '\immediate\closeout\aux' (on many modern computers, the operating system does that automatically).

▶ **Exercise 9.1:** Sometimes, equations should be numbered consecutively throughout the document. In such a case, it is desirable (and easy) to number them automatically. Automatic numbering of equations frees the author from having to renumber the equations each time the document is modified. Write a macro '\EqNum#1' that will automatically increment an equation number (in a \count register) and typeset it in the equation. The macro should have an optional parameter that, when used, would label the equation.

9.2 General Cross-referencing

The problem of cross-referencing is not limited to equation numbers. Imagine a document divided into chapters and sections, and also having figures, equations, examples, and exercises. While writing the document, the author may want to refer to a chapter, an exercise, a figure, or even a page number. In fact, in the case of figures and equations, we can assume that every figure and every numbered equation will be referred to.

We therefore need to develop two macros, \label and \ref. Any item that
the author wants to refer to should first be labeled by saying \label{*name*}, where
name is a unique name identifying the item. Macro \label writes the cross-reference
information on the .aux file, to be input in the second pass. Macro \ref is used
to refer to an item. In the first pass, an expansion such as \ref{*name*} should be
inactive. In the second pass it should use the cross-reference information to typeset
the number of item *name* (a chapter number, figure number, page number, etc.).

Typical labeling may look like this:

```
\chapter 12 {The Mountains}

\label{ch:mntns}The mountains are artificial and atypical...
...
\section {Mount Bloom}

\label{se:bloom}A refreshing exception is this mountain, located ...
...
\figure {filename}{Chalk hill}
...
more and more peculiar\label{pg:peculr} all the time ...
```

Assuming that Chapter 12 will be referred to, we expand \label at the start
of the chapter, with 'ch:mntns' as the argument (the unique name). The name can
be anything, but the prefix 'ch:' is useful, as it will later tell us that the name
is that of a chapter. Macro \label has to write on the .aux file a record of the
form '\def\ch:mntns{{12}{345}}', where 12 is the chapter number and 345 is the
page number (of course, \csname should be used to create a macro name such as
ch:mntns). Similarly, the expansion '\label{se:bloom}' should write the record
'\def\se:bloom{{12.3}{356}}', where 12.3 is the section number, and 356, the
page number.

The labeling example above should thus create an auxiliary file looking something like the following:

```
\def\ch:mntns{{12}{345}}
\def\se:bloom{{12.3}{356}}
\def\fg:Chalk hill{{12.1}{369}}
\def\pg:peculr{{12.1}{387}}
```

(where the details of \csname have been eliminated). The main question is: How
does \label know that the item being labeled is Chapter 12 or section 12.3? The
answer: \label always writes the value of the control sequence \lblitem, as the
first argument, on the .aux file. The author has to make sure that \lblitem always
contains the number of the item to be labeled.

At the start of a chapter, macro \chapter is expanded. In addition to its other
tasks, this macro should define \lblitem as the chapter number, 12. Thus

```
\def\chapter#1 #2{\def\chnum{#1}\def\chname{#2}...
\edef\lblitem{\chnum}...
..}
```

Similarly, at the start of a section, macro `\section` is expanded, which defines `\lblitem` as the current section number. Thus

```
\newcount\secnum
\def\section#1{\advance\secnum by 1...
\edef\lblitem{\chnum.\the\secnum}...
...}
```

Macro `\figure` should also define `\lblitem` as the number of the current figure. If the document contains examples that may have to be referred to, then each should start with an expansion of a macro similar to `\exm` below.

```
\newcount\exno
\def\exm{\advance\exno1\goodbreak\medskip{\bf\the\exno.}\
  \edef\lblitem{\the\exno}}
```

▶ **Exercise 9.2:** Why use `\edef` and not just `\def`?

With this in mind, the definition of `\label` is straightforward:

```
\def\strip#1>{}
\def\label#1{\edef\CrossRef{\write\Aux{\string\expandafter%
  \string\def\string\csname\space#1\string\endcsname%
  {{\expandafter\strip\meaning\lblitem}{\noexpand\folio}}}}%
\CrossRef}
```

If the replacement text of `\lblitem` is 12.3, then `\meaning\lblitem` creates the string 'macro:->12.3', and applying `\strip` leaves the string 12.3.

In the labeling example above, the line
`more and more peculiar\label{pg:peculr} all the time ...`' writes, on the .aux file, a record of the form '`\def\pg:peculr{{12.1}{387}}`' where 12.1 is the current value of `\lblitem` (perhaps the number of the last figure or section). Here we are trying to label the page number where the word peculiar appears. When refereing to pg:peculr, we should use only the page number (the second argument, 387), and not the first argument, 12.1, which would normally be meaningless.

To refer to a cross-reference, we therefore define two similar macros. Macro `\ref` expands the first argument of a cross-reference; macro `\pageref` expands the second argument, the page number. The two macros are similar and illustrate useful applications of `\expandafter` and `\csname`.

```
\def\ref#1{\ifundefined{#1}\message{! No ref. to #1;}\else%
  \expandafter\firstparam\csname #1\endcsname\fi}
\def\firstparam#1{\expandafter\paramone #1}\def\paramone#1#2{#1}
\def\pageref#1{\ifundefined{#1}\message{! No ref. to #1;}\else%
  \expandafter\secondparam\csname #1\endcsname\fi}
\def\secondparam#1{\expandafter\paramtwo #1}\def\paramtwo#1#2{#2}
\def\ifundefined#1{\expandafter\ifx\csname#1\endcsname\relax}
```

Chapter 20 shows how the macros above are used in practice, and how the auxiliary file is declared, opened, written to, and input.

9.3 Creating a Table of Contents

This example shows how a table of contents (TOC) of a document can automatically be created by TeX, using cross-referencing through an auxiliary, .toc, file. The interesting feature here is that three passes are normally necessary, as shown below, to get all the page numbers right.

The main idea in this example is to separate the preparation of the file from the preparation of the TOC itself. Our auxiliary file does not contain the complete TOC but, rather, the minimum amount of information necessary to generate the TOC. This makes it easy for anyone to use the macros presented here, since all the formatting and typesetting of the TOC is done by four macros, which are easy to read and modify. As before, all passes are identical, and TeX does not know what pass it is executing at any time. The entire execution amounts to running the same job three times.

Each pass starts by opening the .toc file. If the file exists, the entire TOC is created by the single command '\input\jobname.toc'. In the first pass, the file normally does not exist, which generates a warning message in the log file. The pass then reopens the same file as output (\write).

The second pass inputs the file and generates the TOC on the first page of the document. This moves the start of the document to page 2, causing all the page numbers in the TOC to be wrong. Pass 2, however, creates another .toc file, this time with the right page numbers, and a third pass is necessary to typeset a TOC with the correct page numbers.

▷ **Exercise 9.3:** Consider the case of a long TOC that occupies more than one page. Will the third pass typeset the TOC with the correct page numbers in such a case?

The third pass may be omitted if the size of the TOC (in pages) is known in advance. This, however, is rare.

The .toc file used here is very simple. It does not contain any formatting information. Instead, for each TOC item, the file contains a simple record with the following fields:

■ One of the codes \ch, \se or \ss, for a chapter, section, and subsection, respectively;

■ the chapter (or section or subsection) number, followed by a ':';

■ the chapter (or ...) name, followed by '\page';

■ the page number, followed by a '\\';

These records are written by macros \chapter, \section, and \ssection, which are expanded at the appropriate points in the document. Examples of these macros are shown ahead.

The entire TOC can be typeset, in the later passes, by the single command '\input\jobname.toc'. This is done by defining macros \ch, \se, \ss, to typeset lines in the TOC. Macro \ch, for example, typesets a chapter line in the TOC. It is expanded automatically during the \input, each time a record starting with a \ch is read off the file. Macros \se, \ss behave similarly. These macros (plus, perhaps, a \tochead, to typeset the heading of the TOC) are the only ones that typeset the TOC and, as a result, the only ones that need to be changed when a different TOC format is required.

Here are the promised examples. The precise way the chapter and section headings are typeset, as well as the fonts used, are easy to change.

```
\font\tochd=cmssdc10 scaled\magstep2
\font\chead=cmssdc10 scaled\magstep2
\font\shead=cmbx12
\font\sshead=cmbx10


\newcount\chnum \chnum=0 \newcount\snum \newcount\ssnum
\def\chapter#1{\advance\chnum by 1 \snum=0
{\chead \bigbreak\bigskip\noindent\the\chnum. #1 \bigskip\nobreak}
\edef\save{\string\ch\the\chnum:#1\string\page\noexpand\folio%
 \string\\}
\write\toc\expandafter{\save}}


\def\section#1{\advance\snum 1 \ssnum=0 {\shead\medbreak\smallskip
\noindent\the\chnum.\the\snum\ #1\smallskip\nobreak}
\edef\save{\string\se\the\chnum.\the\snum:%
#1\string\page\noexpand\folio\string\\}
\write\toc\expandafter{\save}}


\def\ssection#1{\advance\ssnum by 1 {\sshead\medbreak\smallskip
\noindent\the\chnum.\the\snum.\the\ssnum\ #1 \nobreak\smallskip%
 \nobreak}
\edef\save{\string\ss\the\chnum.\the\snum.\the\ssnum:%
#1\string\page\noexpand\folio\string\\}
\write\toc\expandafter{\save}}
```

Macro \chapter is invoked at the start of each chapter. Its (only) argument is the chapter name. It performs the following tasks:

■ It computes the chapter number and typesets the chapter heading, taking care of pagination.

■ It saves, in macro \save, the TOC information for the auxiliary file and executes a delayed \write with an immediate expansion of \save (note the use of \edef and \expandafter, which are explained later).

Macros \section and \ssection operate similarly.

One of the first things to do at the start of a pass is to check for the existence of the .toc file. The commands

```
\newread\toc \immediate\openin\toc=\jobname.toc
\ifeof\toc
 \message{! No file \jobname.toc;}
\else
 \tochead \input\jobname.toc \vfill\eject
\fi
\immediate\closein\toc
```

do the following:

1. Open the file for input.
2. Check for end-of-file.
3. If there is no end-of-file, the file exists, and the TOC is typeset by
   3.1. Macro \tochead typesets any desired heading for the TOC.
   3.2. An \input command reads the entire file and typesets the TOC. Since each record on the file starts with a macro name, that macro is expanded, to typeset one line in the TOC.
   3.3. \vfill\eject is executed to close the TOC.
4. If an end-of-file is sensed, the file does not exist, and a warning is placed in the log file.
5. The .toc file is closed as input.

Following these commands, the document is typeset, and the other macros create the .toc file. If this is pass 1, the .toc file is created. If this is pass 2 or 3, the .toc file is input, and a new one is created. After pass 3, the file should be manually deleted by the user. Once again note that TeX does not know what pass it is doing, and it behaves the same in all passes.

9.4 The Page Number

Two simple problems have to be solved by the macros, namely, writing the correct chapter number and the correct page number on the TOC file.

The page number (\folio) is not known when the TOC record is created. It only becomes known when the OTR is invoked. The \write should therefore be delayed. This is a common problem and is solved simply by saying \write instead of \immediate\write.

The chapter number (\the\chnum), on the other hand, is known and should be expanded immediately. If its expansion is delayed to the OTR, the number expanded will be the chapter number in effect during the OTR. The same problem applies to the section and subsection numbers.

To solve the two problems, macro \chapter contains the lines

```
\edef\save{\string\ch\the\chnum:#1\string\page\noexpand\folio%
  \string\\}
\write\toc\expandafter{\save}}
```

The first line defines macro \save with the necessary information for a single TOC record. We use \edef instead of \def to guarantee that the chapter number (\the\chnum) used by \save will be the current one (the one in effect during the macro definition).

The \noexpand\folio, on the other hand, guarantees that \folio will not be expanded when \save is defined; instead, it will be expanded when \save is expanded (in the OTR).

The \expandafter is used to reverse the order of execution of the two tokens following. When TeX finds the \write\toc command, it expects to find a '{' followed by the material to be written on the file. TeX is ready to save that material (as a whatsit) in the MVL, for a delayed write. However, in our case, TeX finds the \expandafter instead of a '{'. This causes TeX to expand token \save first, and then to place the '{' in front of it. The result is to force TeX to expand \save immediately (to write its replacement text in the MVL as a whatsit), even though the \write is done later. Macro \save is now ready for the next TOC item.

To fully appreciate the details of the macros above, the reader should try them several times, changing the \edef to \def, omitting the \noexpand, and omitting the \expandafter. The resulting errors are very illuminating.

9.5 Formatting the TOC

The only macros involved with the actual typesetting and formatting of the TOC are

```
\def\tochead{...}
\def\ch#1:#2\page#3\\{\bigbreak\medskip
 \line{{\shead#1\ #2}\ \hrulefill\ {\bf#3}}\nobreak}
\def\se#1:#2\page#3\\{\smallskip\noindent{\sshead#1\ #2}\hskip2em#3}
\def\ss#1:#2\page#3\\{\par{\sshead#1\ #2}\hskip2em#3}
```

Macro \tochead typesets the heading at the top of the first TOC page. The other three macros use parameter delimiters to read the correct arguments off the file.

9.6 Remarks

Here are some final remarks on the limitations of our method, as well as suggestions for useful extensions.

■ The size of a record on a file is limited by the operating system of the computer. Each of our lines contains a (chapter or section) name and, as a result, those names cannot be too long. It is interesting to note, in the example below, how the \tt command was expanded, in the .toc file, to '\fam \ttfam \tentt', thereby artificially increasing the chapter name. Note that this limitation has nothing to do with TeX.

■ When a long document, such as a book, is developed, each chapter is normally typeset individually, generating its own .toc file. In such a case, it is better to use the traditional method instead of the one described here. The traditional method creates the TOC by a special job that inputs each of the .toc files, and typesets the TOC using roman numerals for page numbers. The TOC is then manually placed at the beginning of the document.

■ This example can be extended by adding cross-reference capabilities. Any chapter or section could be labeled and then referred to by means of the label (section 9.2). This requires an auxiliary file in addition to the .toc file.

■ Since high quality is what TeX is all about, we have to think of pagination. Bad pagination may occur if the last line of the TOC is typeset on the top of a new page. In such a case, that single line will be *the only thing* on the page. To prevent such a situation, a \nobreak command should manually be inserted in the .toc file, right before the last item. The .toc file should thus end with

```
...
table of contents item
\nobreak
table of contents item
```

▶ **Exercise 9.4:** What's a good way of inserting such a \nobreak automatically?

9.7 A Simple Example

Here is a listing of a hypothetical .toc file created by our macros.

```
\ch1:General\page2\\
\se1.1:Introduction\page2\\
\se1.2:A General Pass\page2\\
\ss1.2.1:Why three passes?\page2\\
\se1.3:The {\tt.toc} file format\page2\\
\ss1.3.1:Writing the file\page2\\
\ch2:Creating the Table of Contents\page3\\
\ch3:Formatting the TOC\page4\\
\ch4:The Special Problems\page4\\
\se4.1:The Solution\page4\\
\ch5:Exercises\page4\\
\ch6:Closing Words\page5\\
\se6.1:Final Remarks\page5\\
\ss6.1.1:Limitations\page5\\
\ss6.1.2:Extensions\page5\\
\ch7:Example of a {\tt.toc} file\page5\\
```

This file results in the following TOC:

Men seldom make passes at girls who wear glasses.

— *Dorothy Parker*

10. Special Topics

Certain features of TeX—notably registers, macros, recursion, token strings and conditionals—make it into a programming language. As a result, it can do more than typesetting. Here we show how TeX can be used for calculations, and how it can interface with other applications.

10.1 Pasting Material in Textures

This is done with the `\special` command [228], an unusual and underutilized feature of TeX. This command simply places its argument in the `.dvi` file, so it can be used and processed by the printer driver. Each implementation may support different arguments, so using this command creates an incompatible source file. The command is typically used to bring pictures from the outside and paste them into a TeX document.

Textures, a TeX implementation for the Macintosh, supports several arguments of `\special`. They are used to pick up pictures in several formats and paste them in. The pictures can be prepared by any Macintosh application and can be in PICT, POSTSCRIPT, or encapsulated POSTSCRIPT formats. The arguments are summarized below; they all have two parts, the first of which is a keyword.

keyword	rest of argument	picture format	picture location
picture	picture name	PICT	in Picture window
pictfile	file name	PICT	in separate file
postscript	PS commands	POSTSCRIPT	rest of argument

postscriptfile	file name	POSTSCRIPT	in separate file
illustration	file name	encapsulated PS	in separate file

The first \special type uses a picture window, which is part of the Textures source file. A picture can be copied into the picture window (and named) from any PICT file, through the clipboard. The window shows the dimensions of the picture. In the source text, a picture can be placed in the text by reserving the necessary space, moving the reference point to the desired location, and saying \special{picture name}. Macros \picture and \pictfile below illustrate one way of achieving that.

```
\def\picture#1 by #2 (#3){\vbox to #2{
\hrule width#1 height0pt depth0pt \vfill
\special{picture #3}}}
\def\pictfile#1 by #2 (#3){\vbox to #2{
\hrule width#1 height0pt depth0pt
\vfill \special{pictfile #3}}}
```

They generate an empty \vbox of the same height as the picture. The box is given the width of the picture by placing an \hrule with height zero. The rest of the box is vfilled, to move the reference point to the bottom, and the \special command is then issued at the left bottom corner of the box.

▶ **Exercise 10.1:** What is the purpose of '\hrule width#1 height0pt depth0pt' in the picture macros?

The pictures below were placed by the construction

```
\smallskip\line{%
\pictfile 1.25in by 1.13in (plant1) \hfil
\vbox{\hsize=2.25in\noindent The first example is two small ..
.. in one hbox}
\hfil\pictfile 1.29in by 1.04in (plant2)}
```

The first example is two small plants. They are about the same size and were typeset as two vboxes, with this text as another vbox in between. The three vboxes were then placed in one hbox.

If the picture is small and can fit on the side of a single paragraph, then room can be left for it on either side of the paragraph by the \hangindent, \hangafter commands [102]. A more general method can be found in (Ref. 27). It can reshape several paragraphs so room is left for a large picture on the side.

10.2 Creating Shaded Boxes with PostScript

The next example shows how \special{postscript ...} can be used, in Textures, to enclose text in ruled boxes. It uses macro \shade, which is listed in Chapter 20 (and also in Ref. 28). A rectangular box is created, with either sharp or rounded corners, around text. The text can be placed, by the user, in either an \hbox or a \vbox, or it can be written explicitly as the macro argument (see second example below). The rectangle is filled with the desired shade of gray and is optionally surrounded by a stroke of any desired thickness. It is also possible to make the shaded area larger than the text by any amount. The macro has five parameters:

■ a real number specifying the percent of white in the shaded background

This is a .87 background.

■ a dimension specifying the extra size of the shaded area (7pt on each side)

■ a count specifying the width of the stroke in points. A zero or empty argument creates no stroke.

■ a count—the radius of the rounded corners, in points. A zero or empty argument produces square corners.

■ the text to be boxed. It is either straight text (if it fits on one line) or is enclosed, by the user, in an \hbox or \vbox.

The arguments are separated by commas, and the last one is delimited by '\\'.

Examples. Here are some simple examples of the use of \shade.

```
\shade0.90,16pt,5,10,\vbox{\hsize=100pt
\noindent A test in vertical mode.
Filled with 90\% white \&
a 5pt stroke}\\
```

A test in vertical mode.
Filled with 90% white
& a 5pt stroke

```
\shade0.97,5pt,1,3,%
straight text, not preboxed\\
```

straight text, not preboxed

```
\shade0.90,5pt,,3,%
\hbox{No stroke at all}\\
```

No stroke at all

```
\noindent A\lower30pt\hbox{%
\shade0.98,10pt,1,5,
\vbox{\hsize=95pt\noindent
A test in horizontal mode. There are
no blank spaces between this box and the
'A' or the 'D' (98\% white)}\\}D
```

A test in horizontal mode. There are no blank spaces between this box and the 'A' or the 'D' (98% white)

```
\shade0.99,8pt,40,,%
\hbox{A very heavy stroke}\\
```

A very heavy stroke

```
\shade0.5,5pt,1,5,\hbox{\shade.99,5pt,1,3,%
A Nested Expansion\\}\\
```

A Nested Expansion

▶ **Exercise 10.2:** Macro \shade has been tested by the author on the Macintosh computer only, using Textures. Try it on your computer and make sure it works!

. . . but always if they wanted to do anything special they
said this was Saturday night, and then they did it.

— *James Matthew Barrie, Peter Pan*

10.3 Arithmetic Calculations

One of the things TEX was never intended to do is arithmetic calculations. Nevertheless, since it has a limited capability for integer arithmetic (it does arithmetic on 32-bit two's-complement numbers [118]), impatient users tax TEX's capacity and do all kinds of calculations with real numbers. Performing real arithmetic in TEX can be done either by using \dimen registers—which can take noninteger values—or by *scaling* integers. If you want to operate on real numbers in the range $0\ldots99$, you can multiply each by, say, 1000, and truncate. The result will be integers in the range $0\ldots99,000$, representing the original real numbers with a five-digit precision. For example, the real number 12.3456 will become the integer 12,345 (and the .0006 is lost).

The \rnd macro below, due to T. Reid (Ref. 23), is a simple example of scaled calculations. It performs integer arithmetic on \count registers, in order to obtain random numbers in the range $0\ldots99$. The algorithm is

```
\rndnum:=\rndnum*371+1; Lines 4, 5
\rndnum:=\rndnum mod 100000; Lines 6-10
\rndval:=\rndnum; Line 12
\rndval:=\rndval mod 100; Line 13
```

and the macro is

```
\newcount\rndnum \rndnum=50 \newcount\rndval \newcount\rndtemp
\newdimen\rnda \newdimen\rndb \newdimen\unit \unit=5pt
\def\rnd{%
  \global\multiply\rndnum by 371
  \global\advance\rndnum by 1
  \ifnum\rndnum>99999
    \rndtemp=\rndnum
    \divide\rndtemp by 100000
    \multiply\rndtemp 100000
    \global\advance\rndnum by-\rndtemp
  \fi
  \global\rndval=\rndnum
  \global\divide\rndval by 1000 \relax}
```

The random numbers are used below to demonstrate how things can be placed at any point on the page. Macro \randbox draws four random numbers, converts them into dimensions, uses two of them as the (x,y)-coordinates of a point, and the other two, as the width and height of a box. The macro then typesets a number, surrounds it by a box, and places it at point (x,y) on the page. A loop is used to repeat the process 35 times.

```
\newdimen\rnda \newdimen\rndb \newdimen\unit \unit=3.5pt
\newdimen\rndc \newdimen\rndd
\def\randbox{%
 \rnd \rndc=\rndval pt \ifdim\rndc<12pt \rndc=12pt\fi
 \rnd \rndd=\rndval pt \ifdim\rndd<12pt \rndd=12pt\fi
```

```
\rnd \rnda=\rndval\unit \rnd \rndb=\rndval\unit
\vbox to 0pt{\kern\rnda
\hbox{\kern\rndb\boxit{\vbox to\rndd{\vss
  \hbox to\rndc{\hss\the\ns\hss}\vss}}0}\vss}%
\nointerlineskip}
```

▶ **Exercise 10.3:** Read the above macros carefully and explain how they make sure that the coordinates calculated from the random numbers do not overflow the page boundaries.

The loop itself is a simple example of the use of \loop–\repeat.

```
\newcount\ns \ns=35
\loop \ifnum\ns>0
  \randbox
\advance\ns by-1 \repeat
```

Note. The Bézier style is perhaps the best known calculating program in TeX, but by browsing through the many TeX archives (and reading TeXhax), the author has found macros to calculate sine and cosine (and has himself written a macro to calculate a square root).

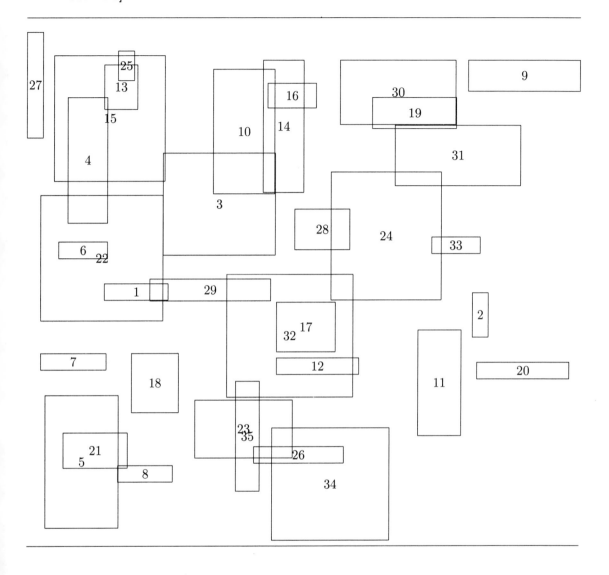

Taking Three as the subject to reason about—A convenient number
to state—We add Seven, and Ten, and then multiply out By One
Thousand diminished by Eight. The result we proceed to divide, as
you see, By Nine Hundred and Ninety Two: Then subtract
Seventeen, and the answer must be Exactly and perfectly true.

— *Lewis Carroll, The Hunting of the Snark*

11. Leaders

A leader is a very common device used by typesetters to lead the eye of the reader across the page. Here is a typical example from a table of contents:

Ch. 1. Introduction 11

In traditional, noncomputerized typesetting, the use of leaders was limited. In TEX, however, both horizontal and vertical leaders can easily be created. Leaders of different shapes can be combined to create special effects (such as a dot–dashed line, or a textured area; see ahead). Leaders can also be staggered to produce attractive patterns.

This has traditionally been one of the underused areas of TEX, and it is the purpose of this chapter to encourage the use of leaders and to demonstrate that leaders are a powerful and easy-to-use tool of typesetting. In fact, after using leaders for a while, the author would like to add the following warning (see Ref. 3, p. viii):   WARNING: Leaders can be hazardous to your other interests.

11.1 Horizontal Leaders

To specify leaders, the user should specify what a single leader is (a dot in our example) and how much space (how wide a window) should be filled by copies of the leader (from the word "Introduction" to the word "11"). These two specifications are the arguments of the \leaders command [223]. Thus, the earlier example was created by

```
\def\dotlead{\leaders\hbox to1em{\hfil.\hfil}\hfil}
\line{\bf Ch.~1. Introduction\rm\dotlead 11}
```

The control sequence `\leaders` has two arguments. The first one must be either a box (containing a single leader) or a rule; the second one should be a glob of glue such as `\hfil`, `\hskip24pt`, `\hskip\wd0`, and so on. It works by repeatedly copying the first argument (the single leader) over the space defined by the second argument. In our example, a single leader was not just a box with a dot but, rather, a `1em` wide box, with a centered dot. Defining `\def\dotlead{\leaders\hbox{.}\hfil}` creates leaders that are too dense:

Ch. 1. Introduction...11

Also note that the size of the leaders (the window in which they fit) was specified as `\hfil`. The size to which this `\hfil` was stretched became the size of the leaders. A natural question at this point is: What if this size is not an integer multiple of the size of a single leader? It turns out that TeX has three commands, `\leaders`, `\cleaders`, and `\xleaders`, to handle such a case in three different ways.

When `\leaders` is used, TeX first locates the innermost box A containing the `\leaders` command. It then fills up A, from the left, with copies of the leader. There may be some space left on the right. In 1 below, A is an `\hbox to\hsize`. In 2, the rule is stretched from the 'n' to the '1'. In 3, some space is left on the right because there was no room for the next copy of the leader.

```
1:  |. . . . . . . . . . . . . . . . . . . . . . .|
2:  |Ch. 1. Introduction_____11|
3:  |Ch. 1. Introduction . . . . . . . . . . . 111|
```

This process has the advantage that, in a multiline leaders application, the dots will be vertically aligned.

Ch. 1. Introduction 11
Ch. 2. Preface 111
Ch. 3. Material 1111

The `\cleaders` command centers the leaders in the leaders window, regardless of the size of the enclosing box A. There is normally some space left on both sides of the window. The `\xleaders` is still different. It distributes the window space evenly between the individual copies of the leader. The three lines below have the same width but are very different. The second line was done with `\cleaders` and the third one, with `\xleaders`.

Of course, sometimes we know how wide the leaders window should be. Here is an example where we want to typeset a string of n asterisks, but the value of n is only known at run time (see also [374]).

```
\setbox0=\hbox{*}
\message{Enter number of asterisks}
\read16to\n
A\hbox{\leaders\copy0\hskip\n\wd0}B
```

There is an important variation of leaders where the single leader is a rule, instead of a box. Defining

```
\def\dotlead{\leaders\hrule\hfil}
\line{\bf Ch.~1. Introduction\rm\dotlead 11}
```

results in

Ch. 1. Introduction_____11

and

```
\def\dotlead{\leaders\hrule\hskip1in}
\line{\bf Ch.~1. Introduction\hfil\rm\dotlead\hfil 11}
```

creates

Ch. 1. Introduction   _____   11

Note that this is a case where \hrule makes sense in horizontal mode! This is why, when you make a mistake such as \hbox{\hrule}, the error message says
! You can't use '\hrule' here except with leaders.

When a rule is used instead of a box, there is no difference between the three leaders commands.

Useful patterns, such as dashed or dot–dashed lines, are easy to create with leaders. Defining

```
\def\hdashfill{\cleaders\hbox{\vrule height.4pt width1.5pt%
  \kern2pt\vrule height.4pt width1.5pt}\hfill}
\def\hdotdashfill{\cleaders\hbox{%
\vrule height.4pt width1.5pt\kern1pt\smash{.}\kern1pt%
\vrule height.4pt width1.5pt}\hfill}
```

we can create _ _ _ _ _ _ _ by means of \hbox to.5in{\hdashfill} and, similarly, for
. _ . _ . _ . .

▸ **Exercise 11.1:** Why is the period smashed?

11.2 Brace Fills

Font cmex has the four characters ⌣⌢ that are combined with leaders to define the plain macros \upbracefill and \downbracefill [354]. Here is one definition:

```
\mathchardef\braceld="37A \mathchardef\bracerd="37B
\mathchardef\bracelu="37C \mathchardef\braceru="37D
\def\upbracefill{$\bracelu\leaders\vrule\hfill\bracerd
\braceld\leaders\vrule\hfill\braceru$}
```

Following this, the command \hbox to45pt{\upbracefill} produces ‿‿‿‿‿.

11.3 Staggered Leaders

We now turn to examples of staggered leaders. Leaders on successive lines can be staggered, leading to interesting effects. The following

```
\newcount\stagger
\def\stagleaders{\global\advance\stagger by 1
\cleaders\hbox to 4pt{\ifodd\stagger\kern2pt\fi
 \vrule height2pt width2pt\hss}}
\vbox{\offinterlineskip\hsize=0.5in\stagger=0
\loop
\line{\stagleaders\hfil}
\ifnum\stagger<10\repeat}
```

creates staggered leaders in a compact \vbox . This can be used, in principle, to create shaded areas. In practice, however, it is slow and may overflow memory.

▸ **Exercise 11.2:** Why is the \global necessary in the definition of \stagleaders?

The same method can be used to create a table of contents, similar to the one of *The TEXbook*, with staggered leaders:

```
\newcount\stagger
\def\stagleaders{\global\advance\stagger by 1
\cleaders\hbox to 20pt{\ifodd\stagger\kern10pt\fi.\hss}}
\def\tocline#1 #2.#3;{\line{\hbox to 1em{\bf\hss#1}\quad#2\thinspace%
\stagleaders\hfil\hbox to 1em{\hss#3}}}
```

```
\tocline1 Name of Chapter.1;
\tocline12 Name of Chapter.93;
\tocline23 Name of Chapter.725;
```

The above results in

Another example of staggered leaders is the following chessboard created by:

```
\boxit{\vbox{\leaders\vbox{\offinterlineskip
\hbox{\leaders\hbox{\kern.25in\vrule width.25in height.25in}%
 \hskip2in}
\hbox{\leaders\hbox{\vrule width.25in height.25in\kern.25in}%
 \hskip2in}
}\vskip2in}}0
```

11.4 Vertical Leaders

As with most other TEX features, leaders also have a vertical parallel. Vertical leaders can be created by specifying a vertical glue instead of a horizontal one as the second argument of \leaders. Thus, the simple construct

```
\def\vdotlead{\leaders\hbox to1em{\hfil.\hfil}\vfil}
\vbox to.3in{\hrule\vdotlead\hrule}
```

creates the I-beam at 1 below. The 'box' at 3 was created by

```
\def\vdotlead{\leaders\hbox to0pt{\hss.\hss}\vfil}
\setbox2=
\vbox{
 \hrule
 \hbox to.3in{\vbox to.3in{\vdotlead}\hfil\vbox to.3in{\vdotlead}}
 \hrule}
```

▶ **Exercise 11.3:** Use vertical leaders with a \vrule to typeset the ruled box in 2 below.

▶ **Exercise 11.4:** Use leaders to typeset the word "lead" as in 4 below.

[Exercise 21.8] shows how interesting patterns can be created by combining both horizontal and vertical leaders. Example 5 below illustrates such a combination. It was created by

```
\def\vrulefill{\leaders\vrule\vfill}
\vbox{\offinterlineskip
\hbox to25pt{\hdashfill}
\hbox to25pt{\vbox to25pt{\vrulefill}\hfil\vbox to25pt{\vrulefill}}
\hbox to25pt{\hdotdashfill}}
```

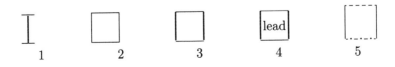

1   2   3   4   5

11.5 Exercises

▶ **Exercise 11.5:** Use macros \leadbox, \stagbox, \crossrules, and \fourdia-
monds to create the patterns below (each is 48pt high and wide).

11.6 Example

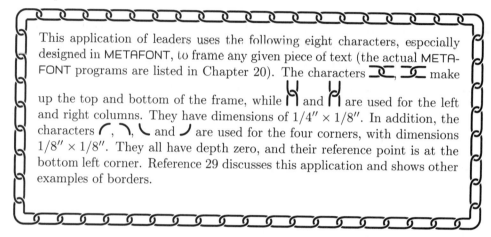

Macro \borderedtext#1\endborderedtext below performs the task in several
steps:

■ The text to be framed (parameter #1) is saved in \toks0 (line 1 below). It is then
typeset in \box0, with a width of 5 inches, and the resulting height of the box is
measured (lines 2–3).

■ Macro \modulo is expanded, to bring the height up to the nearest multiple of
0.25″ (line 4). The text is again typeset in \box0 with the new height (lines 5–6).

■ The border is built in \box2, which is wider and higher than \box0 (lines 7–11).

■ The text is placed inside the border by means of \hbox{\rlap{\box0}\box2}
(line 12).

The eight characters are defined by means of

```
\def\topp{{\pers\char0}} \def\bott{{\pers\char1}}
\def\lft{{\pers\char2}} \def\rt{{\pers\char3}}
\def\NW{{\pers\char5}} \def\NE{{\pers\char4}}
\def\SW{{\pers\char7}} \def\SE{{\pers\char6}}
```

where \pers is the name of the special font.

The modulo calculation is performed by repeated subtraction, instead of by a
multiply and divide. This is slower, but there is no loss of accuracy.

```
\def\modulo{\dimen1=\dimen0 \dimen0=.25in
\loop
```

```
\ifdim\dimen1>.25in
\advance\dimen1by-0.25in
\advance\dimen0by0.25in
\repeat}
```

Macro \borderedtext is straightforward and surprisingly easy to read.

```
1  \def\borderedtext#1\endborderedtext{\toks0={#1}
2  \setbox0=\vbox{\leftskip=.25in\hsize=4.75in\vskip.125in
3   \noindent\the\toks0\vskip.125in}\dimen0=\ht0
4   \modulo \dimen1=\dimen0 \advance\dimen0by.25in
5  \setbox0=\vbox to\dimen1{\leftskip=.25in\hsize=4.75in\vskip.125in
6   \noindent\the\toks0\vfil\vskip.125in}
7  \setbox2=\vbox to\dimen0{\offinterlineskip
8  \hbox to5in{\NW\cleaders\hbox{\topp}\hfill\NE}
9  \hbox to5in{\leaders\vbox to\dimen1{\leaders\hbox to5in
10  {\lft\hfil\rt}\vfil}\hfil}
11  \hbox to5in{\SW\cleaders\hbox{\bott}\hfill\SE}}
12  \centerline{\raise.125in\rlap{\box0}\box2}}
```

This application can easily be modified to have every page framed. The OTR can create the border in \box2, then say '\shipout\hbox{\raise.125in\rlap{\box255}\box2}}'.

> You cannot be a leader, and ask other people to follow you,
> unless you know how to follow, too.
>
> — *Sam Rayburn*

> What well-appointed leader fronts us here?
>
> — *William Shakespeare, Henry IV-2*

12. Tables

"TEX is intended for the creation of beautiful books—and especially for books that contain a lot of mathematics." The purpose of this chapter is to convince the reader that the above quote (from [v]) remains valid after the word "mathematics" is replaced by the word "tables."

It is easy to construct beautiful tables with TEX, so much so that (Ref. 30) has more than 100 examples of tables! Here we try to illustrate the main commands and techniques for constructing tables, using fewer than 20 examples.

Another quote, from (Ref. 31), is also in order: "Tables are the only area in TEX where you *do not* want to write macros." In fact, macros for tables have attracted the attention of many macro writers, and several macro packages exist that make it easy to specify tables of any size and shape.

There are three commands for creating tables, namely, \settabs, \halign, and \valign. Of these, the first [231] is simple but not very powerful, and the third is intended for use in H mode, inside a paragraph, so its use is limited. The \halign command, however, is both simple and powerful, making it the most common tool for tables. It is interesting to note that even the plain macro \matrix is defined [361] in terms of \halign.

12.1 \halign

This command [235, 282] is used to create tables. Since most tables are large, they must be placed between paragraphs. As a result, \halign is a vertical command and, when it is encountered while in H mode, the current paragraph is terminated. It is possible to say \vbox{\halign{...}} and thus, since \vbox is mode neutral, to create tables even inside a paragraph. However, it is better to use \valign, which is a horizontal command, to create a small table inside a paragraph.

The general form of the \halign command is '\halign{⟨preamble⟩⟨data⟩}'. The preamble specifies the format of each column of the table and the separation between the columns; the data items follow the preamble and are specified row by row. Because of the braces, everything done in the table is local. TₑX's first step in setting the table is to read all the data and analyze it to determine what elements go in each column. The widest element in each column determines the width of the column. Once the column widths and the spacings between the columns are known, TₑX can actually set the table.

Each row of the table is set in an \hbox that is then added to the MVL (or, if the \halign is enclosed in a \vbox, to the vertical list in that box) in the normal way. This is why a page break may occur at any place in the table (unless it is enclosed in a \vbox). It is also possible to insert vertical mode material between the lines of a table, in a way similar to \vadjust.

The spacings between the columns are called \tabskip and are illustrated in the diagram below. They are of type ⟨glue⟩ and can thus be flexible. It is clear that the first and last tabskips can be used to position the table horizontally on the page.

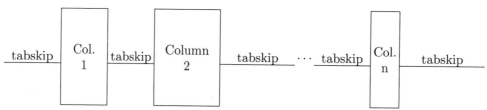

The format of the preamble is '...#...&...#...&... ...\cr', where each '...#...' indicates a template for a column (the # indicates where the data item should go), and the & are column separators. The data, which follows the preamble, is specified in a similar way; individual data columns in a row are separated by &, and the data for a given row is terminated with a \cr.

12.2 Examples of \halign

The following examples illustrate the most important commands and techniques used to construct tables.

The first example has a trivial preamble (just a # for each column). The table

1. onceupon	a	time
there	lived	two	birds
in	a	distant	nest

has been created by

```
\tabskip=0in \halign{#&#\tabskip=0.1in&#&#\cr
    1. once& upon& a      &time \cr
    there&lived&two       &birds\cr
        in&a     &distant&nest \cr}
```

Note the spaces between columns 2 and 3, and between columns 3 and 4.

In the second example all tabskips are 0.2in.

2. once	upon	a	time
there	lived	two	birds
in	a	distant	nest

The commands are

```
\tabskip=0.2in
\halign{\hfil#\hfil&\hfil#&#\hfil&#\cr
    2. once& upon&      a&time \cr
    there&lived&      two&birds\cr
        in&     a&distant&nest \cr}
```

▶ **Exercise 12.1:** The last two columns of table 2 are typeset left justified even though the preamble specifications are different. What is the difference between '&#&' and '&#\hfil&'?

The preamble in example 3 has specifications for column fill and control. Note that the \bf is confined to column 3 even though no braces are used. It is as if the preamble specifications for each column have an implicit pair of braces around them.

3. once	upon	*a*	X_{time}
there__	lived	*two*	X_{birds}
in_____	...a	*distant*	X_{nest}

```
\tabskip=0.3in
\halign{#\hrulefill&\dotfill#&\bf*#*&$X_{#}$\cr
    3. once& upon&      a&time \cr
    there&lived&      two&birds\cr
        in&     a&distant&nest \cr}
```

▶ **Exercise 12.2:** How can one specify an entire table in boldface?

Now comes an example of \omit and \noalign. The \noalign command can be used to insert vertical mode material between table rows. It can only be specified following a \cr and is typically used to insert extra space between lines (by means of \vskip), to suppress the interline space (with a \nointerlineskip), or to place an \hrule. The \omit can be used to change the normal preamble specifications for a certain entry. An '\omit\hfil abc\hfil' will typeset 'abc', centered in the column, regardless of the original preamble specifications for that column. Note the tabskips used here.

	4. once	upon	a		X_{time}
	there	lived	two		gnus
	on	a			X_{nest}

```
\tabskip=0.4in
\halign{\hfil#\tabskip=0.1in&#
&#\tabskip=0.7in&$X_{#}$\tabskip=0.1in\cr
   4. once& upon&      a&time      \cr
\noalign{\vskip 6pt \hrule}
     there&lived&     two&\omit gnus\cr
 \omit on&    a&       &nest       \cr}
```

▶ **Exercise 12.3:** Since \halign is vertical mode material, is it possible to nest a table inside a table by saying \noalign{\halign{...}}?

  Example 5 illustrates \multispan, a useful command that allows one table entry to span more than one column (that's how the heading was generated). Note also the use of \rlap to attach material of zero width to a table entry (a \phantom can also be used for this purpose). This example also illustrates a rule that stretches over part of the table (the reader should study leaders before attempting this).

look!		A HEADING		
5. once	upon	a	time	
there	lived	two	birds	
in	a	distant**	nest	

```
\tabskip=0.2in \halign{#&     #&       #&#\cr
look!&\multispan2 \hfil A HEADING\hfil\cr
\noalign{\vskip-9pt}
     &\multispan2\leaders\hrule\hfil\cr
\noalign{\vskip 1pt\hrule height 2pt\vskip 5pt}
   5. once& upon&      a&time \cr
     there&lived&     two&birds\cr
       in&    a&distant\rlap{**}&nest \cr}
```

▶ **Exercise 12.4:** What would happen in table 5 if \multispan3 was used instead of \multispan2?

  The next step is to add vertical and horizontal rules to the table, such as in:

		A Heading		
6. once	upon	a	time	
there	lived	two	birds	
in	a	distant	nest	

```
\def\aStrut{\vrule height12pt depth4pt width0pt}
\vbox{\offinterlineskip\hrule\tabskip=0pt
```

```
\halign{#\vrule\tabskip=0.1in&#&#&#\vrule&
 #&#&#\vrule\tabskip=0pt\aStrut\cr
&\multispan5\hfil A Heading\hfil&\cr
\noalign{\hrule}
&6. once& upon&      & a     & time&\cr
&  there&lived&\omit&two     &birds&\cr
&     in&   a&       &&distant & nest&\cr}
\hrule}
```

The horizontal rules are easy to create. The ones at the top and bottom are not even part of the table and are created by \vbox{\hrule\halign{...}\hrule}. (The \vbox is necessary so that the rules are stretched to the width of the table instead of to \hsize.)

▶ **Exercise 12.5:** Why are the two extreme tabskips set to zero in example 6?

The horizontal rule inside the table is created by \noalign{\hrule}. The vertical rules are different. They are created in segments such that each row contains the usual data items plus short vertical segments. One approach is to add a dummy column to the table for each vertical rule. In example 6 above we end up with a 7-column table, where columns 1, 4, and 7 contain just vertical rules. To better understand it, we duplicate it below, as example 7, without the \offinterlineskip:

The gaps between the short vertical segments are due to the interline spaces, which are added automatically. Trying to eliminate them by adding the command \offinterlineskip results in 8 below:

Here the rows are squashed together, which makes the entire table hard to read. To open up the rows, a strut is used, which results in example 6 above.

▶ **Exercise 12.6:** How can the effect of the strut be eliminated between two given table rows?

Another approach, illustrated in 9 below, is to create a vertical rule in the same column with a data item, and to use a \quad, to separate the rule and the data.

```
\vbox{\offinterlineskip\hrule\tabskip=0pt
```

```
\halign{\vrule \quad#\tabskip0.1in&#&
\vrule#&#&\hfil#\hfil\quad\vrule \aStrut
\tabskip=0pt\cr
&\multispan3\hfil A Heading\hfil&\cr
\noalign{\hrule}
   9. once& upon&     &  a       &time  \cr
      there&lived&\omit&two       &birds \cr
         in&     a&  &distant &nest  \cr}
\hrule}
```

▶ **Exercise 12.7:** "Compact" tables without any spaces have their uses. Try to duplicate the following "table," without looking at the answer. Note that the table is magnified, for easy reading, but you should eliminate all spacing. ⦂⦂⦂⦂

▶ **Exercise 12.8:** The two \quads add space on the left and right of table 9. How can similar space be added at the top and bottom?

　　　Example 10 is more complicated. It shows how an entry in a table may consist of several paragraphs of text by placing it in a box (a \vtop in our case). Note the use of the plain macro \normalbaselines [351]. The table was given a modern look through the use of rules (both thick and thin) and special spacing. The two lines of the subhead have been moved closer, while the rows are well separated (by means of \vertskip).

```
\def\vertskip{\omit&height1pc&&&\omit\cr}
{\offinterlineskip\halign{
\vtop{\hsize=6pc\normalbaselines\pretolerance=10000% no hyphenation
\rightskip=0pt plus6em\noindent# %raggedright
\hbadness=10000} % dont report underfull boxes [29, 272]
\tabskip=1em&\vrule#&#\hfil&\vrule#&
\vtop{\hsize=12pc\normalbaselines\rightskip=0em plus6em\strut#}\cr
\noalign{\hrule height4pt\smallskip}
\multispan5\strut\bf10. A Modern Table\hfil\cr
\noalign{\smallskip\hrule}
\omit&height3pt&\omit&&\omit\cr
\bf Left&&\bf Center&&\omit\bf Right\hfil\cr
\noalign{\kern-2pt}
\bf Column&&\bf Column&&\omit\bf Column (wider)\hfil\cr
\vertskip
The first column is placed in a raggedright&&
The $2^{\rm nd}$ column has&&
The third column is contained in a wider vtop with a flexible
 right margin,
so each entry may consist of several\cr
\vertskip
vtop, with no hyphenation and no&&
single-line entries&&
paragraphs set in ragged right and, in principle, hyphenated.\cr
```

```
\vertskip
reporting of underfull boxes&&
typeset left justified&&
A reminder: For the individual vrule segments to meet,
 an offinterlineskip
(outside the halign) is needed, and at least one strut
 in each table row.\cr
\vertskip
\noalign{\hrule}}
}% end of group for \offinterlineskip
```

This is a somewhat complex example, but a very practical one. The verbatim listing above should be read carefully and compared to the actual table below. For better understanding, the user should modify the source code in various ways, and then watch for the effect of each modification.

10. A Modern Table

Left Column	Center Column	Right Column (wider)
The first column is placed in a raggedright	The 2nd column has	The third column is contained in a wider vtop with a flexible right margin, so each entry may consist of several
vtop, with no hyphenation and no	single-line entries	paragraphs set in ragged right and, in principle, hyphenated.
reporting of underfull boxes	typeset left justified	A reminder: For the individual vrule segments to meet, an offinterlineskip (outside the halign) is needed, and at least one strut in each table row.

▶ **Exercise 12.9:** The two top rows of example 10 are too close. How did this happen?

12.3 Remarks

We can say '\halign to⟨dimen⟩{...}' and '\halign spread⟨dimen⟩{...}'. The former creates a table whose width is the given ⟨dimen⟩, and the latter spreads the table horizontally to make it wider than its natural width by ⟨dimen⟩. Both work by flexing the tabskips, so the user should make sure that they are not rigid.

There are two ways to center a table. It can be placed in a \vbox that is then centered by means of two $$. Alternatively, the two extreme tabskip values can be made flexible to center the table. The former method has the advantage of inserting the display math glues above and below the table, but the table cannot be broken across pages. If it does not fit on the current page, it may have to be floated to the next one.

▶ **Exercise 12.10:** Why a \vbox and not an \hbox?

The latter method produces a table that will break across pages but has the disadvantage that \halign to\hsize has to be used (see below), making it hard to place an \hrule in the table.

```
\tabskip=0pt plus1fil
\halign to\hsize{\tabskip=12pt<preamble>\tabskip=0pt plus1fil\cr
<data>}
```

Note: The construct $$\halign{...}$$ is not a display math formula and is not typeset in math mode at all. It is called an *alignment display* [190] and is used to typeset tables shifted horizontally on the page. Such an alignment gets the \abovedisplayskip and \belowdisplayskip glues above and below it, and each row is then shifted to the right by the current value of \displayindent. Thus $$\displayindent=1in\halign{...}$$ will typeset a table shifted 1 inch to the right. The plain macro \eqalignno uses an alignment display.

Placing tabskips: The tabskip placed on the left of a table is the one in force when the '\halign{' is read. The tabskip used to separate columns is the one in force when the '&' separator is read. The tabskip placed on the right of the table is the one in force when the \cr of the preamble is read. Any tabskips defined within the \halign are local. Example:

```
\tabskip10pt
\halign{\tabskip5pt#\tabskip25pt&#&\tabskip15pt#\cr
a&b&c\cr
d&e&f\cr}}
```

The tabskip on the left of the table is 10pt. The \tabskip5pt is read and stays in force until overriden by the \tabskip25pt. The first and second columns are thus separated by 25pt, as are the second and the third ones. The tabskip in force when the \cr is read is 15pt, and this is placed on the right of the table. Outside the table, the tabskip in force is still the 10pt.

▶ **Exercise 12.11:** How can one tell what tabskips are really used in a given table?

▶ **Exercise 12.12:** (Easy) What is the effect of '\tabskip1pt\tabskip2pt'?

The commands and features discussed above are the most important ones in connection with tables. In addition, there are a few more that are less commonly used:

■ A table can be indented to the right by '$$\displayindent=3in\halign{...}$$'. This also places the math display glues above and below the table.

■ \crcr [249] is useful when the \halign is created by a macro.

■ \everycr [275] can be used when, say, a \noalign{\hrule} should be placed after every \cr in the table.

■ \hidewidth [243] can be used to allow a table entry to overflow into the next column.

■ \vcenter [242] can be used to center a table vertically with respect to a display math formula. Thus

```
$$[a+b]\left[\vcenter{\tabskip=0pt
 \halign{$#$\cr a\cr-\cr b\cr}}\right]=a^2-b^2.$$
```

creates

$$[a+b]\begin{bmatrix} a \\ - \\ b \end{bmatrix} = a^2 - b^2.$$

■ A long periodic preamble can be made shorter by the use of && [241]. The preamble
     \hfil#\hfil& \hfil#&#\hfil&\hfil#&#\hfil&\hfil#&#\hfil&...\cr
can be written \hfil#\hfil&& \hfil#&#\hfil\cr. Preamble fields following the && will be used repeatedly.

■ Nested tables are possible. An \halign may be nested in another \halign by placing it in a \vbox. Similarly, an \halign entry may be a paragraph, enclosed in a \vbox. Thus

```
\halign{\tabskip=12pt\hfil#\hfil&\hfil#\hfil\cr
A&\vbox{\hsize=.1in\halign{#\tabskip=6pt&#\cr1&2\cr3&4\cr}}\cr
\vbox{\hsize=1in\noindent Be it as it may, go do as I say}&B\cr}
```

produces

1 2
3 4

A
Be it as it may,
go do as I say

                  B

(without the surrounding rules).

■ Spaces preceding an & in the preamble are typeset in the final table; spaces following an & are discarded. This can be verified by simple experiments such as \halign{# &#\cr1&2\cr}.

▶ **Exercise 12.13:** Produce the table $\begin{aligned} x &= & 12345 \\ f(x) &= & 11235 \end{aligned}$. (the main problem is to align the two x's vertically).

A common problem is to create a column of numbers aligned by the decimal point. We illustrate three solutions. The first one uses the fact that all ten digits in font cmr have the same width. We make the '@' an active character and assign it the width of a digit. When the table is prepared, we substitute an '@' for each missing digit. Thus the table on the right was obtained by

$$\begin{aligned} &123.45 \\ &6.78 \\ &10. \\ &.7 \\ &918.18 \end{aligned}$$

```
\catcode'\@=13 \def@{\hphantom{0}}
\halign{#\cr 123.45\cr @@6.78\cr @10.@@\cr @@@.7@\cr 918.18\cr}
```

The second solution is to type each number in two parts, to become two columns in the table. The example above can be typed as \halign{\hfil#.&#\hfil\cr 123&45\cr 6&78\cr 10&\cr &7\cr 918&18\cr}.

The third approach is to declare the period an active character and use it to separate the number into two table columns.

```
\def\per{.}\catcode'\.=13 \def.{&}
\halign{\hfil#&\per#\hfil\cr 123.45\cr-10.980\cr0.\cr.6\cr67\cr}
```

▶ **Exercise 12.14:** The decimal point is inserted by macro `\per` at the start of the second column. Alternatively, the point can be inserted at the end of the first column by the preamble '`\hfil#\per&#\hfil\cr`'. Is there a significant difference between the two preambles?

There are three ways to open up the rows of a table:

1. Place a strut together with `\offinterlineskip`, as in the examples above.

2. Say `\noalign{\vskip...}` between rows.

3. Use the `\openup` command before starting the table. This command is a plain macro [364] that advances `\baselineskip` (and also the two parameters `\lineskip` and `\lineskiplimit`). It should normally be used inside a group, thus {\openup 4pt\halign{...}}.

▶ **Exercise 12.15:** What is the effect of `\offinterlineskip\openup1pt`?

12.4 \valign

This command [249] is the opposite of `\halign` but is much less useful and also confusing. This is why [249] says, "People usually work with TEX at least a year before they find their first application for `\valign` ...."

When should `\valign` be used? Most tables are meant to be read row by row, so it is natural to use `\halign` to set them. If a table is meant to be read column by column, the user should consider setting it with `\valign`. Imagine a table such as *a* below; it can be set with an `\halign`, but adding one item, such as in *b*, would require a lot of work.

allow	bed	climax	defray
allowable	bedabble	climb	defrayal
allowance	bedash	climable	defrayer
allower	bedaub	climber	defrock
⋮	⋮	⋮	⋮

allow	bed	climax	defray
allowable	bedabble	climb	defrayal
allowance	*bedaggle*	climable	defrayer
allower	bedash	climber	defrock
	bedaub	⋮	⋮
⋮			

        a   *b*

Using `\valign`, the original table is set by

```
\valign{<preamble>
allow&allowable&allowance&allower&$\vdots$\cr
...}
```

which makes it easy to add items to any column. In general, \valign is the opposite of \halign, so the important differences between them are the following:

- In a \valign, preamble entries refer to rows.
- The entire \valign is H mode material and can be part of a paragraph.
- Each entry in a \valign is typeset in V mode, so its width is \hsize. As a result, each entry should normally be enclosed in a box.
- In an \halign, entries in the same row are aligned by their baselines. In a \valign, on the other hand, entries in the same column are aligned on the left. The result is that a column can be centered or right justified only if its width is known in advance.
- \tabskip now separates *rows*, \noalign can be used to insert (horizontal) material between *columns*, and \span spans *rows*.
- No default glues, such as \baselineskip, are inserted between rows of a \valign. Struts should normally be used to control row separation.
- Vrules between columns are easy in a \valign; just use \noalign{\vrule} following the \cr. An \omit\vrulefill can be used to get vrules of different heights that don't extend all the way down.
- Hrules can be placed in a \valign by adding dummy rows, such as \hrule#&, in the preamble.

The following example is both simple and practical.

```
\valign{\tolerance=9000\hbadness=10000\vfil#\vfil\cr
\hsize=1.5in\noindent This alignment ... (without the rules).\cr
\hbox to6pt{}\cr
\hsize=2in\noindent Each set of data spec ...columns of data.\cr
\hbox to6pt{}\cr
\hsize=1.5in\noindent Exercise (easy): ... at the top.\cr}
```

This alignment has one preamble spec, so it produces a table with one row (without the rules).	Each set of data specifications corresponds to a column and, because of the two vfils, the columns will be vertically centered. This is thus a method for vertically centering several boxes or columns of data.	Homework (easy): Change this to produce columns aligned at the top.

13. Advanced Math

In general, mathematical typesetting with TeX is easy to learn, and the results far surpass anything created by commercial word processors or page layout software. As a result, this chapter treats only those topics that may present problems to the average user or that are not treated in detail in *The TeXbook*.

13.1 General Background

Perhaps the most striking feature of math mode is the automatic spacing. Even a beginner knows that spaces don't count in math mode, and spacing is done automatically according to the classes (section 13.6) of the entities involved. Thus the expressions `$a + b$`, `$a+ b$` and `$a +b$` are all equivalent, resulting in $a + b$.

There are two math modes, math inline (delimited by a pair of single '`$`') and display math (delimited by a pair of double '`$$`'). The former is used for math expressions that fit on the same line with text; the latter is used for expressions that are either too long or too tall and should thus occupy their own vertical space on the page. Note that math display formulas can be part of the current paragraph even though they are preceded and followed by large, flexible chunks of vertical glue.

It should be noted that the double '`$$`' must be tokens of catcode 3 and should be consecutive. In an expression such as '`$ $a^2 $$`' the first '`$`' is followed by a space, so it starts math inline mode. The second one terminates it. When the '`a^2`' is read, TeX is outside math mode, so the '`^`' causes an error. The same thing happens when we try to supply one '`$`' from a macro. In '`\def\tlp{$}`

'$\tlp a^2...\tlp$', the first '$' is followed by \tlp, which doesn't have catcode 3 (doesn't have any catcode), so it starts math inline mode. The macro expansion generates another '$', which terminates this mode. The '^' is again seen outside of math mode, which causes the same error as above. Also, since a display should be centered, it cannot be generated in an \hbox. If we say \hbox{$$...$$}, the first pair would start and end math inline mode (resulting in an empty expression), and the second pair would do the same.

▶ **Exercise 13.1:** What would be the result of \def\tlp{$$}, followed by the expansion \tlp a^2\tlp ?

There are also the two token parameters \everymath and \everydisplay. They are inserted at the start of every inline and display expression, respectively.

A display formula is preceded by a glob of \abovedisplayskip glue and is followed by the \belowdisplayskip glue. Both have plain values of 12pt plus 3pt minus 9pt; very flexible, which makes it easier for the page builder to find a good page break. As a proof of the attention to detail paid by TeX, it should be mentioned that there are two more glues, \abovedisplayshortskip and \belowdisplayshortskip (with plain values of '0pt plus 3pt' and '7pt plus 3pt minus 4pt', respectively). They are used when the line above (or below) the display formula is too short and does not reach the left edge of the formula, as in the example following this line:

$$e^{\pi i} + 1 = 0.$$

Centering displays. Even beginners know that display formulas are centered, but that's not the entire truth. A display is centered in a box whose width is the value of \displaywidth. The box is then typeset \displayindent units from the left margin. After reading the first '$$' of a display, \displaywidth is set to \hsize and \displayindent, to zero. This has the effect of centering the formula with respect to the page. To typeset noncentered display formulas, the values of the two parameters can be changed inside the display, just before the closing '$$'. Example: '$$⟨formula⟩ \displaywidth=\hsize \displayindent=-1in$$' will shift the display one inch to the left.

This does not work very well, since we don't know the size of the formula beforehand, so here is another approach to the problem.

13.2 Noncentered Displays

To typeset noncentered displays, a macro \indentDisplay#1#2 is developed, where '#1' is the amount by which the display should be shifted to the right, and '#2' is the formula itself. A typical expansion is '\indentDisplay{1in}{x^2...}'.

The first idea is to place the math expression in an \hbox and indent the box by any desired amount. The expression should be placed between single '$' (see exercise 13.2). To make TeX treat it as a display, rather than inline math, a \displaystyle should be used. The math display glues should also be added above and below the \hbox. The \hbox is placed *inside the paragraph*, between lines, by means of a \vadjust. (The \leavevmode is needed in case the macro is invoked in v mode.) The first version is

```
\def\indentDisplay#1#2{\leavevmode\vadjust{\vskip\abovedisplayskip
\hbox{\kern#1$\displaystyle#2$}\vskip\belowdisplayskip}\hfil\break}
```

▸ **Exercise 13.2:** Why single and not double '\$'?

▸ **Exercise 13.3:** Why the `\hfil\break` ?

This is simple but has the disadvantage that `\eqno` cannot be used (since it is invalid in math inline mode). The next version uses delimited parameters to extract the equation number, if present, and to typeset it, separate from the formula, at the right margin.

```
\def\indentDisplay#1#2{\setdisplay{#1}#2\eqno\eqno\endd}
\def\setdisplay#1#2\eqno#3\eqno#4\endd{\leavevmode%
\vadjust{\vskip\abovedisplayskip
\line{\kern#1$\displaystyle#2$\hfil#3}
\vskip\belowdisplayskip}\hfil\break}
```

Readers who would like to gain experience in macro writing should study this example in detail, since it is simple and, at the same time, demonstrates the power of delimited parameters.

▸ **Exercise 13.4:** How is parameter '#4' above used?

13.3 Fonts in Math Mode

When typesetting text there is, at any given time, a single current font, and characters are taken from this font. The current font can be changed by means of the `\font` command. When typesetting mathematics, the `\font` command is ignored, a single current font is not enough, and 16 font families are used, each consisting of 3 fonts. A font family can be selected by saying, for example, `\fam7` (an assignment, short for `\fam=7`), but this makes sense only in math mode. The `\fam` command is ignored outside of math mode.

The total number of math characters available at any time is therefore $16 \times 3 \times 256 = 12,288$! New families can be defined inside groups, temporarily increasing even this large number.

The three members of a family are used for text size, script size, and scriptscript size as in, for instance, A^{B^C}. Fonts can be associated with the three members by the commands `\textfont`, `\scriptfont`, and `\scriptscriptfont`.

In general, any font can be assigned to any family, except that families 2 and 3 are treated in a special way and should contain the math symbols and math extension fonts, respectively. (The math extension font contains large symbols and pieces that are used to build even larger ones.) To get normal math typesetting, a roman font should be assigned to family 0, and a math italics font, to family 1. The `plain` format defines family 0 [351] by
`\textfont0=\tenrm \scriptfont0=\sevenrm \scriptscriptfont0=\fiverm`
following which family 0 can be selected by `\fam0`. Since this command is ignored outside of math mode, the `plain` format says `\def\rm{\fam0\tenrm}`. When `\rm` is

used in math mode, the `\fam0` command is executed and `\tenrm` is ignored. Outside math mode, the effect is to select `\tenrm` and ignore `\fam0`. As an example, the code `$\rm A+B^{C^D}$` produces $A + B^{C^D}$ (all in roman type).

Note that it is not necessary to select all three members of a family (except for family 3, which should have all three members selected). The empty `\nullfont` is substituted for any fonts that have not been selected. Also, an `\hbox` in math mode temporarily changes the mode to H, where the current text font is used, instead of any family.

Family 1 selects the italic fonts that are so common in math mode.

```
\textfont1=\teni \scriptfont1=\seveni \scriptscriptfont1=\fivei
\def\mit{\fam1} \def\oldstyle{\fam1\teni}
```

The `\mit` command is meant to be used in math mode only and is ignored outside it. It can be used for italic Greek capitals. For example, `$\mit\Sigma\dots\Pi$` produces $\mathit{\Sigma} \dots \mathit{\Pi}$. The `\oldstyle` command can be used in or out of math mode. It produces the digits 0123456789.

Family 2 selects the math symbols, and family 3, the extensible symbols.

```
\textfont2=\tensy \scriptfont2=\sevensy \scriptscriptfont2=\fivesy
\def\cal{\fam2}
\textfont3=\tenex \scriptfont3=\tenex \scriptscriptfont3=\tenex
```

The `\cal` command can be used to typeset the capital letters in calligraphic form. '`\cal XYZ`' produces '$\mathcal{XYZ}$'. Family 4 selects text italics. The definition is `\textfont4=\tenit \def\it{\fam4\tenit}`. Thus when we say `\it` in math mode, we get `\fam4`, but when we say the same thing outside math mode, we get the `\tenit` font. Note that the `plain` format selects only a text font for this family. If we try to use the script font of family 4 (by, e.g., `$\it X_i$`), we get the message '`! \scriptfont 4 is undefined (character i).`' Obviously, the `plain` format assumes that text italics should be used in math mode just for text, not for the mathematics itself.

The alert reader, looking at [351], will discover that the actual definition is `\newfam\itfam \textfont\itfam=\tenit \def\it{\fam\itfam\tenit}`

The `\newfam` command is used to select a fresh, unused, family number, so the user won't have to memorize those numbers. The selection of the font families for slanted, boldface, and typewriter is now obvious and should be guessed by the reader before looking at the commands below.

```
\newfam\slfam \def\sl{\fam\slfam\tensl} % \sl is family 5
\textfont\slfam=\tensl
\newfam\bffam \def\bf{\fam\bffam\tenbf} % \bf is family 6
\textfont\bffam=\tenbf \scriptfont\bffam=\sevenbf
\scriptscriptfont\bffam=\fivebf
\newfam\ttfam \def\tt{\fam\ttfam\tentt} % \tt is family 7
\textfont\ttfam=\tentt
```

Note that the `plain` format allows subscripts in boldface but not in slanted or typewriter fonts.

Armed with this knowledge, it is now possible to select fonts other than CM for use in math mode. The Lucida font, for instance, is widely available, and contains the necessary math symbols (although not a slanted or a typewriter version). To use it in a document for both text and math, the following declarations are necessary:

```
                                       \newfam\lucit \newfam\lubf
                                       \def\Lucida{\l
\font\l=lucida at 10pt                 \textfont0=\l \scriptfont0=\ls
\font\ls=lucida at 7pt                 \scriptscriptfont0=\lss
\font\lss=lucida at 5pt                \def\rm{\fam0\l}%
\font\lsy=lucidamatsym at 10pt         \textfont1=\lmi \scriptfont1=\lmis
\font\lsys=lucidamatsym at 7pt         \scriptscriptfont1=\lmiss
\font\lsyss=lucidamatsym at 5pt        \textfont2=\lsy \scriptfont2=\lsys
\font\lmi=lucidamatita at 10pt         \scriptscriptfont2=\lsyss
\font\lmis=lucidamatita at 7pt         \textfont3=\lex
\font\lmiss=lucidamatita at 5pt        \textfont\lucit=\tenlit
\font\lex=lucidamatext at 10pt         \def\it{\fam\lucit\tenlit}%
\font\tenlit=lucidaI at 10pt           \textfont\lubf=\tenlbf
\font\tenlbf=lucidaB at 10pt           \scriptfont\lubf=\sevenlbf
\font\sevenlbf=lucidaB at 7pt          \scriptscriptfont\lubf=\fivelbf
\font\fivelbf=lucidaB at 5pt           \def\bf{\fam\lubf\tenlbf}}
```

following which the single command \Lucida will switch to the Lucida font for both text and math. To toggle between CM and Lucida, just say

```
...
CM used here
...
\begingroup\Lucida
...
Lucida used here
...
\endgroup
...
CM used here
...
```

Figure 13.1 is an example that shows text and math in CM and Lucida side by side:

At the time of writing (late 1995) there are two more fonts containing the necessary math symbols, which can be used instead of CM. One is the Euler font,

$$\int_0^\infty f(x)\,dx = \sum_{i=0}^{n} P_i B_i.$$

$$B_{n,i}(t) = \binom{n}{i} t^i (1-t)^{n-i}. \quad (1)$$

Definition. The mediation operator $t[\![P_0, P_1]\!]$ between two points P_0, P_1 is defined as

$$t[\![P_0, P_1]\!] = tP_1 + (1-t)P_0$$
$$= t(P_1 - P_0) + P_0.$$

It is shown below that this operator can be applied to *any number of points* and that it creates curves that interpolate between the points. It has the advantages of being a simple mathematical function and thus fast to calculate.

$$\int_0^\infty f(x)\,dx = \sum_{i=0}^{n} P_i B_i.$$

$$B_{n,i}(t) = \binom{n}{i} t^i (1-t)^{n-i}. \quad (1)$$

Definition. The mediation operator $t[\![P_0, P_1]\!]$ between two points P_0, P_1 is defined as

$$t[\![P_0, P_1]\!] = tP_1 + (1-t)P_0$$
$$= t(P_1 - P_0) + P_0.$$

It is shown below that this operator can be applied to *any number of points* and that it creates curves that interpolate between the points. It has the advantages of being a simple mathematical function and thus fast to calculate.

Figure 13.1. CM and Lucida.

used with the computer concrete metafont in (Ref. 32). The other is the Times-math font, (Ref. 33), designed to be used with the well-known Times font.

An important point to realize is that TeX only looks at the actual fonts assigned to the various families *at the end of the math expression.* Each expression can use all 16 families, but not more. If we try, $...\textfont5=abc...$, then abc would be used as the textfont of family 5 *for the entire expression*, not just the latter part.

13.4 The Four Styles

One of the first things that catches the eye, when typeseting mathematics, is the use of different character sizes for large operators and for superscripts. To fully understand the size changes, the user should be familiar with the concept of *style*, described below.

When entering display math mode, TeX goes into *display* style. When in this style, it picks up all characters from the textfont of the current family, so normally they come out in the same size as the surrounding text, except that large operators come out bigger, and super- and subscripts change the style to *script*. When TeX is in script style, it picks up all characters from the scriptfont of the current family. Any superscript or subscript encountered while in script style changes the style to *scriptscript*, where characters are picked up from the scriptscript font of the current family. There is no scriptscriptscript style, as that would use characters too small for comfortable reading.

When entering inline math mode, TEX switches to *text* style, where it picks up characters from the textfont of the current font, so they come out the same size as the surrounding text; TEX switches to script and scriptscript styles as before. Thus in the two examples `$\sum x^{n^2}$` and `$$\sum x^{n^2}$$` the n is typeset in script style, the 2, in scriptscript style, and the x, in text style (in the former) and display style (in the latter). The results are $\sum x^{n^2}$ and $\sum x^{n^2}$.

In both display and text styles, characters are picked up from the textfont of the current family. The differences between the styles are as follows:

1. In display style, the large operators come out larger.

2. In text style, TEX tries to limit the vertical size of the expression. This is reflected in:

2a. the placement of limits (see section 13.8);

2b. the placement of superscripts and subscripts. Look carefully at the two expressions A^2, A^2. The latter places the superscript a little higher and is what you get in display style.

▶ **Exercise 13.5:** How was the latter expression above typeset in inline mode?

2c. When TEX is in text style, it switches to script style whenever it has to stack things vertically. Compare $a + \frac{z^2+1}{z}$ and $a + \dfrac{z^2 + 1}{z}$. Both were created by the expression `a+{z^2+1\over z}`. In the former, which was typeset in inline mode, the z was typeset in script style, so the 2 came out in scriptscript style. In the latter (a display), the z came out in display style, and the 2, in script style.

In cases where TEX's automatic style selection is judged wrong, the user can manually select a style with one of the commands `\displaystyle`, `\textstyle`, `\scriptstyle`, and `\scriptscriptstyle` [141]. A common example is $A^{\frac{m+1}{n}}$. Trying `$A^{\textstyle{m+1\over n}}$` produces $A^{\frac{m+1}{n}}$, which most people consider preferable.

The above commands apply until the end of the current subformula or until the style is changed by another such command. A good example is the four expressions

`$A^B+8\over9$` `$\textstyle A^B+8\over9$`

`$\scriptstyle A^B+8\over9$` `$\scriptscriptstyle A^B+8\over9$`

that result in (1) $\frac{A^B+8}{9}$, (2) $\frac{A^B+8}{9}$, (3) $\frac{A^B+8}{9}$, and (4) $\frac{A^B+8}{9}$, respectively.

In 1, the A, the 8 and the 9 are in script style, while the B is in scriptscript style. In 2, the A and the 8 are in text style and the B is in script style. The 9 is in script style, as before, not affected by the `\textstyle` command. Expression 3 is identical to 1, and in 4, the A, the B, and the 8 are in scriptscript style, while the 9, as before, is in script style.

There are also four more "cramped" styles (cramped display, etc.), that are used in places where room is scarce, such as under a square root sign. The main feature of the cramped styles is that superscripts are not raised as much as in the normal styles.

▶ **Exercise 13.6:** Devise a simple math expression that will illustrate the differences between text and cramped text, and between display and cramped display styles.

13.5 Mathematical Units of Dimension

Few users realize that commands such as \hfil, \hskip and their relatives can also be used in math mode. Here is the complete list: \hfil, \hfill, \hskip, \hss, \hfilneg, \leaders, \mkern, and \mskip. However, the last two must have their glue expressed in *mathematical units* (mu). A mathematical unit of glue (muglue [168]) equals 1/18 the width of the em of family 2 (the math symbols fonts). Incidentally, the size of the em is determined by the font designer, and it becomes the value of \fontdimen6 of the font. Thus, when in display style or text style, the em of font \textfont2 is used; when in script style, the em of \scriptfont2 is used. Thus in $a\mskip6mu b^{a\mskip6mu b}$, the two skips are different. The plain format defines [349, 357] a thin skip as '\mskip3mu', a medium skip as '\mskip 4mu plus 2mu minus 4mu', and a thick skip as '\mskip 5mu plus 5mu'.

Note that mu is a keyword, whereas \mu is a control sequence, producing the Greek letter μ.

There are also 256 \muskip registers, each capable of storing a glob of muglue. They can be allocated by either a \newmuskip or \muskipdef.

For daring users, here is an example of leaders in math mode: '$a\_\_\_b$', created by $a\leaders\hrule \mskip18mu b$. There are others, more complicated, on [357].

▶ **Exercise 13.7:** Use leaders to create the following, 0.5 inch wide, patterns:

$$\equiv\equiv\equiv\equiv \qquad\qquad \text{ΤΙΤΙΤΙ}$$

13.6 Spacing in Math Mode

Spacing is very important in math expressions and is more complex to handle than in text. The spacing between elements in a math expression depends on their mathematical meaning. A good look at $a = (-x, y)(1 \times 2) - b/c$ shows that there is much more space on both sides of the '=' and the (binary) '−' than around the '/'. Also the unary '−' has no space on the right. Parentheses are treated in a special way, as are punctuations.

As a result, each character appearing in a math expression belongs to a class, and the class determines the spacing that will appear around the character. There are eight classes, shown below.

Class	Meaning	Example	Class	Meaning	Example
0	Ordinary	/	4	Opening	(
1	Large operator	\sum	5	Closing	)
2	Binary operator	+	6	Punctuation	,
3	Relation	=	7	Variable family	x

When a math expression is read, it is converted into a math list made up of atoms of eight types. The first seven types are identical to the first seven classes above, and the eighth one is an *inner atom*, a special type of subexpression (page 257). The spacing between adjacent atoms is then determined by the following table [170]:

Left atom	Right atom							
	Ord	Op	Bin	Rel	Open	Close	Punct	Inner
Ord	0	1	(2)	(3)	0	0	0	(1)
Op	1	1	*	(3)	0	0	0	(1)
Bin	(2)	(2)	*	*	(2)	*	*	(2)
Rel	(3)	(3)	*	0	(3)	0	0	(3)
Open	0	0	*	0	0	0	0	0
Close	0	1	(2)	(3)	0	0	0	(1)
Punct	(1)	(1)	*	(1)	(1)	(1)	(1)	(1)
Inner	(1)	1	(2)	(3)	(1)	0	(1)	(1)

where 0, 1, 2, and 3 stand for no space, thin space ($\,$), medium space, ($\>$) and thick space ($\;$), respectively. Parentheses indicate that the space should be inserted only in display and text styles, not in script and scriptscript styles. Asterisks correspond to combinations that should not occur in a typical math expression. Normally, a binary operation should not follow another one, and a closing should not follow a relation. The mathematical meaning of $(a++b=)c$, for example, is not clear, so TeX converts the '=' and the first '+' to Ord.

Here are some examples of spacing in math expressions:

1. $a(b+c)$ is converted into a math list of six atoms of types Ord, Open, Ord, Bin, Ord, Close. According to the table, the spaces will be   Ord Open Ord\>Bin\>Ord Close. The final result is $a(b + c)$.

2. $<a,8>=-x^2$. The x^2 is considered a single atom, so there are eight atoms in the list. The unary minus is considered of type Ord, so the spacing becomes Rel\;Ord Punct\,Ord\;Rel Rel\;Ord Ord. The final result is $< a,8 >= -x^2$.

In this example, the notation $< a, 8 >$ stands for a vector, so the angle brackets '<' and '>' should be considered Open and Close atoms. TeX, of course, does not know that, and treats the brackets as Rel (because they are usually used for the relations 'less than' and 'greater than'). As a result, there is too much space (a thick space) between '<' and 'a' and between '8 'and '>'. This is an example where the simple spacing rules do not work, and the user should correct the spacing manually.

To see the contents of a math list, a \showlists command should be used in math mode. Just say, for example, $<a,8>=-x^2\showlists$. To also see the actual spacing used, say $<a,8>=-x^2\showlists$\showlists (the second \showlists is outside math mode). Alternatively, the math expression can be placed in a box register '\setbox0=\hbox{$<a,8>=-x^2$}', which is then shown '\showboxbreadth=... \showbox0'.

There are actually 13 types of atoms [158]. The types not shown in the table above are treated, for spacing purposes, as type Ord, except fractions, which are treated as type Inner.

The \mathchar command in math mode is analogous to the \char command outside math mode. It makes it possible to select a character from any of the current font families and declare it as belonging to any class for the purpose of spacing. Thus \mathchar"603A selects the character at position 3A (hexadecimal) of family 0, and declares it a punctuation (class 6). If this character is used often, it is possible to assign it a name by \def\colon{\mathchar"603A} or, even simpler, \mathchardef\colon="603A (analogous to \chardef).

There is also the related command \mathcode. This command associates a character input from the source file with a character from one of the font families. Thus \mathcode'\:="303A means that a colon read in math mode should be replaced with the character at position 3A of family 0 and should be treated as a relation. The plain format contains [344] a large group of characters defined by \mathcode.

When reading a math expression, TeX looks at the \mathcode of each character to decide how to handle it. Note that the mathcode does not say anything about which of the three fonts in the family to select. This is determined by the current style and by the vertical structure of the math expression.

The different mathcodes of ':' and \colon are the reason why $(a:b)(a\colon b)$ result in $(a : b)(a\!:\!b)$ (different spacing).

If a character is assigned a mathcode of "8000, it becomes an active character [289] in math mode only! Saying \mathcode'\x="8000 makes the 'x' an active character in math mode. To actually define it, the user should say something like {\catcode'\x=13 \gdef x{2}} (the 'x' should temporarily become an active character outside math mode and then be defined). Following this, the string 'abxy$x=y$' would result in 'abxy2 = y'. The \plain format assigns this mathcode to the space, underscore, and prime.

▸ **Exercise 13.8:** What is the definition of \prime?

One of the functions of INITEX is to assign mathcodes to all the characters. All the letters get mathcodes of "71xx, and all digits, a mathcode of "70xx, where xx is the hexadecimal position of the character in the font. As a result, lettters are initially picked from family 1 (math italics) and digits, from family 0 (roman). They are considered class 7 (see below). All other characters receive mathcodes of "00xx, where xx is as before. All other characters are therefore considered Ord, and are picked up from family 0.

Class 7, variable, deserves special attention. A character with a mathcode of "7yxx is normally picked up from family \fam but, if \fam is outside the range 0–15, the character is picked up from family y.

As mentioned earlier, all letters have mathcodes of "71xx, and all digits, mathcodes of "70xx. They all belong to class 7. When entering math mode, the family is always set to −1, so letters are picked up from family 1 (math italics) and digits, from family 0 (roman). All other characters are assigned mathcodes of "00xx, so they are considered Ord (class 0) and are picked up from family 0.

13.7 Ordinary Symbols

Symbols of type Ord (ordinary characters) include the Greek characters, the calligraphic capitals, the decimal point, and some math symbols that don't belong in any other class.

These are the lowercase Greek letters (note that there is no \omicron, since it is identical to the letter 'o'):

α	\alpha	ι	\iota	ϱ	\varrho
β	\beta	κ	\kappa	σ	\sigma
γ	\gamma	λ	\lambda	ς	\varsigma
δ	\delta	μ	\mu	τ	\tau
ϵ	\epsilon	ν	\nu	υ	\upsilon
ε	\varepsilon	ξ	\xi	ϕ	\phi
ζ	\zeta	o	o	φ	\varphi
η	\eta	π	\pi	χ	\chi
θ	\theta	ϖ	\varpi	ψ	\psi
ϑ	\vartheta	ρ	\rho	ω	\omega

Here are the uppercase Greek letters (there are only 11. The other ones look like certain roman letters.):

Γ	\Gamma	Ξ	\Xi	Φ	\Phi
Δ	\Delta	Π	\Pi	Ψ	\Psi
Θ	\Theta	Σ	\Sigma	Ω	\Omega
Λ	\Lambda	Υ	\Upsilon		

Calligraphic capitals: To get $\mathcal{A}\ldots\mathcal{Z}$, just say $\cal A\ldots Z$. The reader should look at [351] to see how the control sequence \cal is defined as font family 2.

The following are math symbols of type Ord:

$\aleph$	\aleph	$\prime$	\prime	$\forall$	\forall
$\hbar$	\hbar	$\emptyset$	\emptyset	$\exists$	\exists
$\imath$	\imath	∇	\nabla	$\neg$	\neg
$\jmath$	\jmath	$\surd$	\surd	$\flat$	\flat
ℓ	\ell	$\top$	\top	$\natural$	\natural
$\wp$	\wp	$\bot$	\bot	$\sharp$	\sharp
$\Re$	\Re	$\Vert$	\Vert	$\clubsuit$	\clubsuit
$\Im$	\Im	$\angle$	\angle	$\diamondsuit$	\diamondsuit
∂	\partial	$\triangle$	\triangle	$\heartsuit$	\heartsuit
∞	\infty	$\backslash$	\backslash	$\spadesuit$	\spadesuit

It is important to understand that any math symbol (or even an expression) can be treated as type Ord by enclosing it in braces. Thus in the expression $a+b$, the + is a binary operation, with spaces on both sides, whereas in $a{+}b$, the + is considered an Ord. The results are $a + b$ and $a+b$. Some people also prefer $a\times b$ over $a \times b$.

The \mathord command declares a subexpression to be an ordinary symbol (Ord). In $(a\mathord+b)$, the '+' is considered an Ord (rather than the usual Bin), so the spacing is $(a+b)$ rather than $(a + b)$.

▶ **Exercise 13.9:** What is the difference between '$a{+}b$' and '$a\mathord+b$'?

13.8 Large Operators

The "Large" operators come in two sizes, for text and display styles. Note that some have names that start with \big. Those have smaller versions that are binary operators. Compare, say, \bigcap to \cap '∩∩'. A binary operator has spaces on both sides. A large operator gets different spacing, and it may have limits:

$\sum\sum$ \sum		$\cap\cap$ \bigcap		$\odot\odot$ \bigodot	
$\prod\prod$ \prod		$\cup\cup$ \bigcup		$\otimes\otimes$ \bigotimes	
$\coprod\coprod$ \coprod		$\sqcup\sqcup$ \bigsqcup		$\oplus\oplus$ \bigoplus	
$\int\int$ \int		$\vee\vee$ \bigvee		$\uplus\uplus$ \biguplus	
$\oint\oint$ \oint		$\wedge\wedge$ \bigwedge			

The math operators that are considered "large" (of class Op) have three important properties:

1. The positions of their limits (super- and subscripts) can be controlled. In display style, the limits of an Op are normally displayed above and below it. In text style, they are placed to the right. This behavior can be changed by the \limits, \nolimits commands; it can be reset to the default values by the \displaylimits command. Note that the three commands apply only to an atom of type Op. The expression

```
$\sum_0^\infty\quad\sum\nolimits_0^\infty\quad
\sum\limits_0^\infty\quad\sum\displaylimits_0^\infty$
```

produces $\sum_0^\infty \quad \sum\nolimits_0^\infty \quad \sum\limits_0^\infty \quad \sum_0^\infty$, whereas

```
$$\sum_0^\infty\quad\sum\limits_0^\infty\quad
\sum\nolimits_0^\infty\quad\sum\displaylimits_0^\infty$$
```

results in

$$\sum_0^\infty \quad \sum_0^\infty \quad \sum\nolimits_0^\infty \quad \sum_0^\infty$$

It is easy to create new large operators with the \mathop command. A new Op can be a character in a font, or it can be a subformula. Assuming that \beegI is the name of a big font, the code \def\bigF{\mathop{\hbox{\beegI f}}} defines the large operator $\int$. It can be used, for example, in '$A+\bigF_0^2x$' to create $A+\int_0^2 x$, and in '$$A+\bigF_0^2x$$' to create

$$A + \int_0^2 x.$$

▸ **Exercise 13.10:** Why is the \hbox necessary in the definition above?

It is also possible to define a \mathop\limits or \mathop\nolimits.

2. If the large operator consists of a single character, it is vertically centered with respect to the axis, and its italic correction is taken into account when positioning limits. If the large operator consists of anything else, then its baseline is aligned with the axis. The following: $$-\!\!\!\int\limits_{-\pi}^\pi\!\!\!-\mathop{gg}_{-\pi}^\pi-\mathop{g}-\mathop{\kern0pt g}-$$ results in

$$-\int_{-\pi}^{\pi} - \underset{-\pi}{\overset{\pi}{gg}} - g - g -$$

It is easy to see that the integral is centered with respect to the minus signs (which are located on the axis), but the gg is not. Also, the limits are simply centered above and below the gg, not taking into account the slant. The single g following the gg has its baseline aligned with the axis, but the next g is centered because the \kern0pt creates a \mathop with more than a single character.

▸ **Exercise 13.11:** Why is the big $\int$ above not centered around the axis?

The big 'f' above always comes out at the same size, regardless of the current style. To get different sizes, depending on the style, the \mathchoice command may be used. It takes four arguments and typesets one of them, depending on the current style. After the definition:

```
\def\bigF{{\mathchoice{\mathop{\hbox{\beegI f}}}
{\mathop{\hbox{\tenrm f}}}
{\mathop{\hbox{\sevenrm f}}}
{\mathop{\hbox{\fiverm f}}}}}
```

the expression ${\displaystyle\bigF}\bigF^{\bigF_\bigF}$ results in $\int f^{f_f}$.

3. Some large operators can "grow." A good example is $\sqrt{\sqrt{\sqrt{a}}}$, created by $\sqrt{\sqrt{\sqrt a}}$.

As usual, placing a large operator in braces turns it into type Ord.

13.9 Binary Operations

They get thick spaces on both sides (except when preceded by an opening or followed by a closing). If desired, the spaces can be eliminated by enclosing the binary operation in braces.

+	+	−	-	÷	\div
±	\pm	∩	\cap	∨	\vee
∓	\mp	∪	\cup	∧	\wedge
\	\setminus	⊎	\uplus	⊕	\oplus
·	\cdot	⊓	\sqcap	⊖	\ominus
×	\times	⊔	\sqcup	⊗	\otimes
*	\ast	◁	\triangleleft	⊘	\oslash
⋆	\star	▷	\triangleright	⊙	\odot
⋄	\diamond	≀	\wr	†	\dagger
∘	\circ	◯	\bigcirc	‡	\ddagger
•	\bullet	△	\bigtriangleup	∐	\amalg
		▽	\bigtriangledown		

Note that the \setminus operation is identical in shape to the \backslash, which is of type Ord. The difference between them is, of course, the spacing.

The \mathbin command declares a subexpression to be a binary operator. This way the proper spacing is placed around it. Compare, for example, (a,b) to $(a \, b)$. The former was done by (a,b), and the latter by $(a\mathbin,b)$.

13.10 Relations

This is a large class including negations and arrows.

<	<	>	>	=	=
≤	\leq	≥	\geq	≡	\equiv
≺	\prec	≻	\succ	∼	\sim
⪯	\preceq	⪰	\succeq	≃	\simeq
≪	\ll	≫	\gg	≍	\asymp
⊂	\subset	⊃	\supset	≈	\approx
⊆	\subseteq	⊇	\supseteq	≅	\cong
⊑	\sqsubseteq	⊒	\sqsupseteq	⋈	\bowtie
∈	\in	∋	\ni	∝	\propto
⊢	\vdash	⊣	\dashv	⊨	\models
⌣	\smile	|	\mid	≐	\doteq
⌢	\frown	‖	\parallel	⊥	\perp

Negated relations. The prefix \not may precede many of the relations above, inverting their meaning:

≮	\not<	≯	\not>	≠	\not=
≰	\not\leq	≱	\not\geq	≢	\not\equiv
⊀	\not\prec	⊁	\not\succ	≁	\not\sim
⋠	\not\preceq	⋡	\not\succeq	≄	\not\simeq
⊄	\not\subset	⊅	\not\supset	≉	\not\approx
⊈	\not\subseteq	⊉	\not\supseteq	≇	\not\cong

⋢ \not\sqsubseteq   ⋣ \not\sqsupseteq   ≭ \not\asymp

The \not operation is simply defined as character "36 of family 2 (\math-char"3236) whose width is zero.

Arrows. Note that some arrow names start with a capital latter. Others start with the prefix long.

←	\leftarrow	⟵	\longleftarrow	↑	\uparrow
⇐	\Leftarrow	⟸	\Longleftarrow	⇑	\Uparrow
→	\rightarrow	⟶	\longrightarrow	↓	\downarrow
⇒	\Rightarrow	⟹	\Longrightarrow	⇓	\Downarrow
↔	\leftrightarrow	⟷	\longleftrightarrow	↕	\updownarrow
⇔	\Leftrightarrow	⟺	\Longleftrightarrow	⇕	\Updownarrow
↦	\mapsto	⟼	\longmapsto	↗	\nearrow
↩	\hookleftarrow	↪	\hookrightarrow	↘	\searrow
↼	\leftharpoonup	⇀	\rightharpoonup	↙	\swarrow
↽	\leftharpoondown	⇁	\rightharpoondown	↖	\nwarrow
⇌	\rightleftharpoons				

The \mathrel command declares a subexpression to be a relation. This way the proper spacing is placed around the subexpression. Compare, for instance, (a, b) to (a , b). The former was done by (a,b), and the latter by $(a\mathrel,b)$.

13.11 Combining Relations

New math symbols can sometimes be generated by combining existing symbols, but this should be done carefully, keeping in mind the classes of the symbols being combined and the class of the newly created symbol. If we want, say, to combine the three symbols ∘, ≀ and •, it may be wrong to say $\circ\wr\bullet$, as this would produce ∘ ≀ •. The three symbols are binary relations and so get spaces on both sides. To concatenate them without any gaps, we first have to declare them relations (remember that no spaces are placed around relations). This is done by $\mathrel\circ \mathrel\wr \mathrel\bullet$, which produces ∘≀•.

The new symbol resulting from this combination is of class Rel by default, but it's easy to change its class. The expression $a ∘≀• a∘≀• a$ (look carefully at the spacing) was created by

```
\def\newsym{\mathbin{\mathrel\circ \mathrel\wr \mathrel\bullet}}
\def\Newsym{\mathpunct{\mathrel\circ \mathrel\wr \mathrel\bullet}}
$a\newsym a\Newsym a$
```

The combination ⟷ illustrates another problem. Both arrows are relations, so \mathrel is not needed. However, if we simply say $\leftarrow\rightarrow$, we get ←→. This is because the arrows are narrower than their boxes. To connect them smoothly, we should use negative math kern. We define \def\joinrel{\mkern-3mu}, and now $\leftarrow\joinrel\rightarrow$ produces the desired result.

The \buildrel macro [361] is a good example of how to combine existing symbols into a new Rel. Its definition is

```
\def\buildrel#1\over#2{{\mathrel{\mathop{\kern0pt #2}\limits^{#1}}}}
```

and it creates a new symbol of class Rel by placing two existing symbols one above the other. Note the use of `\kern0pt`. It's a "do nothing" operation whose purpose is to make sure that #2 is not a single character. Examples are

$\xrightarrow{\infty}$ `\buildrel\infty\over\longrightarrow`

$\overset{\text{def}}{\equiv}$ `\buildrel\rm def\over\equiv`

▸ **Exercise 13.12:** What is the result of `$\buildrel\top\over\bot$`?

▸ **Exercise 13.13:** Use `\buildrel` to generate $\overset{\circ}{\angle}{}^{\bullet}$

13.12 Openings and Closings

All the left delimiters can "grow" when preceded by `\bigl` or its relatives. A left delimiter is normally preceded by some space but is not followed by any. Six left delimiters are available:

(	(	{	\lbrace	⟨	\langle
[	\lbrack	⌊	\lfloor	⌈	\lceil

Each left delimiter has a corresponding right delimiter:

)	)	}	\rbrace	⟩	\rangle
]	\rbrack	⌋	\rfloor	⌉	\rceil

The `\mathopen`, `\mathclose` commands declare a subexpression to be an opening or a closing, respectively. This way the proper spacing is placed before or after the subexpression. Compare $< a, b >$ to $<a, b>$. The former was done by `$<a,b>$`, and the latter by `$\mathopen<a,b\mathclose>$`.

13.13 Punctuations

This is a small class, containing just the comma, semicolon, and the symbol `\colon`. These symbols are followed by a thin space. (Note that the regular colon, obtained by typing ':', is a relation, so it normally gets thick spaces on both sides.)

The `\mathpunct` command declares a subexpression to be a punctuation. This way the proper spacing is placed around it. Compare $(a!b)$ to $(a! b)$. The former was done by `$(a!b)$`, and the latter by `$(a\mathpunct!b)$`.

13.14 Inner Subformulas

A math expression may be divided into subexpressions. Anything enclosed in braces is considered a subexpression, as is anything placed in an `\hbox` or a `\vcenter`. The argument of `\overline` is also a subexpression. There is also the concept of an *inner* subexpression, which is surrounded by extra space. Examples of inner subexpressions (or inner formulas) are fractions and `\left`...`\right` constructions.

▸ **Exercise 13.14:** Devise a simple test to "show" an inner subformula.

The `\mathinner` command can be used to declare a subexpression as inner. A typical example is the `\ldots` macro, defined [359] as `\def\ldots{\mathinner{\ldotp\ldotp\ldotp}}`. There is a distinct difference between the results of `$a\ldots b$` and those of `$a\ldotp\ldotp\ldotp b$`.

13.15 Delimiters

A delimiter in math mode is any character that can follow \left or \right, or one of the withdelims commands. Everything between a \left and its matching \right becomes a subexpression whose height and depth are measured carefully so that the right size delimiters can be assigned.

An important feature of delimiters is that they should be able to "grow," which is done in three ways:

1. Two variants, a "small" and a "large," can be specified by the user for each delimiter. This is done using the concept of a delimiter code. The relevant commands are \delcode and \delimiter.

A delimiter code consists of six hexadecimal digits, where the first three give the family number and position of the "small" variant of the character, and the last three, the same for the "large" variant. There is a \delcode command for this but, in practice, the \delimiter command is used to assign both a class and a \delcode to a character. For example, a left brace is defined [359] as \def\lbrace{\delimiter"4266308}. The left brace is thus defined as a class 4 (opening) character, whose small variant is character "66 of family 2, and whose large variant is character "08 of family 3.

A variant may be missing, such as in \def\lmoustache{\delimiter"4000340} (no small variant). A negative \delcode means that the character should not act as a delimiter. When TeX starts, all delcodes are −1. A delcode of zero means no variants. The command \delcode`.=0 is especially useful. It makes it possible to use the period following \left or \right, with nothing typeset. (Actually, an empty box is typeset, of width \nulldelimiterspace, which has a plain value of 1.2pt [348].)

2. Several sizes of a character may be included in family 3 and may be linked by the charlist command. Note the absence of a '\', since this is a METAFONT command. This command is written by the font designer, and it creates a list in the font's .tfm file. The list is read by TeX when the font is loaded, and is used each time a math expression is being constructed.

In the case of a left brace, family 3 contains (in file 'bigdel.mf') the command
charlist oct"010": oct"156": oct"032": oct"050": oct"070";
meaning that the characters at (octal) positions 10, 156, 32, 50, and 70 of family 3 are to be linked in a list of bigger and bigger left braces. The diligent reader will notice that the character at position '70 is ' $\big\{$ ' and not a left brace, which brings us to the next point.

3. Even bigger characters can be constructed by stacking two or three characters (building blocks) vertically and connecting them with copies of a fourth character.

To continue with the example of left braces, the same family 3 contains four special characters that are building blocks, from which even bigger left braces can be constructed. This again is specified by the font designer, by means of the extensible command. In the case of a left brace, the command is: extensible oct"070": oct"070",oct"074",oct"072",oct"076";. It means that a left brace

is to be assigned position 70 (octal) in family 3 and is to be built from the characters at (octal) positions 70 ' ⌈ ', 74 ' ⎱ ' and 72 ' ⌊ ' stacked vertically, with copies of
character 72 ' ⌊ ' inserted in between.

Another illuminating example is the left angle bracket. The compact command
\langle{\delimiter"426830A} defines it as class 4 (openings) with a small variant
taken from "268 and a large one, from "30A. The command charlist oct"012":
oct"104": oct"034": oct"052"; links '12 (which equals hexadecimal "0A) to
three bigger left angles, for a total of five variants.

The five sizes ⟨ ⟨ ⟨ ⟨ ⟨ are obtained by

\langle \bigl\langle \Bigl\langle \biggl\langle and \Biggl\langle.
Because of its special shape, the left angle bracket cannot be extended by stacking
parts vertically. Therefore, only the five sizes above are available, and there is no
extensible command for this character.

To summarize, the rules for selecting a delimiter are as follows:

1. Depending on the current style, either the scriptscriptfont, or the scriptfont,
or the textfont of the family is used.

2. If the character found is large enough, it is used for the delimiter.

3. Otherwise, if the character is extensible, it is extended to the right size, and
used.

4. Otherwise, if the character is part of a charlist, its successor is used.

5. If it is not, then the character with greatest height plus depth found so far
is selected and used.

The four plain macros \big, \Big, \bigg, and \Bigg are useful in cases where
the user wants a delimiter of a specific size. These macros select delimiters of size
8.5pt, 11.5pt, 14.5pt, and 17.5pt, respectively. They are defined using \left. The
definition [360] of \big, for example, is essentially
\def\big#1{\hbox{$\left#1\vbox to8.5pt{}\right.$}}.

For each of the four \big macros, the plain format defines three more, with
specific classes. Thus for \big we have the definitions: \def\bigl{\mathopen\big}
\def\bigm{\mathrel\big} \def\bigr{\mathclose\big}.

They are used, of course, as an opening, a relation, and a closing, respectively.

13.16 Radicals

The \radical command is used to create square root signs that can grow but,
as is usual with TeX, it is a general command that can be used to create new
symbols. Any math character (or several characters) can be selected instead of the
'√', and \radical appends a rule to the right of the character, and extends the
rule to cover the width of the subexpression that follows. The thickness of the rule
equals the height of the character, and the bottom of the rule is aligned with the
baseline of the character, so special characters should be designed for this purpose.

The first example of a radical is the square root sign √. To construct it at the right size, both the \radical and the extensible commands are used. Here are the details:

1. Square roots are defined by \def\sqrt{\radical"270370 }, so small square roots are created by character "70 of font family 2, and bigger ones, by character "70 of font family 3. However, the latter character is the start of a charlist of five square root signs so, in all, six such signs are available to \radical to choose from.

2. For even bigger radicals, the extensible command below is used (in file 'bigdel.mf'). It stacks the two parts '√' '⌐' vertically, with copies of ' | ' in between. The actual command is
extensible oct"164": oct"166",0,oct"164",oct"165";

An example of a non square root radical is \def\weird{\radical"36B325 }. It selects the two characters at positions "6B and "25 of family 3 as the radical signs. The former is used for small radicals, and the latter for big ones. The display

$$\weird a\;\weird{\int x\over\sum y}$$ now creates $\rfloor \overline{a} \left\lfloor \frac{\int x}{\sum y}\right.$.

▸ **Exercise 13.15:** Why the gap between the weird sign and the rule?

13.17 \mathaccent

This command is the basis for macros that create math mode accents that can grow. It should be followed by a math character code and by a subexpression. It uses the mathchar to select an accent, then centers the accent on top of the subexpression. An example is the **plain** definition [359] \def\widehat{\mathaccent"362}. Character "62 of \textfont3 is a hat, followed by two larger hats in positions "63 and "64. The hat that best fits the width of the subexpression is selected.

▸ **Exercise 13.16:** How does TEX know that there are three and only three hats?

An unusual example is \def\wideweird{\mathaccent"034A}. The reader should examine the effects of $\wideweird{u}$, $\wideweird{uw}$, $\wideweird{uwx}$, $\wideweird{uwxyz}$.

13.18 Horizontally Extensible Symbols

There is a variety of symbols that can be extended horizontally. Those that are not primitives use either leaders or a linked list (charlist) to become extended.

The two primitives \overline and \underline extend a rule over or under their argument. (A related macro is \underbar, which is defined [353] by means of \underline, but can only be used outside of math mode. It places a rule under text but ignores the depth of the characters.)

The **plain** macros \overrightarrow and \overleftarrow are defined on [359] in terms of macros \rightarrowfill and \leftarrowfill, which themselves are defined [357] by means of leaders. Similarly, \overbrace and \underbrace are defined [359] by means of \downbracefill and \upbracefill, respectively. The latter are defined [357] by means of leaders.

The \widehat and \widetilde symbols are defined [359] in terms of the primitive \mathaccent. This primitive searches the list of available tilde (or hat) symbols for the one that best fits over its argument. There are three tilde and three hat signs in font cmex10 (family 3), so this method cannot produce very wide symbols.

The following examples $\overbrace{x_1 + \cdots + x_n}^{n>1}$, $\overrightarrow{\mathbf{V}_i \mathbf{V}_j}$, $\overline{a + \overline{a} + \widehat{ab}}$ were produced by

```
$\overbrace{x_1+\cdots+x_n}^{n>1}$,
$\overrightarrow{{\bf V}_i{\bf V}_j}$,
$\overline{a+\overline a+\widehat ab}$
```

▶ **Exercise 13.17:** Use \overbrace to typeset $\overbrace{a + b + \cdots = n}^{n>0}\ x_a + x_b + x_c + \cdots$.

13.19 Super- and Subscripts

In the expression $A_i^j + \cdots$, the A_i^j is called an *atom*. A is the nucleus, j is the superscript, and i is the subscript of the atom.

The following properties of super- and subscripts are worth mentioning:

1. Superscripts and subscripts are normally vertically aligned but can be staggered by grouping. Thus $x^2_{34}+{y^2}_{34}$ produces $x^2_{34} + y^2{}_{34}$.

2. The vertical positions of super- and subscripts depend on the vertical sizes of subexpressions. The following: $((x^2)^3)^4$, $\left((x^2)^3\right)^4$, ${({({(x^2)}^3)}^4$ generates $((x^2)^3)^4$, $\left((x^2)^3\right)^4$, and $\left((x^2)^3\right)^4$, respectively. The '2' is the superscript of the x, and the '3' and '4', the superscripts of the two right parentheses. In the first expression all the superscripts are at the same height, because the x and the parentheses are typeset in the same style. In the second expression, the parenthesis preceding the '4' is bigger, so the '4' is raised. In the third expression all the parentheses are the same size, but the braces define subexpressions. The '3' is raised because it is a superscript of a tall subexpression (tall because of the '2'). The '4' is raised for the same reason; it is a superscript of the tall subexpression $((x^2)^3)$.

3. A super- or subscript can be placed in an \hbox, which can be raised or lowered. The following x^2, $x^{\raise3pt\hbox{$\scriptstyle2$}}$ results in x^2, x^2.

4. The vertical positions also depend on certain fontdimen parameters assigned to the math symbols font \textfont2. The following three parameters control the vertical positions of superscripts. They specify the minimum distance from the baseline of the nucleus to the baseline of the superscript. \fontdimen13 is used for expressions in display style. \fontdimen14 is used for expressions in noncramped text style (or smaller). \fontdimen15 is used for expressions in cramped text style (or smaller).

The following two parameters control the vertical positions of subscripts. They specify the minimum distance from the baseline of the nucleus to the baseline of

the subscript. \fontdimen16 is used when there is no superscript. \fontdimen17
is used when there is a superscript.

The following is a simple example where both scripts are moved away from the
nucleus:

```
$a^2_2$,
\fontdimen14\textfont2=5pt
\fontdimen17\textfont2=5pt
$a^2_2$
```

The results are a_2^2, $a_{\,2}^2$.

In addition, \fontdimen18 specifies the maximum distance of the superscript
baseline below the top of the nucleus. Similarly, \fontdimen19 specifies the minimum distance of the subscript baseline below the bottom of the nucleus.

5. The \scriptspace parameter determines the horizontal space following
super- and subscripts. Its **plain** value is 0.5pt. It is only looked at once, just before the closing '$', so in $a^2 \scriptspace=5pt b^2 \scriptspace=15pt a^2$,
each superscript is followed by 15pt of space.

13.20 Ellipses

There are several commands to create an ellipsis. The \ddots, \ldots, \cdots,
and \vdots commands are **plain** macros [359] that can only be used in math mode.
They work by typesetting a period, moving it in the desired direction, and repeating
three times. The \dots command (another **plain** macro [356]) can be used inside
or outside math mode. It is easy to use the definition of \ddots to define an \adots
(for *ascending dots*) macro.

```
\def\adots{\mathinner{\mkern2mu\raise1pt\hbox{.}\mkern2mu
   \raise4pt\hbox{.}\mkern2mu\raise7pt\hbox{.}\mkern1mu}}
```

which may be useful in serious mathematics and also for typesetting fun patterns
such as ⋅⋅⋅⋅⋅⋅⋅⋅⋅⋅⋅⋅⋅⋅ . It is possible to create a wave pattern by

```
\setbox0=\hbox{$\adots\ddots$}
\setbox1=\hbox{$\ddots\adots$}
\def\wave{\leavevmode\copy0\lower9pt\copy1}
```

such that \wave\wave produces ⋅⋅⋅⋅⋅⋅⋅⋅⋅⋅⋅⋅⋅⋅⋅⋅⋅⋅⋅⋅

13.21 The Vertical Bar

Vertical bars in the math modes can be created by '\vert', by a '|', and by \mid. Here is a short discussion of the three.

\vert is declared by plain as \delimiter"26A30C. It is thus selected as either character "6A of family 2, or character "0C of family 3. It is used by \left, \right and their relatives to create big vertical bars.

Many modern keyboards have a vertical bar '|' key. In the math modes this is identical to \vert and can grow as necessary. Outside math mode, this simply typesets character "7C from the current font. In fonts cmr, cmbx, and cmti this character happens to be an em-dash. In cmtt it is a short vertical bar '|'.

\mid is defined as a plain \mathchar"326A. It thus picks character "6A of family 2 and declares it a relation (class 3). Using \mid in an expression typesets the same character as (the small version of) \vert, but with different spacing around it and without large versions.

13.22 \mathsurround

Last but not least, \mathsurround [162] should be mentioned. It specifies the (discardable) glue to be placed before and after any inline math expression. Its \plain value is 0. Saying 'step$a=b$by\mathsurround=6pt$a=b$step' creates 'stepa = bby a = b step'.

13.23 Exercises

▸ **Exercise 13.18:** Write math expressions to produce $\begin{smallmatrix} & 2 & 3 & 4 \\ 1 & & O & & 5 \\ & 8 & 7 & 6 \end{smallmatrix}$ in both inline and display modes. This requires a sharp eye, a little thinking, and a good knowledge of several math features.

▸ **Exercise 13.19:** Use boxes and your knowledge of spacing in math mode to create the structure $\dashv\!\!\top\!\!\vdash$ out of the four symbols ⊣, ⊥, ⊤, and ⊢.

▸ **Exercise 13.20:** (tongue in cheek) In what sense is the expression '$ab\sqrt{}$' absurd?

▸ **Exercise 13.21:** Create the simple pattern $a^{2^{2^{2^2}2}}a$.

14. Line & Page Breaks

The concepts of badness and penalties are central to these algorithms, which themselves are of interest to the advanced user. The concept of badness was described in section 3.16. Penalties were discussed in the Introduction (page 5. The reader should review these before reading any further, but here is a short recap.

The badness of a box is the amount by which it has to be stretched (or shrunk), divided by the maximum allowed stretch (or shrink). (The actual definition is slightly different.) The badness is nonnegative.

A penalty is an integer that expresses the desirability of breaking a line or a page at a certain point. Positive penalty discourages a break; negative penalty encourages it. To affect line breaks, the penalty should be placed inside a paragraph by saying, e.g., \penalty500, in H or RH mode. Saying the same thing in one of the vertical modes would place the penalty between the lines of a paragraph, thereby affecting the page break decisions.

This chapter discusses the line-break and page-break algorithms in detail, using many examples. The reader is encouraged, as usual, to do the examples independently and to try to solve all the exercises.

Before delving into the algorithms, two more things need to be discussed; the meaning of *tolerance*, and the concept of *demerits*.

14.1 The Tolerance

The concept of tolerance is intuitively easy to understand. We already know that if a box does not have enough material to fill it up, TEX will stretch its globs of glue even beyond their maximum stretchability and will classify the box as either loose or underfull. Similarly, if we place too much material in a box, its glues will shrink, but not below their minimum size, and the box will be classified as either tight or overfull. The classification depends on the values of parameters \hbadness and \hfuzz.

It is important to realize that the value of \hbadness does not affect the line-break decisions and thus has no effect on the way the document is typeset. A line of text may be considered overfull because of \hbadness, yet that line may become part of the final paragraph because the user has specified low tolerance. A low tolerance value means fewer choices for the paragraph builder in determining line breaks. The paragraph builder may be forced, in such a case, to include overfull lines in the paragraph. When the user increases the tolerance, the paragraph builder has more choices in selecting line breaks and thus can avoid breaking at overfull lines.

The value of parameter \tolerance does not affect the way the glue is flexed, nor does it affect the classification of a line of text. It only affects the way line breaks are chosen and thus the way the entire document is typeset. The tolerance is used only when a paragraph is being constructed (either in the main body of the text or in a \vbox), and it tells the paragraph builder how to evaluate the badness of lines when line breaks are being considered. The greater the tolerance, the more badness will be tolerated in lines of text.

The tolerance is measured in units of badness, and the rule is that lines with badness up to the current value of \tolerance will be tolerated. Lines with badness exceeding the tolerance will not be tolerated and so will not become lines in the final paragraph.

As a simple example, we set \hbox to52pt{A bit of text} and follow with \showthe\badness. The result is 94. We now typeset

\vbox{\pretolerance=94\tolerance=94\hsize=52pt\noindent A bit of text
A bit of text}

(\pretolerance is discussed on page 277) and since the badness does not exceed the tolerance, we get

A bit of text
A bit of text

However, when we reduce the tolerance to 93, TEX is no longer allowed to tolerate lines with badness of 94, and as a result, it **must** try different line breaks. The best that it can do is

A bit of text A■
bit of text

with an "overfull box" message. The reason for this bad result is that both passes of the line-break algorithm fail (see page 277 for a discussion of pass failure).

The plain value of \tolerance is 200, and that of \pretolerance, 100 [348].

14.2 The Definition of Demerits

The *demerits* of a line are calculated each time a breakpoint a is considered for the line. The definition [98] is $d = b^2 \pm p^2$, where p is the penalty found at point a, and b is the badness of the line created by breaking at a. The minus sign is used if p is negative. This definition achieves two goals.

■ The demerits are measured in units of badness squared and are therefore very sensitive to any small changes in either the badness or the penalties. As a result, the demerits are a large number. They are usually in the range of 10^4–10^5 but, in extreme cases, can reach even 10^8.

■ The penalty either reinforces the badness or works against it. Consider a line with a badness of 1000. If the penalty found at the end of this line is 1000, then the demerits of the line will be 2×10^6 units. However, if the penalty is -1000, then the demerits will be zero (the -1000 units of penalty cancel the 1000 units of badness).

Another 10,000 units (the `plain` value of parameter `\adjdemerits`) are added to the demerits of a line if the line is *visually incompatible* with its predecessor (see below). More units are added in certain cases of lines ending with a hyphenation. The actual definition of demerits includes a parameter, `\linepenalty` [98], that can be set by the user to effectively increase or decrease the value of the badness, such that 1000 units of penalty would cancel either more or less than 1000 units of badness.

▸ **Exercise 14.1:** Since badness and penalties are arithmetically added in the definition above, they must have the same units (this is also implied by saying that they cancel each other or work together). What are those units?

The document has a more pleasing appearance if we can avoid having a loose line next to a tight one. This is why visual compatibility is introduced [97]. A line of text is classified as either tight, decent, loose, or very loose, and two adjacent lines are said to be visually incompatible if their classes are not adjacent. Thus a loose line next to a very loose one is okay, but a loose line following a tight one form a visually incompatible pair.

A line with badness < 13 is decent. A line with badness ≥ 13 is tight if it has to shrink and loose if it has to stretch. A line is classified as very loose if its badness ≥ 100 and it has to stretch.

▸ **Exercise 14.2:** What's special about badness 13?

14.3 The Line-break Algorithm

This algorithm [98–99] is also described in detail in Ref. 34. It is implemented in §813–890.

We start with an outline of a typical line-breaking algorithm used by a good word processor. The algorithm uses three values for the *normal, minimum,* and *maximum* spacings between words. It proceeds by appending words to the current line, assuming normal spacing. If, at the end of a certain word x, the line is too long, the word processor tries to shrink the line. If that is successful, the next word y will start the next line, and the current line is broken after word x. Otherwise, the word processor backspaces over word x and tries to stretch the line. If that is successful, x becomes the first word of the next line.

If neither shrinking nor stretching works (both exceed the preset parameters), a good word processor should try to hyphenate the offending word, placing as much of it on the current line as will fit. The word processor may use one of three methods for hyphenation. It may have sophisticated built-in hyphenation rules, it may have a hyphenation dictionary, or it may use a few simple rules and ask for the user's confirmation. The rest of the hyphenated word is placed at the start of the next line. A word processor that does not hyphenate has to resort to stretching and may generate very loose lines.

The important feature of all such methods is that once a breakpoint has been determined, the algorithm *does not* memorize it but starts afresh with the next line. We can call such an algorithm "first fit," and its main feature is that it does not look at the paragraph as a whole. Such an algorithm produces reasonably good results. For a high-quality typesetting job, however, it is not fully satisfactory. For such a job, an algorithm is needed which considers the paragraph as a whole. Such an algorithm starts by making only tentative decisions about line breaks. If something goes bad toward the end of the paragraph, the algorithm may go back to the first lines and reconsider their original, tentative, breakpoints. Such an algorithm can be called "best fit."

TeX's algorithm determines several feasible breakpoints for each line and calculates the badness and the demerits of the line for each such breakpoint. After doing this for the entire paragraph, the final breakpoints are determined in a way that minimizes the demerits of the entire paragraph. Mathematically the problem is: (1) to set up a graph where the nodes are the feasible breakpoints, and each edge connects two breakpoints; (2) to find the shortest path in this (acyclic) graph.

Before discussing the details of the algorithm, we show an example of an actual paragraph. The paragraph was taken from the short story of [Ch. 6] and was typeset in a narrow \vbox. The text is

```
Mr.~Drofnats---or ''R. J.,'' as he pre ferred to be called---was
happiest when he was at work typesetting  beautiful documents.
```

The command

```
\vbox{\hsize=2.5in \tolerance=1000 \tracingparagraphs=1
Mr.~Drofnats---or ''R. J.,'' as ... beautiful documents.}
```

produces a detailed tracing (shown below) in the log file, and the short paragraph:

> Mr. Drofnats—or "R. J.," as he pre-|
> ferred to be called—was happiest when| he|
> was at work typesetting beautiful docu-|
> ments.|

The small vertical dashes indicate the feasible breakpoints. These points are also shown, as circled nodes, in the graph below, which summarizes all the possible ways of breaking the paragraph into lines. The graph also shows the demerits associated with each feasible breakpoint. A quick look at the graph shows that there was only one feasible breakpoint for the first line (the point following 'pre-'), but two such points for the second line (points 2, 3 in the diagram). There were two feasible breakpoints for the third line. The first one (point 4) can be a breakpoint if line two is broken at 'when' or at 'he'. The second one (point 5) can only be a breakpoint if line two is broken at 'he'. Finally, line four can only be broken at 'ments.' (the end of the paragraph), and both points 4 and 5 lead to 6.

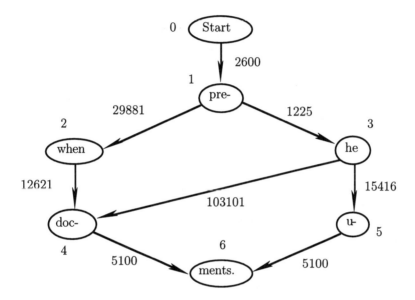

The graph shows that there are three ways of breaking the paragraph into lines. The first is to go 0-1-2-4-6 (with total demerits 50,202), the second is 0-1-3-4-6 (with t=112,026), and the third, 0-1-3-5-6, with the smallest total of 24,341. Notice that there is no path from feasible breakpoint 2 to 5. That would correspond to breaking the second line at 'when' and the third line at 'u-', making that line too tight. The results of going 0-1-2-4-6 and 0-1-3-4-6 are shown below for comparison.

Mr. Drofnats—or "R. J.," as he pre-
ferred to be called—was happiest when
he was at work typesetting beautiful doc-
uments.

Mr. Drofnats—or "R. J.," as he pre-
ferred to be called—was happiest when he
was at work typesetting beautiful doc-
uments.

The demerit values in the graph were taken from the tracing in the log file. The same values are also listed on [98–99] but it is interesting to note that the 1984, edition of *The TEXbook* lists slightly different values. For example, the total demerits of the shortest path is 24,305 instead of 24,341, a difference of 36 units. A careful comparison of the old and new (1990) values shows that the reason for the differences is very small changes in the values of certain glue components. The shrink component of the interword glue of the cmr10 font was changed from 0.1111pt to 0.11111pt [75] (a difference of $\approx 1.38 \times 10^{-7}$ inches, a microscopic size). Similarly, the shrink component of the glue that follows a comma in the same font was changed from 0.88887pt to 0.88889pt.

The fact that such minute changes in the values of glue components cause a difference of 36 units in the total demerits of our short paragraph shows that the definitions of badness and demerits make sense. The two quantities are very sensitive to any small changes in the dimensions of the boxes and glues involved.

Here is the listing created by \tracingparagraphs=1:

```
@firstpass
@secondpass
[]\tenrm Mr. Drofnats---or ''R. J.,'' as he pre-
@\discretionary via @@0 b=0 p=50 d=2600
@@1: line 1.2- t=2600 -> @@0
ferred to be called---was hap-pi-est when
@ via @@1 b=131 p=0 d=29881
@@2: line 2.0 t=32481 -> @@1
he
@ via @@1 b=25 p=0 d=1225
@@3: line 2.3 t=3825 -> @@1
was at work type-set-ting beau-ti-ful doc-
@\discretionary via @@2 b=1 p=50 d=12621
@\discretionary via @@3 b=291 p=50 d=103101
@@4: line 3.2- t=45102 -> @@2
u-
@\discretionary via @@3 b=44 p=50 d=15416
@@5: line 3.1- t=19241 -> @@3
ments.
@\par via @@4 b=0 p=-10000 d=5100
@\par via @@5 b=0 p=-10000 d=5100
@@6: line 4.2- t=24341 -> @@5
```

Lines starting with a single @ summarize the badness, penalty, and demerits (b, p, and d, respectively) associated with an active breakpoint. Lines starting with @@k show the total demerits accumulated when going from active breakpoint 0 to active breakpoint k. (End of example.)

We now turn to a general description of the line-break algorithm. The central structure is a list of "active breakpoints," which is created by the algorithm and is used by it all the time. Initially the list contains one breakpoint, namely the start of the paragraph. Text is read until there is a little more than enough material for one line. The paragraph builder then considers several feasible breakpoints for the line, at the end of several consecutive words. For each such feasible breakpoint a, the badness of the resulting line is calculated. If it exceeds the tolerance, point a is dropped; otherwise a becomes an active breakpoint and is appended to the list. Thus, at the end of a line, there typically are several active breakpoints a in the list.

It is now clear that higher values of the tolerance allow the paragraph builder to select more active breakpoints. If we don't mind having "badder" lines, we can increase the tolerance (the exercises in [Ch. 6] should be reviewed at this point).

Toward the end of the next line of text, the paragraph builder starts looking for feasible breakpoints b for that line. For each such point, the paragraph builder considers all the active breakpoints a of the previous line. It considers all the lines obtained by connecting every breakpoint a to the current breakpoint b. If the badness of such a line is acceptable—that is, it does not exceed the tolerance—point b becomes an active one and is appended to the list. Note that the same point b can be an active breakpoint for several points a. In such a case, b will go into the list several times, each time for a different a.

This process is repeated for every line, appending more and more active breakpoints to the list. At the end of the paragraph, the active breakpoints become nodes in an acyclic graph where each edge corresponds to a line of text generated by selecting two breakpoints. The edges are weighted by the demerits, and the final step is to select the shortest path from the top of the graph (the node representing the start of the paragraph) to the bottom (the node representing the end of the last line of the paragraph).

The algorithm above is first applied to the text without any hyphenations (pass 1). If pass 1 fails, the paragraph builder hyphenates *every* word in the paragraph and starts a second pass in which it applies the same algorithm, this time looking for feasible breakpoints inside hyphenated words. However, each hyphenation point receives a penalty \hyphenpenalty (with a plain value of 50) when it is generated. If an active breakpoint is chosen at a hyphenation, this penalty increases the demerits of the candidate line, which serves to discourage breaks at hyphenations. Experience shows [96] that only about two lines per paragraph, on the average, are broken at a hyphen.

Here is the algorithm in more detail: 1. The paragraph builder starts with the first pass, attempting to set the entire paragraph without any hyphenations. If it is possible to break the paragraph into lines with badnesses $\leq$ \pretolerance, the first pass succeeds. Otherwise, the first time a line is encountered whose badness exceeds \pretolerance for every feasible breakpoint, pass 1 is declared a failure.

2. If the first pass fails, the paragraph builder hyphenates every word in the paragraph, places penalty \hyphenpenalty at every hyphenation point, and starts pass 2, where it uses \tolerance instead of \pretolerance. Pass 2 may fail for the same reason as pass 1. (The existence of pass 2 implies that TeX hyphenates only when necessary.)

3. If pass 2 fails, the paragraph builder checks the value of parameter \emergencystretch. If it is zero, pass 2 is completed in spite of the bad lines, and the paragraph is set with overfull/underfull lines and error messages.

4. If \emergencystretch has a positive value, the paragraph builder pretends that every line has additional stretchability equal to \emergencystretch, and starts pass 3, the emergency pass, using \tolerance as in pass 2. Note that the additional stretchability isn't really added to the lines, so any underfull boxes will remain underfull.

5. If pass 3 is performed, the paragraph will be set according to the results

of that pass, even if it fails. If pass 3 fails, the paragraph will be set with over-full/underfull lines and error messages.

When \tolerance is set to large values, "badder" lines are tolerated. In particular, when \tolerance is set to 10,000, the paragraph builder becomes infinitely tolerant and will tolerate any lines, no matter how bad. Also, the algorithm is implemented such that it will not produce overfull boxes when \tolerance=10000, only underfull ones. This is why typesetting the same paragraph twice, with \tolerance values of 9999 and 10,000, can lead to very different results (see experiments 2 and 3 ahead).

When \pretolerance is set to a large value, the first pass will have a better chance of success. This implies fewer hyphenations in the final paragraph. When \pretolerance=10000, the first pass will always succeed (unless it runs into an unhyphenatable word that's wider than an entire line). When \pretolerance=0, the first pass will always fail. When \pretolerance=-1, the first pass will be skipped, resulting in a paragraph with more hyphenations than necessary.

When \tolerance is set to a small value, such as 10, only lines with badnesses ≤ 10 will be tolerated. Experience shows that when English text is set in a 6-inch margin (or wider), using font cmr10, the paragraph builder can still find enough active breakpoints and set most paragraphs without any overfull lines. When \tolerance=0, only lines with zero badness would be tolerated. It is impossible always to have lines with badness zero, so all three passes usually fail in such a case, producing many overfull lines. When \tolerance=-1, no line can be tolerated, no matter what the badness. The three passes fail, and every line will be overfull.

The effect of \emergencystretch is to allow each line of text more stretch and thus to reduce its badness. Thus a line that was originally overfull may not be so when \emergencystretch becomes large. This results in a larger number of active breakpoints, making it easier for the paragraph builder to set the paragraph without overfull lines. The plain value of \emergencystretch is 0pt, and increasing it is similar to increasing the tolerance and has the effect of reducing the number of overfull lines.

There is a difference between increasing the tolerance and increasing \emergencystretch. The former case may result in a graph where every path has infinite demerits. In such a case any path is as good as any other one, and this tends to create a paragraph where one line is very bad and the others are relatively good. The latter case decreases the badnesses of all lines, so lines that used to have infinite badness may now have a finite one. The graph may now have one or several paths with finite demerits, resulting in a paragraph with several bad lines, but no horrible ones. Here is an example.

These identical paragraphs were typeset in a 1-inch wide column. Paragraph 1 was typeset without any special settings; it is bad, because of the narrow margin. Paragraph 2 was typeset with \tolerance=9999, so it only has one overfull line (and a few underfull ones). Paragraph 3 was done with \tolerance=10000, so it only has underfull lines. Note how one line came out horribly looking. Paragraph 4 was typeset with \tolerance=200, but with \emergencystretch=10pt. There are no horrible lines, but, in general, the lines look worse than those of paragraph 3.

(Paragraph 5.) The serious reader should typeset the same paragraph with

This paragraph is■ typeset several times,■ with a 1in wide margin, to com- pare the results of■ a large tolerance to those of large emergencystretch.■	This paragraph is typeset several times, with a 1in wide margin, to compare the re- sults of a large tolerance to those■ of large emer- gencystretch.	This paragraph is typeset several times, with a 1in wide margin, to compare the re- sults of a large tol- erance to those of large emer- gencystretch.	This paragraph is typeset sev- eral times, with a 1in wide mar- gin, to compare the results of a large toler- ance to those of large emergen- cystretch.
1	2	3	4

\raggedright. The line breaks will be identical to those of 4 but with much better interword spacing and with no errors. This is also a good place to review the definition of \raggedright (page 94), which may be a good solution in cases of many overfull/underfull lines.

In the next experiment, the same paragraph is typeset four more times, all with \tolerance=9999 (and thus with the same line breaks). Paragraph 6 was set with \hbadness=0, and thus produces messages about underfull, loose, tight, and overfull lines. It illustrates the fact that when a narrow margin and a low \hbadness are combined, practically every line is judged bad.

Paragraph 7 was set with \hbadness=10000. Even though the line breaks are the same, only the one overfull line is reported. Paragraph 8 is like 7, but with \hfuzz=3.4pt, to eliminate the message about the overfull line (note that the line still sticks out on the right). Paragraph 9 was set with \hbadness=-1, so *every* line is judged bad. This is a good method to display the badness of lines.

▶ **Exercise 14.3:** Normally, the badness reflects the amount of stretch or shrink of the interword glue. With \raggedright, however, that glue is rigid, and the badness reflects the amount of space at the right margin. Can the method above be used to display the badness of *every* line?

Note. This nine-paragraph example illuminates many aspects of the line-break algorithm and should be performed by the reader (end of example).

This paragraph is typeset several times, with a 1in wide margin, to compare the re- sults of a large tolerance to those■ of large emer- gencystretch.	This paragraph is typeset several times, with a 1in wide margin, to compare the re- sults of a large tolerance to those■ of large emer- gencystretch.	This paragraph is typeset several times, with a 1in wide margin, to compare the re- sults of a large tolerance to those of large emer- gencystretch.	This paragraph is typeset several times, with a 1in wide margin, to compare the re- sults of a large tolerance to those■ of large emer- gencystretch.
6	7	8	9

The line-break algorithm is affected by other parameters. Things such as \leftskip, \parshape, and \hangindent affect the sizes of individual lines and are thus important to the paragraph builder. Another such parameter is \loose-ness.

One way for the user to select better page breaks is to add lines to, or delete lines from, certain paragraphs. This can be done either by rewriting the text, or by setting the \looseness parameter to the right value. If \looseness is set to 2, the paragraph builder tries to add two lines to the paragraph by doing the following:

- It goes through the line-break algorithm as usual, to find out the number of lines, n, that the paragraph would have in the optimal case.

- Now that it knows the value of n, it artificially increases the stretch of the inter-word glue and tries different line breaks that will result in $n + 2$ lines.

- If that fails (there are lines with badness exceeding the tolerance), it tries for $n+1$ lines. It that also fails, the paragraph would be set optimally, with n lines.

Similarly, when \looseness=-2, the paragraph builder will shrink glues, trying to end up with a paragraph of $n-2$ or, at least, $n-1$ lines. Note that \looseness is set back to zero at the end of a paragraph, and is thus only active for one paragraph.

▸ **Exercise 14.4:** What is the effect of setting \looseness to a very large negative value, such as -1000?

▸ **Exercise 14.5:** What other parameters are reset after each paragraph?

As a result of the discussion above, it is now clear that when wide columns (about 6 inches or wider) and a 10pt or 12pt proportional font are used, there is no problem in setting paragraphs, and overfull lines are rare.

When narrow columns are specified, or when a fixed-space font is used, the paragraph builder has fewer choices of selecting active breakpoints, which results in overfull lines. In such a case, the user has the following options:

1. Rewrite the text until the overfull lines disappear. This may be time consuming or even, if the column is very narrow, impossible. This solution may be worth the time if the document absolutely requires the highest typesetting quality.

2. Increase the tolerance. This will reduce the number of overfull boxes, but the lines will look awful. This is a good option since it shows the author which lines are bad and thus provides hints as to how the text should be rewritten.

3. Increase the value of \emergencystretch. This will reduce the badness of lines and, consequently, the number of overfull boxes. If the author is not planning to rewrite the text, then this may be a better choice than increasing the tolerance, for the reason discussed earlier.

4. Increase the value of \hbadness and/or \hfuzz. This has the effect of declaring fewer lines as overfull/underfull. There will be fewer error messages, but the bad lines would still stick on the right margin as before. Note that \hbadness does not affect the choice of line breaks, only the criterion used in classifying a line as overfull, tight,and so on. This may be a good option if high-quality typesetting is not required and, at the same time, the author does not care to know which lines are bad. This, of course, runs against the spirit of TEX, so let's hope that this option is rarely used.

5. Execute \overfullrule=0pt. This will suppress the thick rules indicating overfull lines. There will still be error messages, and the lines will still stick out on the right.

6. Use \raggedright. This improves the interword spacing and may be the best solution for certain documents. In fact, some authors (Ref. 35) claim that raggedright facilitates rapid reading and is the best way to set text.

7. Fiddle with \looseness in order to artificially stretch or shrink lines. This is not recommended, since \looseness was designed to improve page breaks, not the appearance of paragraphs.

8. Change the space factor codes of the punctuation marks to get more stretch (or more shrink) than normal in the interword glue following them.

9. If all else fails, try to increase \hsize to widen the paragraph. This decision, of course, is not always up to the typist and may require a higher authority.

We now turn to another experiment that summarizes many properties of the line-breaking algorithm. The following six short paragraphs are typeset, without indentation, in a short, 60pt margin. We also set \hbadness=-1, so that no line is considered decent. The text

```
\parindent=0pt\hsize=60pt\hbadness=-1
be it as it may\par
x be it as it may\par
xx be it as it may\par
xxx be it as it may\par
xxxx be it as it may\par
xxxxx be it as it may\par
```

is typeset to produce

be it as it may	x be it as it may	xx be it as it may
1	2	3
xxx be it as it may	xxxx be it as it may	xxxxx be it as it may
4	5	6

Analyzing these experiments can add much to the understanding of the line-break algorithm. The first experiment is simple. The 60pt size results in a single, tight line with badness 86. Since the badness is less than \pretolerance, the first pass succeeds.

The second experiment is more interesting. The first pass tries the feasible breakpoint after the second 'it'. This, however, leads to the line 'x be it as it,' which, at 60pt, is underfull with badness 232. This badness exceeds \pretolerance, so the first pass tries to break after 'may'. This leads to the line 'x be it as it may', which is overfull. As a result, the first pass fails. The second pass is unable to find any hyphenation points for 'may' and so fails for the same reason (recall that badness 232 also exceeds \tolerance). The result is that the paragraph is set as a single, overfull line.

In the third experiment, pass 1 succeeds. The line 'xx be it as it' is loose, with badness 15. Adding another 'x', in experiment 4, makes the line tight, with badness 6.

▶ **Exercise 14.6:** Typeset the six paragraphs as above, and analyze the results of experiments 5 and 6.

▶ **Exercise 14.7:** What would be the results of the six experiments when \toler- ance=232?

14.4 Hyphenation

The hyphenation algorithm is described in [App. H] and in Ref. 36 and will not be discussed here. Instead, certain commands and parameters that affect hyphenation will be described.

The \- command [455] inserts a 'discretionary hyphen' into a word. This tells TeX that it can insert a hyphen and break the word at the point. This command is used when the hyphenation algorithm cannot find an otherwise valid hyphenation point.

The \discretionary command [95] also specifies a discretionary hyphen, but it does more than that. It also tells TeX what text to place on either side of the break and what text to use when no break is necessary. A common example is 'di\discretionary{f-}{fi}{ffi}cult' which means that it is permissible to hyphenate between the 'di' and the 'cult'. If no hyphenation is necessary, an 'ffi' should be inserted; otherwise, an 'f-' should precede, and an 'fi' should follow, the break. Note that \- is equivalent to \discretionary{-}{}{}.

The \hyphenation command [452] adds words to a hyphenation exception dictionary. The dictionary overrides the normal hyphenation process and is used for words that are not properly hyphenated by TeX. Thus \hyphenation{acad-e-my acad-e-mies af-ter-thought macintosh} adds the four words to the dictionary. From that point on, the word 'macintosh' (or 'Macintosh') will not be hyphenated, and the three other words will only be broken at the points indicated. Note that upper- and lowercase letters are considered identically by this command.

The \uchyph parameter [454] has to do with words that start with an uppercase letter. Such words should sometimes be hyphenated as usual and sometimes not hyphenated at all. If \uchyph is set to a positive value, such words will be hyphenated; if it is set to a nonpositive value, they will not.

The \showhyphens command [364] is a plain macro. It shows the hyphenations that TeX has found for a word (or several words). Its definition is tricky:

```
\def\showhyphens#1{\setbox0=\vbox{\parfillskip=0pt
\hsize=\maxdimen\tenrm \pretolerance=-1 \tolerance=-1
\hbadness=0 \showboxdepth=0\ #1}}
```

The argument is placed in \box0 with both \tolerance and \pretolerance set to −1. This guarantees failure of all three passes. The setting \parfillskip=0pt \hsize=\maxdimen makes the box extremely wide and without any infinite flexibility. The \hbadness=0 guarantees that the box will be declared underfull (or at least loose), and thus reported in the log file (with the hyphenation points, since pass 2 was performed). The setting \showboxdepth=0 makes sure that only the relevant information is shown.

▶ **Exercise 14.8:** Why is \box0 set to a \vbox and not to an \hbox?

The two parameters \lefthyphenmin and \righthyphenmin [454] specify the minimum number of characters that are allowed to the left and right of a hyphenation point. Their plain values are 2 and 3, respectively. These parameters, of course, must have positive values. Any nonpositive values are replaced with a 1.

The \hyphenchar command [454] specifies the hyphenation character for a given font. The command \hyphenchar\tenrm=': specifies ':' as the hyphen in the cmr10 font. Also, \hyphenchar\ninett=-1 suppresses hyphenation in the \ninett font.

Since in most fonts the hyphenchar is '-' (ASCII code 45), it is convenient to use the command \defaulthyphenchar=45. When a font is loaded, its hyphen character will be set to that value, which can later be changed to anything else with a \hyphenchar.

TeX is being used to typeset many different languages, and they have different hyphenation rules. The \language and \setlanguage commands [455] produce whatsits each time the language is switched in the midst of a paragraph (e.g., from Japanese to Javanese). These whatsits are stored in memory with the rest of the current paragraph, while the paragraph text is being read in H mode. When the paragraph builder is invoked to determine the line breaks, the whatsits are used to select the set of hyphenation rules appropriate for each language. The hyphenation rules are discussed in [App. H].

14.5 The Page-break Algorithm

TeX accumulates boxes in its main vertical list until it has more than enough material for a page. It then selects a point for a page break using an algorithm [110–114] similar to the one for line breaks. TeX determines all the points in the main vertical list that are feasible breakpoints for a page break. It calculates the "cost" of breaking at each of those points and selects the breakpoint with the minimum cost.

The main difference between line breaks and page breaks is that once a page break has been determined, the page is sent to the output routine, which normally ships it out. Thus there is no going back and changing the page break. In principle, the best method would be to read the entire document, save it in memory, determine all the feasible page breakpoints, construct a graph, and select the path of minimum total cost. However, because of time and memory space considerations, TeX does not do that. We say that TeX chooses page breaks by a "local," rather than a "global," optimization method.

A "global" page-break algorithm is described in Ref. 37, which itself was typeset by that algorithm.

The last paragraph of [114] discusses another, minor, difference between page breaking and line breaking.

The cost c of breaking a page at a certain point x is defined as $c = b + p$, where b is the badness of the page that would be formed if the break were at point x, and p is the penalty found at point x. The actual formula [111] is more complex and involves a quantity q that is used to increase the page cost in case of insertions. Again the penalty is used to add either to the badness or to cancel some of it.

A few more points should be briefly mentioned:

1. When a page is broken between two lines, the vertical glue at the breakpoint is discarded, so it does not appear at the top of the next page. The same is true for kern and penalties found at a breakpoint.

2. The entire process of determining feasible page breakpoints and calculating their costs can be displayed in the log file by specifying \tracingpages=1 [112–113].

3. The interline glue (\baselineskip) is normally made rigid. This is done in order to achieve uniform appearance of the pages. However, the vertical glue in a page should have some flexibility, since otherwise TeX would have only one possible breakpoint for the page, and that breakpoint may have a high—even an infinite—cost associated with it. The solution is to add flexibility to the glue between paragraphs (\parskip) and around formulas in math display mode. By stretching or shrinking those glues, TeX may find many feasible breakpoints, resulting in better page breaks. In special cases, such as a one-page document, \baselineskip can be made flexible.

Tolerance is another word for indifference.

— *W. Somerset Maugham*

Not for their own demerits, but for mine, fell slaughter on their souls: heaven rest them now!

— *William Shakespeare, Macbeth*

15. Handling Errors

The novice user should become familiar with [Ch. 6, Ch. 27] before reading any farther. The notation [§xxx] is heavily used here, for the benefit of advanced readers who want a more detailed understanding of error handling. However, there is no need to consult the actual WEB code in order to understand error messages and respond to them.

When something goes wrong, either a warning, an error, or a fatal error is issued. In either case, a message is displayed on the screen (and in the log file). In the case of an error, TeX stops and waits for a user's response. In the case of a fatal error, the program quits, since no user response can correct the situation.

15.1 Warnings

Warnings are mostly concerned with "bad" boxes. They don't require immediate response, but they normally deserve careful consideration after the document has been typeset. They give important clues on how to improve the appearance of the document by changing the tolerance, increasing the value of \emergencys-tretch, or by rewriting parts of the text. The following is a (probably incomplete) list of warnings:

1. Missing Character: There is no xx in font yy! [§581].
2. Underfull or loose box reported [§660, §663, §675].
3. Overfull or tight box reported [§663, §666, §667, §675].

15.2 Errors

The main error procedure is [§82]. Error messages are intended to give the user enough information about the error and its precise location. The user's response should depend on the error, the point in the document where TEX has located the error, and on the user's experience. Many times it is possible to correct the error "on the fly," without aborting. Sometimes it is only possible to convince TEX to ignore the error and continue, so it can find more errors. The document will have to be retypeset. In other cases the only response possible is to stop the typesetting, correct the error in the input file, and restart.

All error messages have the same general format [§72]. They start with a line describing the error, followed by two lines showing the point in the document where the error was discovered, followed by a line with a '?' prompt.

1. Perhaps the most common example is a misspelled control sequence name, such as \hship typed instead of \hskip. The input 'pre error \hship1in post error', will cause the following error message:

```
Undefined control sequence.
l.123 pre error \hship
                      1in post error
?
```

The error message in this case is clear, 'Undefined control sequence'. The following two lines show that TEX has read the text up to and including \hship, but hasn't read '1in post error' yet. The '1.123' means that the error was detected on line 123 of the current input file [§313]. The '?' is a prompt for the user's response. A first-time user should respond with a '?', which will produce [§85] the following list of options:

```
Type <return> to proceed, S to scroll future error messages,
R to run without stopping, Q to run quietly,
I to insert something, E to edit your file,
1 or ... or 9 to ignore the next 1 to 9 tokens of input,
H for help, X to quit.
```

where the letters 'S', 'R', 'Q', 'I', 'E', 'H', and 'X' can be either upper or lower case. Here is an explanation of the options:

▪ A ⟨return⟩ means to ignore the error and proceed.

▪ An 'S' means to enter scroll mode. Error messages will be displayed on the screen and in the log file, but TEX will ignore them, as if the user has pressed ⟨return⟩, and proceed as best it can. It will only stop when an input file cannot be found.

▪ An 'R' (run mode) is the same as 'S', except that TEX will not stop even for a missing input file.

▪ A 'Q' is like 'R' and also runs quietly. Errors will go in the log file but will not be displayed on the screen.

▪ An 'I' followed by some material means to insert the material at the point where the error was discovered (see examples below).

- An 'E' causes TEX to stop typesetting and to invoke the editor. This option should
be selected when the error cannot be ignored and cannot be corrected by any of the
other options. The 'E' option is implementation dependent and is not available on
all computers.
- An 'H' produces a more detailed explanation of the error message. In our case it
will result in

```
The control sequence at the end of the top line
of your error message was never \def'ed. If you have
misspelled it (e.g., '\hobx'), type 'I' and the correct
spelling (e.g., 'I\hbox'). Otherwise just continue,
and I'll forget about whatever was undefined.
```

A frustrated user sometimes tries another 'H', hoping to get even more error
information. However, the second 'H' only produces [§89]

```
Sorry, I already gave what help I could...
Maybe you should try asking a human?
An error might have occurred before I noticed any problems.
``If all else fails, read the instructions.''
```

- An 'X' means to go back to the operating system. TEX will discard the current
page and quit. The .dvi file, with previous pages, is saved and can be viewed or
printed.
- A positive number n means to skip over the n input tokens following the error
point. An input token is either a control sequence, a single character, or a space.
The value of n can be up to 99 [§88] but, in practice, only small values are used.
- There is also a secret 'D' option, for debugging [§84, §1338]. However, it doesn't
seem to be implemented on any system known to this author.

If you want to scroll through error messages in just part of your document, you
can enclose that part with the pair \scrollmode and \errorstopmode. There are
also \nonstopmode and \batchmode commands that correspond to the 'R' and 'Q'
options, respectively.

A ⟨return⟩ is sometimes the simplest option. In our case, the bad command
\hship would be ignored when we hit ⟨return⟩, and the '1in' would simply be
typeset. The user can later edit the input file and correct the error. Another
possibility is to type '3', to skip over the next three tokens '1in'. This produces the
response

```
l.123 pre error \hship1in
                        post error
?
```

TEX is showing what tokens have been skipped over, and waits for a response.
A ⟨return⟩ is the simplest response at this point.

A ⟨return⟩, of course, is not always the best response. In many cases, ignoring
an error causes more, spurious, errors. A simple example is 'pre error \def post
error'. This is a serious error, and ignoring it produces more errors, while TEX is
looking for the missing macro name and definition.

To correct the \hship, it is best to insert an \hskip by typing 'i\hskip'. The inserted material replaces the last token read.

2. Here is an example of an incomplete command. The text 'pre error \hskip post error' causes the error message:

```
 Missing number, treated as zero.
<to be read again>
                          p
l.1 pre error \hskip p
                          ost error
 ?
```

TEX expects the \hskip to be followed by a ⟨glue⟩; instead, it finds the 'p' of 'post'. It backspaces over the 'p' and creates a zero. The zero should be followed by a valid unit of glue, and a good choice is to insert one by typing, e.g., 'iin' (note that the 'i' for insert does not have to be followed by a '\'). Now the 'p' is read again and normal execution resumes.

3. Again an incomplete command. 'pre error \hskip1 post error' produces

```
 Illegal unit of measure (pt inserted).
<to be read again>
                          p
<to be read again>
                          o
l.1 pre error \hskip1 po
                          st error
 ?
```

The \hskip1 is followed by a 'p', which could be the beginning of 'pt'. The 'p', however, is followed by an 'o', which causes this specific message. The unit 'pt' is automatically inserted, and both the 'o' and 'p' are backspaced over. TEX is now positioned between 'pre error \hskip1' and 'pt post error'. The simplest response is ⟨return⟩, which causes the execution of \hskip1pt. However, we can also type 'i\hskip1in', which will cause the execution of \hskip1pt\hskip1in. Another possibility is to type 'i\unskip\hskip1in', which will undo the \hskip1pt and execute \hskip1in.

▶ **Exercise 15.1:** What's a good response to the following error 'pre error \hskip1p1 post error'?

4. Another common error is missing '$' signs around math formulas. The fragment of text 'pre error text 2^x post error text' produces the message:

```
 Missing $ inserted.
<inserted text>
                          $
<to be read again>
                          ^
```

```
1.1 pre error text 2^
                      x post error text
?
```

After reading the '^', TₑX realizes that a '$' is missing. It inserts one and backspaces over it. Unfortunately, it is too late, since the beginning of the formula (the '2') has already been read. In our case, since the formula is so short, we can type '2', which will skip over the '$^'. The response will be

```
1.1 pre error text 2^
                      x post error text
?
```

and now we can hit ⟨return⟩, to continue. The resulting document will be wrong, and will have to be redone, but this gives us a chance to discover more errors.

    If the formula is long, it is better to hit ⟨return⟩. TₑX will enter math mode, will typeset the formula and the rest of the paragraph in math mode, and will stop at the start of the next paragraph, inserting another '$'. We can then hit ⟨return⟩ again, to continue.

▶ **Exercise 15.2:** The text 'The correct name for '^' is ''circumflex.''' causes an error. What is it and what is the best response?

    5. If an error is found inside a long command, such as '\setbox0=hbox{abc}', the start of the command is sometimes ignored. The above error causes the message

```
 A <box> was supposed to be here.
<to be read again>
                    h
1.1 \setbox0=h
              box{abc}
?
```

and also ignores the '\setbox0='. If we try ⟨return⟩, TₑX will simply typeset 'hboxabc'. Can we correct the situation such that the command will be executed? The best response is a '4', to delete the four characters 'hbox', followed by 'i\setbox0=\hbox'. See also [ex. 27.2] for a similar error.

    6. In a long document, a user may accidentally redefine a primitive or a `plain` macro, use it for a while, and later forget about the new definition and try to use it in its original meaning. A simple example is a document where the word LEFT needs to be typeset very often in chapter 1. The author may define \def\left{{\sc Left}}. In later chapters, the same author may write $\left a+b\right$ and, since \left no longer has its original meaning, strange messages will be issued.

    The best response in such a case is to insert the original definition of \left. If this is impossible (it is unknown, is too long, or is a primitive), an 'E' may be the next best choice.

7. Subtle errors may lie dormant for a long time before being discovered. A common case is a macro ending with a ⟨glue⟩. Something like `\def\abc{...` `\hskip1pt}` seems innocent enough. It may work for a long time without giving any trouble. However, if it is used in the context `...\abc plush carpets...`, it produces the error message

```
 Missing number, treated as zero.
 <to be read again>
                     h
 1.2 text\abc plush
                    carpets... \bye
 ?
```

The correct definition is, of course, `\def\abc{... \hskip1pt\relax}`.

▸ **Exercise 15.3:** Who is liable to make such an error?

8. A similar example is the declaration `\newcount\c`. It looks innocent, but `\c` is a `plain` macro (which one?). A later attempt to use it in its original meaning, such as 'a\c b', would cause a '`Missing number, treated as zero`', error message.

9. When an error is detected inside a macro, the resulting message is longer. It contains information about the specific point in the macro where the error wss detected, as well as the point in the document where the macro was expanded. Here is a simple `\chapter` macro with an error:

```
\newcount\chapnum
\def\chapter{\advance\chapnum by1 {\bf\chapnum:}}
```

The expansion 'pre error text \chapter post error text' will result in the message:

```
 Missing number, treated as zero.
 <to be read again>
                     :
 \chapter ... by1 {\bf \chapnum :
                                  }
 1.123 pre error text \chapter
                               post error text
 ?
```

This means that `\chapter` was expanded on line 123, and the error was discovered after the ':' in the macro was scanned. The error, of course, is not a missing number but a missing `\the`.

This also shows that TEX does not always correctly identify the real cause of an error.

In case of nested macros, an error in an inner macro causes several pairs of lines to be displayed, each showing the point where a macro was expanded by another macro. The error '`\footline{\count0\hfil}`' is a good example. The resulting message consists of several pairs of lines showing the points where `\footline` was typeset by \plainoutput, where \plainoutput was expanded by the document's output routine, and where the routine was expanded in the document.

If such listings occur often and are unnecessarily long, their size can be limited by setting the parameter \errorcontextlines to a small value [34]. The rule is that the top and bottom pairs are always shown, along with additional \errorcontextlines pairs in between.

10. The text 'pre error text \footline{left \smallskip right} post error text' is bad since it places a vertical command inside \footline. We assume that the document is short, so the output routine is invoked by the \bye at the end of the document. If we set \errorcontextlines=0, we get the short error message

```
 Missing } inserted.
 <inserted text>
                 }
 ...
 l.3 \bye
 ?
```

where the dots indicate that some of the message was suppressed. Setting \errorcontextlines to a larger value, however, produces a much more detailed message:

```
1   Missing } inserted.
2   <inserted text>
3                   }
4   <to be read again>
5                         \vskip
6   \smallskip ->\vskip
7                         \smallskipamount
8   <inserted text> left \smallskip
9                           right
10  \makefootline ...\the \footline
11                        }
12  \plainoutput ...y \makefootline
13                          }\advancepageno \ifnum \outpu...
14  <output> ...\Hsize \plainoutput
15                          }
16  \supereject ...r \penalty -\@MM
17
18  \bye ->\par \vfill \supereject
19                                \end
20  l.3 \bye
21  ?
```

Line 20 shows that the error was discovered when \bye was read on input line 3. Lines 18–19 show the definition of \bye and indicate that the error was discovered after the \supereject, but before the \end, was read. Line 16 shows part of the definition of \supereject and the point in \supereject where the error was discovered.

After reading the entire message from bottom to top, we finally realize that the error was discovered when \bye caused the expansion of the output routine, which, in turn, expanded \plainoutput, which itself expanded \makefootline, which typeset \the\footline, which expanded \smallskip.

11. Line 14 in the previous example contains the notation <output>. This indicates that an error was found while the output routine was executed. Another indicator, <write>, becomes part of the error message if the error occurred while writing on an output file. The following:

```
\newwrite\abc
\immediate\write\abc{\xyz}
```

results in

```
 Undefined control sequence.
<write> \xyz
             <inserted text>
                   }\endwrite
1.2 \immediate\write\abc{\xyz}
?
```

▶ **Exercise 15.4:** Is it possible to have both indicators in one message?

A third indicator, <insert>, is shown when an error is found in material that has been inserted in response to another error.

Here is a list of a few less common errors.

1. Use of macro doesn't match its definition [§398].
2. Runaway argument (paragraph ended before argument was complete) [§396].
3. Huge page cannot be shipped out [§641].
4. The following box has been deleted [§992].
5. Insertions can only be added to a \vbox [§993].
6. Improper discretionary list [§1121].
7. Improper alphabetic constant [§442].

15.3 Fatal Errors

No user response can help TEX recover from such errors. It is barely able to mutter a few words before it succumbs [§93].

Examples of fatal errors (see also [299]) are

1. One hundred errors have been discovered since last paragraph [§82].
2. End-of-file on the terminal [§71].
3. TEX capacity exceeded [§94]. This happens when the capacity of any of the many data structures in memory [300] is exceeded. Many times this simply indicates infinite recursion.
4. A \dump is attempted inside a group [§1304].
5. A hardware problem or a bad copy of the software is used [§95]. The message is '! This can't happen' or 'I can't go on meeting you like this'.

6. TeX is running in a nonstop mode and no \end is found at the end of the input file [§360].

7. TeX is running in a nonstop mode and an attempt is made to \read from the terminal [§484].

8. TeX is running in a nonstop mode and the document tries to open a non-existent file [§530].

9. Two alignment preambles are interwoven [§789].

15.4 General Notes

■ The user should realize that certain errors may be detected by a preview program or by the printer driver, not by TeX. These programs may not always issue good error messages and may not provide for any user response. A common example is a missing font. The bitmap font files are only used by the printer driver, not by TeX. If such a file is missing, the printer driver may substitute another font without issuing as much as a warning. Another example is a bad \special. Something like \special{Psotscript ...}, where 'Postscript' is misspelled, will not be understood by the printer driver and may cause any kind of message, or none at all.

■ Sometimes, on long or slow runs, we want to know what TeX is doing at a certain point. One way of doing this is to place \message commands at strategic points (e.g., the start of each chapter). Another way is to interrupt the program. This should be possible on any interactive TeX implementation. Once interrupted, the program displays the current line number and waits for the usual user response.

■ When things go wrong and no error message is issued, it is advisable to use the tracing capabilities of TeX. They are described mostly on [301–303] and can be very useful.

■ If you suspect that your input file may be "dirty," you may want to set \pausing=1 [303]. TeX will stop after reading each input line and will give you a chance to edit it. (This editing will not affect the original file, however.)

A man whose errors take ten years
to correct is quite a man.
— *J. Robert Oppenheimer*

Make us adore our errors; laugh at's, while we strut to our confusion.
— *William Shakespeare, Anthony and Cleopatra*

16. Output Routines

The notation OTR is used for output routine. It is the subject of the next several chapters.

"It would be possible to write an entire book about TeX output routines; but the present appendix is already too long." This quotation, from [400], best describes these chapters. Its aim is to do justice to OTRs, the justice denied them in *The TeXbook* because of lack of space.

These chapters are an improved version of the four OTR articles that appeared in *TUGboat* in 1991 and 1994. The present chapter is an introduction to OTRs and related concepts, followed by examples. The examples are mostly simple OTRs, useful for common applications. No advanced techniques are used. In Chapter 17, various techniques are developed for communicating with the OTR by means of marks, special penalty values, kerns, and special boxes inserted in the text. Some methods require several passes, saving information between the passes either on a file or in memory. Some techniques examine the contents of \box255, going as far as breaking it (or a copy) up into individual components. Whenever possible, the techniques are applied to practical cases. Chapter 18 illustrates techniques for communicating with the OTR from H mode. Chapter 19 treats insertions. OTRs with insertions introduce more complexities and deserve detailed treatment. Specifically, the `plain` format OTR is introduced in section 19.11 since it supports insertions.

Advanced TeX users hardly need be convinced that an understanding of OTRs is important, since OTRs must be used whenever special output is desired. However, the entire topic of OTRs has traditionally been considered complex, and it is! The reasons are as follows: (1) OTRs are asynchronous with the rest of TEX (this is explained in the Introduction) and involve difficult concepts such as splitting boxes and insertions. (2) Certain features that could be very useful in OTRs are not supported by TEX. Specifically, there are no commands to identify marks, rules, and whatsits in a box, and to break up a line of text into individual characters. One can only hope that in the future, with more interest in TEX and more demand from users, typesetting programs will be developed, upwardly compatible with TEX, with the missing features included.

Finally, it is my pleasant duty to thank those people who helped improve these chapters by suggesting OTR problems or approaches to solutions, by carefully reviewing the manuscript, and by pointing out errors and omissions. They are Robert F. Batzinger, Ronald F. Whitney, Victor Eijkhout, Dominik Wujastyk, Malcolm Clark, Jôrgen Pind, Mary McClure, Lothar Meyer-Lerbs, Amy Hendrickson, Chris Carruthers, Daniel Comenetz, and Chris Impens.

⏵ Readers are encouraged to send me problems related to OTRs, suggestions for new techniques, and errors found here.

16.1 Introduction

A few introductory concepts are introduced in this section for the benefit of the inexperienced reader.

Boxes. Definition: A box is an indivisible unit of material, inside which TEX switches to one of the internal modes. Thus in an \hbox, TEX is in restricted horizontal (RH) mode, which means that items placed in such a box will be positioned side by side. If an \hbox is typeset, it is not broken into lines, but is typeset as one line. Similarly, items placed in a \vbox are stacked from top to bottom. When such a box is typeset, the entire box is placed on the same page. More information on the properties of boxes can be found in [Chs. 11–12].

Lists. A *list* is best thought of as a bunch of boxes (and other items) positioned either horizontally or vertically. Thus TEX supports horizontal and vertical lists. The specific items that may appear in a horizontal list can be found on [95]; those that may appear in a vertical list can be found on [110]. An example of a horizontal list is the TEX logo. Its components are the two letters 'T', 'X', a box with the letter 'E', and two pieces of \kern. An example of a vertical list is a page of text. Its components are the individual lines of text and the penalties, glue, and other items between them.

The main loop. This is also known as the *inner loop*, the *chief executive*, or the *main control* (see [§1029, §1030, §1035]). This is where TEX spends most of its time, preparing pages of text that are eventually sent to the OTR. This loop consists of reading characters from the source file, scanning them and converting them into tokens, using the tokens to build boxes, and combining the boxes into lists.

16.2 The MVL and the Page Builder

▸ This section introduces a simple picture of a structure called the *main vertical list* (MVL). The precise way the MVL is organized and used is presented in section 16.3.

When TeX reads the first character of a paragraph, it switches to horizontal mode (H mode), where it reads the rest of the paragraph. It then switches back to vertical mode (V mode) and invokes the line-breaking algorithm. The resulting lines are saved in the MVL. Gradually, more and more lines of text are appended to the MVL and, at a certain point, when the MVL contains more than enough material for a page, TeX invokes (exercises) the *page builder*. The page builder decides on a good point to break the page, cuts a chunk of material off the MVL, places it in \box255, and invokes the OTR. The OTR usually adds things such as a page number, a header, or footnotes, and ships out the page (i.e., writes the page to the .dvi file).

The OTR does not have to ship the page out. It can ship out just part of it, and save or discard the rest. It can also return the rest to the MVL. The rule is that any material left over by the OTR (i.e., any material placed on the OTR's vertical list) is returned to the MVL. This is a very useful feature.

An important point, which should be emphasized at this early stage, is the *asynchronicity* of the OTR. \box255 is *not* gradually filled up with text. It is the MVL that's filled up with lines of text until there is more than enough material for a page. The page builder then cuts a chunk, the size of a page, off the MVL and places it in \box255. At that point, there is normally some material left in the MVL, for the next page. This means that when the OTR is invoked, TeX has already read text from the source file past the end of the current page. Thus it creates the asynchronous nature of the OTR. It is not synchronized with the main loop but is invoked from time to time, as the need arises. Another way to express this idea is to say that the OTR lags behind TeX's main loop.

Note that TeX does not know the size of the paper that eventually comes out of the printer. There are only four parameters that relate to the size and position of the page, namely: \hsize, \vsize, \hoffset and \voffset. The first two become the width and height of \box255. The two "offset" parameters are the amounts by which the left and top margins differ from the default offsets of the printer that typesets the .dvi file. Many drivers default to offsets of 1in, and \hoffset and \voffset give displacements from that point. Thus setting \hoffset=0.5in, \voffset=-0.2in creates a left margin of 1.5in and a top margin of 0.8in. The right and bottom margins are unknown to TeX and are normally only used by the printer driver. In the rare cases where TeX needs to know those quantities, they have to be entered by the user or computed. This is demonstrated in one of the examples (for printer registration marks) shown in section 16.8.

Figure 16.1 shows how those quantities are related. It also shows that the depth of \box255 is the depth of the bottom line of text. As a result, the vertical size of the printed area is the height plus the depth of \box255 and is slightly larger than \vsize.

▸ \pagetotal **and** \pagegoal. Two \dimen registers are maintained by the page

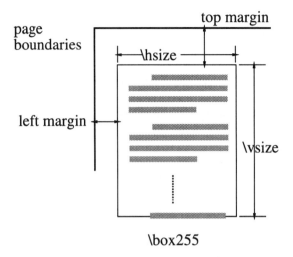

Figure 16.1. Page Boundaries and the Printed Area

builder and are used in the page-breaking algorithm. They are also used for insertions.

\pagetotal [114], is the vertical height of the MVL. We will denote it by t. This parameter starts at zero and is incremented by the page builder each time something with a height (i.e., a box or some glue) is appended to the MVL. Things like marks, penalties, and whatsits do not have any dimensions and do not affect t. Since some vertical glues have flexibility, this parameter is generally flexible. Initially, when the MVL is empty, t is set to zero.

\pagegoal [114], which we denote by g, is the desired height of the page. Generally, it is equal to \vsize but, when insertions are generated, g is decremented. Initially, when the MVL is empty, g is set to \maxdimen. When the first item is placed in the MVL, g is set to \vsize ("...TEX salts away the value of \vsize ..." [114]). When footnotes or other material are generated, to be inserted in the page, their heights are subtracted from g [123].

These two parameters can be "shown," either in the document itself (using \the) or in the log file (using either \showthe or \message). They can even be modified, but this should be done very carefully. It is even possible to get the two values displayed after each line by setting \tracingpages=1 [112]. This option exists mainly to show the feasible points for page breaks and how the page-breaking algorithm works. It is a good idea, however, to try it once, on a short page that also has footnotes, and to follow the changes in the values of the two quantities.

As an example, the two parameters will be used to determine how much space is left on the current page. If t is zero, the space left on the page is the entire page (\vsize). Otherwise, it is the difference $g - t$. Macro \pagespace calculates that difference.

\newdimen\spaceleft

```
\def\pagespace{%
  \ifdim\pagetotal=0pt
    \spaceleft=\vsize
  \else
    \spaceleft=\pagegoal
    \advance\spaceleft by -\pagetotal
  \fi}
```

16.3 The Current Page and LRC

The discussion so far has been general, ignoring certain important details. This section presents a more accurate picture of the MVL and the way the page builder operates.

The MVL consists of two parts, the *current page* and, below it, the *list of recent contributions*. The current page holds the material that will become \box255. The recent contributions are used to temporarily hold recently read material. After an entire paragraph has been read, it is typeset, and the lines of text appended to the recent contributions. At that point, the *page builder* is invoked (exercised). Its job is to move lines, one by one, from the recent contributions to the current page. For each line, the page builder calculates the cost of breaking the page after that line. For the first couple of lines the cost is very high because breaking there would result in an extremely stretched page. Thus, for those lines, the badness b becomes 10,000 and the cost c, 100,000 (see formula on [111]).

At a certain point, when there are enough lines in the current page for a resonably looking page, b (and, as a result, c) starts getting smaller. A while later, there may be too many lines of text in the current page, and it has to be shrunk, increasing b and c again. The entire process can be seen, in real time, by setting \tracingpages=1 [112]. If the page has to be shrunk more than its maximum shrinkability, both b and c become infinite. When c becomes infinite (or when a penalty $\leq -10,000$ is found, see page 300) the page builder goes back to the line of text where the cost was lowest, breaks the top of the current page at that point, and places it in \box255. The bottom part of the current page is then returned to the recent contributions, and the page builder invokes the OTR.

The page builder is exercised at the end of a paragraph, at the end of a display equation within a paragraph, at the end of an \halign, and in a few other cases [122, 286]. The OTR is invoked only by the page builder [§1025], which is why it is never invoked in the middle of a paragraph (unless the paragraph contains display math material).

The advanced reader might want to glance at [§980–1028] for the actual code of the page builder.

Since the page builder is exercised quite often, the list of recent contributions is usually small or empty, and the current page gets larger and larger. When the OTR is invoked, the current page is empty. The \showlists command can always be used to display the two parts of the MVL in the log file.

The quantity t (\pagetotal) mentioned earlier as the height of the MVL is, actually, the height of the current page. It is updated by the page builder each time a line (or glue) is added to the current page.

A better understanding of the page builder and the MVL must include glue and penalties. When a paragraph is typeset, the lines of text are appended to the recent contributions, with glue and penalty items between them. These items are moved to the current page with the lines of text. If the current page is empty, all glues, kerns, and penalties moved to it are discarded. This is how discardable items disappear at the top of a page. When the first box is moved to the current page, a special \topskip glue is placed above it, to keep its baseline 10pt (the value in plain format) below the top of the page. Following that, glue, kern, and penalties are moved, with the text, from the recent contributions to the current page.

When a penalty $\leq -10,000$ is encountered, the page builder breaks a page. If the resulting page does not have enough text lines, it may be underfull. Such penalty values can be used to eject a page (say, by \vfill\penalty-10000), or to communicate with the OTR (which knows the penalty associated with the point at which a page was broken).

It should be stressed, however, that a penalty of $-10,000$ does not invoke the OTR *immediately*. If such a penalty is created inside a paragraph, between lines of text, it is saved in the recent contributions with the lines and is only recognized as special when it is moved, by the page builder, to the current page. As a result, if a paragraph contains

```
...\dimen0=2pt...\vadjust{\penalty-10000}
...\dimen0=1pt...\par
```

the OTR will be invoked after the entire paragraph has been read and broken into lines, and it will find \dimen0 to be 1pt.

A page can only be broken at (i.e., just above) a glue, kern, or penalty. If a page is broken at a glue or kern, the glue stays in the recent contributions (to be discarded when moved to the top of the next page). If the page is broken at a penalty, the penalty is saved in parameter \outputpenalty. This parameter can be used to communicate with the OTR. Also, if the OTR decides to return some material from \box255 to the current page, it may want to place back the original penalty at the breakpoint, by saying \penalty\outputpenalty.

If a page is broken at a point other than a penalty, \outputpenalty is set to 10,000. This is true for TeX version 3.0 and higher [125]. In the older versions, \outputpenalty was set to zero in such a case. The reason for the change is that if a page is broken at a penalty, the penalty must be $< 10,000$, so \outputpenalty cannot normally be set to 10,000. This change, incidentally, means that macros that try to replace the penalty at a page breakpoint should say '\ifnum\outputpenalty=10000\else\penalty\outputpenalty\fi' instead of just \penalty\outputpenalty. (This point was communicated to the author by Daniel Comenetz.)

The precise rules of where a page can be broken are listed on [110]. One of them says, "A page break may occur at glue, provided that this glue is immediately preceded by a non-discardable item." If a set of successive glues is moved to the current page, a page break can occur either before or after that set, but not inside it. If the page builder decides to break the page before the set, the entire set is returned to the recent contributions, to disappear at the top of the next page. If the page is broken after the set, it becomes glue at the bottom of the page. This information will be used in section 17.11, when we try to communicate with the OTR by means of \kern.

The depth of the current page. The discussion so far has been kept simple (although some readers may disagree) by ignoring certain features that have to do with the depths of the boxes involved. These features are important since, normally in a document, successive pages should have the same (or almost the same) vertical size. The height of a page is controlled by (actually, it is equal to) \vsize. The depth of a page should also be under the user's control since, in certain situations, it may spoil the uniform appearance of the document. This is why it is important to consider TeX features that have to do with the depth of a page, and we start by introducing certain quantities that have to do with the depth of vboxes in general.

Consider a large \vbox with lines of text, separated by glue and penalties. The depth of this \vbox [80] is the depth of the bottom component. If that component is a glue or penalty, the depth is zero. If it is a box, then its depth becomes the depth of the entire \vbox, except that it is limited to the value of parameter \boxmaxdepth. If \boxmaxdepth=1pt and the depth of the bottom box is 1.94444pt, then the depth of the entire \vbox will be 1pt and its height will be incremented by .94444pt. This is equivalent to lowering the reference point (or, equivalently, the baseline) of the \vbox by .94444pt. In the plain format, \boxmaxdepth=\maxdimen [348], so it has no effect on the depths of boxes. However, \boxmaxdepth can always be changed by the user.

The *current page* is that part of the MVL that contains the material for \box255 Its current height is t, its goal height is g, but what is its depth? It is, of course, the depth of the bottom line of text—normally a small dimension that may vary a little from page to page. This results in pages with slightly different vertical sizes (i.e., height + depth). However, if the bottom line of text contains a large symbol with a depth of, say, 35pt, the vertical size of the page will be \vsize + 35pt. The page will be much taller than its neighbors, spoiling the uniform appearance of the document. To avoid this, the page builder uses another parameter, \maxdepth, when it appends lines to the current page [125]. The plain format sets [348] \maxdepth=4pt. In our example, when the line with depth = 35pt is added to the current page, the depth of the current page is set to 4pt and the difference of 31pt is added to its height t. A good way to visualize this situation is to say that the baseline of the current page no longer coincides with the baseline of the bottom line, but is located 31pt below it.

The internal quantity \pagedepth (d) contains the depth of the current page, and is updated each time a line (or glue) is added to the current page. d is a "relative" of t, the height of the current page. It should be noted that t has a few more "relatives" [114], the most important of which are \pagestretch and \pageshrink, the amounts of stretchability and shrinkability in the current page. t and its relatives are used by the page builder to determine a page break (some of them are also used for insertions).

A simple experiment can show how those quantities are updated. First, change \parskip to some flexible value such as 1pt plus2pt minus1pt, then typeset text in small pages and place a command such as

```
\message{(total: \the\pagetotal;
 depth: \the\pagedepth;
 shrink: \the\pageshrink;
 stretch: \the\pagestretch)}
```

at the start of each paragraph. It will show the values of the four parameters at the end of the preceding paragraph.

The \message commands will show the value of t growing from paragraph to paragraph, until a page is shipped out by the OTR. The value of d is usually 1.94444pt (the depth of many letters in cmr10) but is different when anything other than cmr10 is used. If the last line of a paragraph happens to contain letters without any depth, d will be zero at the end of that paragraph. Note that d can only be displayed while in v mode, between paragraphs (within paragraphs, \showthe\pagedepth only shows 0pt, not the depth of the last line of the previous paragraph). Each time another \parskip is inserted, between paragraphs, \pageshrink is incremented by 1pt, and \pagestretch, by 2pt.

Controlling the depth of the current page. We have already mentioned that d is limited to \maxdepth. If a text line with a depth $>$\maxdepth is moved to the current page, the depth of the current page (\pagedepth) is set to \maxdepth, and its height t is incremented by the difference (the baseline of the current page is lowered and is now located below the baseline of the bottom line of text).

Here is a simple experiment to clear up this point. First, set

```
\hsize=3.5in \vsize=2in
\tracingpages=1
```

and typeset some text in font cmr10. Most lines of text will have a height of $250/36 \approx 6.94444$pt and a depth of $70/36 \approx 1.94444$pt. The \baselineskip glue between the lines is thus set to 3.1112pt, to achieve a separation of 12pt between consecutive baselines. The last three % lines of TeX's tracing report should be

```
% t=130.0 g=144.54 b=10000 p=0 c=100000#
% t=142.0 g=144.54 b=204 p=0 c=204#
% t=154.0 g=144.54 b=* p=0 c=*
```

This shows that t is incremented by 12pt between lines of text. The last line results in infinite cost, so the page is broken after the line with c=204.

Now repeat the experiment with the depth of the second line increased to 7pt by placing the following box in it \hbox{\vrule depth7pt}. Typesetting the same material now results in different % lines:

```
% t=130.0 g=144.54 b=10000 p=0 c=100000#
% t=145.0 g=144.54 b=10 p=0 c=10#
% t=156.94444 g=144.54 b=* p=0 c=*
```

Figure 16.2 shows the layout of the last three text lines in both cases. In part **a**, the lines all have the same height and depth and are separated with the same size glue. In part **b**, however, the situation is more complex. A \baselineskip of 3.1112pt is inserted between the first and second lines, so their baselines are

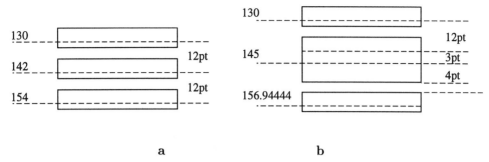

a   b

Figure 16.2. Layout of the Last Three Text Lines

separated, as usual, by 12pt. Since the second line has a depth of 7pt, the value of \pagedepth is set to \maxdepth (= 4pt), and the difference of 3pt is added to the height t. The baseline of the entire page is thus lowered 3pt below the baseline of the second line. Normally, t would be set to $130 + 12 = 142$pt. Instead, its value now is 145pt.

Because of the large depth of the second line, it is separated from the third line by \lineskip [78], which has a plain format value of 1pt. The baseline of the third line is set $4 + 1 + 6.94444 = 11.94444$pt below the baseline of the page, and t is incremented by that amount, to 156.94444pt. d is reset to the depth of the third line, namely 1.94444pt.

Again, appending the third line to the current page has resulted in infinite cost, and it is eventually returned to the recent contributions.

This is part of the overall task of the page builder while constructing the current page.

The height of a box of text. We denote the value of \baselineskip (normally 12pt) by b. A large \vbox with text consists mainly of lines of text, each an \hbox, separated by globs of glue, normally in the (varying) amounts necessary to separate baselines by exactly b, but sometimes just the amount \lineskip. We assume a simple case where no large characters or equations are used. In such a case, all lines of text are separated by b. The height of the box is thus

$$b(n - 1) + \text{the height of the first line}$$

where n is the number of text lines.

The height of \box255. In the case of \box255, enough glue is placed above the first line of text to reach to \topskip from the first baseline. We denote the value of \topskip by h (10pt in plain). So if the baseline of the first line is now h below the top of the page, the height H of \box255 should be $b(n - 1) + h$ (Figure 16.3). However, the height of \box255 is always set, by the page builder, to \vsize. The difference between the two heights is usually supplied by the flexible glues on the page, the most common of which is \parskip.

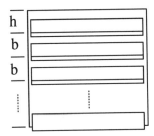

Figure 16.3. The Height of a Page Box

Example: `\vsize=180pt` creates 15-line pages, since $12(15-1)+10 = 178$. The `\parskip` glues on the page are stretched by a total of 2pt. Similarly, `\vsize=189pt` creates 15-line pages, but each page has to be stretched by 11pt.

What if there isn't enough stretchability? In such a case, the bottom of the page remains empty. This is an unusual situation where the height of a box is greater than the sum of the vertical dimensions of its components. Normally such a case is considered an "underfull box" but, in the case of the page builder and `\box255`, "underfull and overfull boxes are not reported when `\box255` is packaged for use by the OTR" [400].

A simple experiment is recommended to clear up this point (note that this experiment uses `\output`, which hasn't been introduced yet; nevertheless it is a useful experiment to perform at this point).

```
\vsize=189pt \parskip=0pt
\output={\setbox0=\vbox{\unvcopy255}
\message{[\the\ht0, \the\ht255;]}
\setbox1=\vbox to\vsize{\hrule width3in
 \vfil\hrule width3in} \wd1=0pt
\shipout\hbox{\box1 \box255}
 \advancepageno}
```

`\vrule height10pt` text to be typeset...

The `\parskip` glue is now rigid, so there is no flexibility on the page at all. The messages will be [178pt, 189pt;]. `\box1` is set to two rules with a 'fil' in between and is superimposed on `\box255`. A look at the typeset page will show that the top rule is placed exactly 10pt above the baseline of the top line, and the bottom rule is well below the bottom line of text.

A better understanding of `\box255` is gained by trying
                    `\setbox0=\vbox to\vsize{\unvbox255}`
Even though both `\box255` and `\box0` have the same height namely, `\vsize`, they don't have the same status. `\box255` may be created underfull without an error message but, when transferred to `\box0`, the destination box becomes underfull, *with* an error message.

The most natural thing to do, in such a case, is to try to fill up \box0 by saying \setbox0=\vbox to\vsize{\unvbox255 \vfill}. Surprise! This may, sometimes, cause an "overfull box." The explanation has to do with the depth of \box0.
Without the \vfill, it is the depth of the bottom line. With the \vfill, it is zero,
and the depth of the bottom line is added to the *height* of \box0, which may cause
it to be overfull.

16.4 Examples

The following sections show how to write an OTR, and they illustrate typical OTRs for common applications. It should again be stressed that the examples
are kept simple and, therefore, are not completely general. They should be read,
understood, and modified for specific needs, rather than copied and used verbatim.

An OTR is simply a sequence of TEX commands assigned to the token-register
\output. Thus \output={...} would cause TEX to execute the commands ...
whenever it decides to invoke the OTR.

The simplest OTR is \output={}. When TEX sees this OTR, it substitutes
the *default* OTR, which is \output={\shipout\box255}. This default OTR is the
simplest one that ships out a page.

\shipout is the TEX primitive that creates a page in the .dvi file. That
page reflects the contents of whatever box follows the call to \shipout. An interesting feature is that \shipout can be invoked anytime, not just from an OTR.
This is demonstrated in the example in section 16.7. Another interesting point is
that \shipout can be redefined (which, of course, is true for any control sequence,
whether a macro or a primitive). One of the methods shown for double-column
pages (page 319) does just that.

The OTR itself can be redefined during a TEX run. The new OTR will be used
when the page builder next invokes the OTR. It is possible to define a macro such as
\def\newotr{...} and assign \output={\newotr} at any time. This will redefine
the OTR. It is even possible to write an OTR that redefines itself! Here is an example:

```
\output={
 \shipout\box255 \advancepageno
 \global\output={
   \shipout\vbox{
     \box255
     \bigskip
     \centerline{\folio}}
   \advancepageno}
 }
```

This OTR typesets the first page without a page number. It then (globally)
redefines itself to typeset the rest of the document with the page number centered
below the text. The reason for the \global is the local nature of the OTR, which
is explained in section 16.4.

It should be noted that the OTR is expected to empty \box255; it can ship
it out, move it to some other box, or return it to the MVL. The latter is done

simply by saying \box255 or \unvbox255 inside the OTR. An OTR that does not do anything with \box255 will cause the error message "unvoid \box255."

The following OTR just empties \box255. \output={\setbox0=\box255}. This does not cause an immediate error message but is probably not what you want to do. It amounts to tossing away the entire document, page by page.

The next example is \output={\uvbox255}. This OTR always returns the page to the MVL, which causes the page builder to immediately find a new page break and invoke the OTR again. Incidentally, the new page break may not be the same as the original one, because of the penalty at the breakpoint. When the breakpoint is chosen, the page builder places the penalty found at the point in parameter \outputpenalty, not in \box255. If the breakpoint has no penalty, \outputpenalty is set to 10,000. The penalty can be returned to its original place by \unvbox255\ifnum\outputpenalty=10000\else\penalty\outputpenalty\fi. This guarantees that the page builder will find the same breakpoint.

An execution of the OTR which does not ship out anything is called a *dead cycle*. Dead cycles have their uses and are illustrated by some of the examples shown later (see, e.g., Approach 2 on page 314). However, many consecutive dead cycles normally indicate an error. This is why TeX counts the number of consecutive dead cycles (in register \deadcycles) and stops the run if \deadcycles $\geq$ \maxdeadcycles. The plain format value of \maxdeadcycles is 25, and it can be changed at any time. Each time \shipout is invoked, it clears \deadcycles.

The page number. In most documents, pages should be numbered. The numbering can only be done by the OTR, since it is only there that the page number is known. The page number can come from any source. Here is an example where the OTR typesets a page number, from a \count register, centered below the printed area:

```
\newcount\pageNum
\output={
  \shipout\vbox{
    \box255\smallskip
    \centerline{\tenrm\the\pageNum}}
  \global\advance\pageNum by1}
```

Note the \tenrm in the preceding example. It is necessary because of the asynchronous nature of the OTR. When the OTR is invoked, TeX can be anywhere on the next page. Specifically, it could be inside a group where a different font is used. Without the \tenrm, that font (the current font) would be used in the OTR.

In the plain format, the \count0 register serves as the page number, and the following two macros are especially useful.

■ \folio typesets \count0 as the page number. However, if \count0 is negative, \folio typesets roman numerals (see section 9.6 for why roman numerals are sometimes used for page numbers).

■ \advancepageno advances the page number by one. This is done by either incrementing or decrementing \count0, according to its sign.

Any \count register can be used to typeset the page number. However, the main advantage of using \count0 is that TeX writes its value on the .dvi file, so a page preview program can easily display the page number with the page. Stated more negatively, driver programs may *only* understand the values of \count0 written into the .dvi file as page numbers and may not be able to selectively print pages whose numbers correspond to some other counter.

Actually, the ten registers \count0...\count9 are used as a (composite) page number. Any nonzero register in this group is written on the .dvi file. Typesetting and advancing any of them must be done explicitly by the user. Macros \folio, \advancepageno only handle \count0.

Grouping and the OTR. The OTR, as mentioned earlier, is a list of tokens. It is also implicitly surrounded by a pair of braces, making it into a group. This means that anything done inside the OTR is *local*, unless preceded by \global. This is a useful feature since, normally, operations within the OTR should be local (i.e., hidden from TeX's usual operations of making paragraphs and putting together math formulas).

16.5 A "Boxed" Page

Here is an OTR for a "boxed" page. It surrounds the page with double rules on all sides, and centers the page number below the double box. Note that the page shipped out is wider and taller than \box255. The value of \hsize in this case is, therefore, not the width of the final page shipped out, but the width of the text lines in \box255.

Macro \boxit typesets text and surrounds it with four rules (see [ex. 21.3]). Parameter #2 is the space between the rules and the text. #1 is a box containing the text.

```
\def\boxit#1#2{%
  \vbox{\hrule
    \hbox{%
      \vrule \kern#2pt
      \vbox{\kern#2pt #1
        \kern#2pt}%
      \kern#2pt\vrule}
    \hrule}}
```

```
\output={
  \shipout\vbox{
    \boxit{\boxit{\box255}9}3
    \medskip
    \centerline{\tenrm\folio}}
  \advancepageno}
```

Figure 16.4 is an example of a small, doubly boxed page.

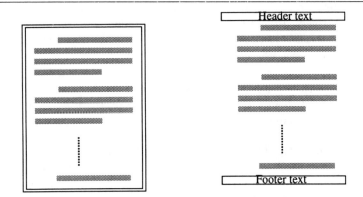

Figure 16.4. A Boxed Page   Figure 16.5. Header and Footer

16.6 Header and Footer

This OTR typesets a header and a footer, both token lists supplied by the user. Typically one of them contains the page number.

```
\output={
  \shipout\vbox{
    \offinterlineskip
    \vbox to3pc{
      \line{\the\headline}
      \vss}
    \box255
    \vbox to3pc{
      \vss
      \line{\the\footline}}}
  \advancepageno}
```

The headline and footline occupy 3pc each, enough for about four lines of text. If the user places more than 3pc worth of material in any of them, there will be no error message (because of the \vss), and the extra material will be typeset on top of the main body of the page. Both \headline and \footline are token lists and may contain \if commands. Examples are

```
\headline={%
  \ifodd\pageno
    \line{\hfil\bf Header\hfil
      \llap{\tenrm\folio}}%
  \else
    \line{\rlap{\tenrm\folio}\hfil
      \bf Header\hfil}%
  \fi}

\footline={\it footer text\hfil}
```

The vertical size of the box shipped out (the printed page) is 6pc plus \ht255 (which is \vsize) plus \dp255 (which is limited to \maxdepth and thus can be kept small).

Figure 16.5 is an example of such a page.

16.7 A Title Page

Sometimes, the book designer specifies a separate title page at the start of each chapter. Here is an example of a \chapter macro that typesets such a page *outside* the OTR. It starts with an \eject, to eject the last page of the previous chapter, then invokes \shipout to ship out a page with the chapter number and name. Note that even though the page number isn't typeset, \chapter still has to advance it.

```
\def\chapter#1 #2;{%
  \vfill\eject
  \shipout\vbox to\vsize{
    \line{\bf Chapter\hfil#1}
    \vfil
    \vbox{\raggedcenter\bf#2}}
  \advancepageno}
```

There seem to be two problems with our simple macro:

1. If the first chapter starts on the first page of the document, our macro will eject a blank page before any text has been typeset.

2. If a page was ejected just before the new chapter started, our macro will eject a blank page.

It turns out that neither of these is a problem. An \eject is essentially a \penalty-10000, and if the very first thing in the document is a penalty, it gets discarded. Also, if the first thing on a new page is such a penalty, it gets discarded. Here is a relevant quote (from [114]), "If you say \eject\eject, the second \eject is ignored, because it is equivalent to \penalty-10000 and penalties are discarded after a page break."

16.8 Printer Crop Marks

We now turn to an OTR that optionally typesets the registration marks, also known as *crop marks*, which are used to align the pages for photography prior to printing and to indicate the size of the final page. The registration marks should be positioned at the four corners of the page, not at the corners of the printed area. The top left mark, for instance, should be located 1in+\voffset above the top of the printed area, and 1in+\hoffset, to the left. Similarly, the top right mark should be placed up and to the right, but by how much?

The problem is that TeX has no idea how wide and tall the paper is. All it knows is the left and top offsets and the dimensions of the printed area (\hsize and \vsize). To place the registration marks properly, the user should specify the dimensions of the paper.

The document should thus start by specifying

```
\newdimen\paperheight
\newdimen\paperwidth
```

```
\paperheight=..in \paperwidth=..in
```

It is also possible, although less desirable, to prompt the user to enter the two dimensions.

```
\newdimen\paperheight
\newdimen\paperwidth
\message{Enter paper height }
 \read-1to\tmp \paperheight=\tmp
\message{Enter paper width }
 \read-1to\tmp \paperwidth=\tmp
```

The next step is to create a \vbox of these dimensions, with the marks at the corners.

```
\newif\iffinalrun \finalruntrue
\newdimen\ruleht \ruleht=.5pt
\newdimen\gap \gap=2pt
\def\verrules{%
  \hbox to\paperwidth{%
    \vrule height1pc width\ruleht depth0pt
    \hfil \vrule width\ruleht depth0pt}
            }
\def\horrules{%
  \hbox to\paperwidth{%
    \llap{\vrule width1pc height\ruleht\kern\gap}
    \hfil
    \rlap{\kern\gap\vrule width1pc height\ruleht}
            }}

\newbox\rmarks \setbox\rmarks=
  \vbox to\paperheight{\offinterlineskip
    \vbox to0pt{\vss\verrules\kern\gap\horrules}
    \vfil
    \vbox to0pt{\horrules\kern\gap\verrules\vss}
                }
```

The dimensions of the box are then set to zero, so it will be superimposed on the printed page '\ht\rmarks=0pt \wd\rmarks=0pt', and the OTR typesets the box (which does not move the reference point), followed by the printed page. This causes the top left corner of the printed page to coincide with the top left registration mark. To center the printed page, it should be lowered and moved to the right.

```
\newdimen\Mdown \Mdown=\paperheight
\advance\Mdown by-\vsize
\advance\Mdown by-6pc \divide\Mdown by 2
\newdimen\Mright \Mright=\paperwidth
\advance\Mright by-\hsize
\divide\Mright by 2
```

```
\output={\shipout\vbox{\offinterlineskip
  \iffinalrun\copy\rmarks\kern\Mdown\moveright\Mright\fi
  \vbox{
    \vbox to3pc{\line{\the\headline}\vss}\box255
    \vbox to3pc{\vss\line{\the\footline}}
        }                }
  \advancepageno}
```

Sometimes the book designer specifies off-center pages. Even-numbered pages should be moved to the right and odd-numbered ones, to the left. This brings facing pages closer together and leaves extra room on the outside margins.

```
...
\divide\Mright by 2
\newdimen\Mleft \Mleft=\Mright
\advance\Mright.5in \advance\Mleft-.5in
\output={
      ...
    \moveright
      \ifodd\pageno\Mleft
      \else\Mright\fi
      ...
```

This off-centering is only done on the final run.

Figure 16.6 shows a small page with registration marks.

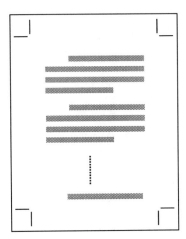

Figure 16.6. Printer Registration Marks

The reader should now realize that three sets of dimensions should be specified when a book is designed and published. (1) The size of the sheet of paper that actually goes into the printer (specified by the human printer). After coming out of the printer, this sheet of paper is trimmed, at the crop marks, to (2) the size of the final page (\paperheight and \paperwidth). Finally, there is (3) the size of the printed area (\hsize and \vsize) on the page.

16.9 A Bordered Page

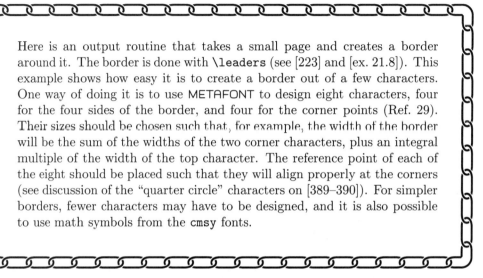

Here is an output routine that takes a small page and creates a border around it. The border is done with \leaders (see [223] and [ex. 21.8]). This example shows how easy it is to create a border out of a few characters. One way of doing it is to use METAFONT to design eight characters, four for the four sides of the border, and four for the corner points (Ref. 29). Their sizes should be chosen such that, for example, the width of the border will be the sum of the widths of the two corner characters, plus an integral multiple of the width of the top character. The reference point of each of the eight should be placed such that they will align properly at the corners (see discussion of the "quarter circle" characters on [389–390]). For simpler borders, fewer characters may have to be designed, and it is also possible to use math symbols from the cmsy fonts.

The border is built in three steps:

1. Macro bordr sets \box2 to a \vbox containing the border.

```
\font\bord=filename
\def\topp{\bord\char0} \def\bott{\bord\char1}
\def\lft{\bord\char2}  \def\rt{\bord\char3}
\def\NW{\bord\char4}    \def\NE{\bord\char5}
\def\SW{\bord\char6}    \def\SE{\bord\char7}}
\newdimen\vtmp

\def\bordr#1,#2;{\setbox0=\hbox{\topp}%
\vtmp=#2 \advance\vtmp-2\ht0
\setbox2=
\vbox to#2{\offinterlineskip
\hbox to#1{\NW\cleaders\hbox{\topp}\hfill\NE}
\hbox to#1{%
\vbox to\vtmp{\leaders\hbox{\lft}\vfil}
\hfil
\vbox to\vtmp{\leaders\hbox{\lft}\vfil}%
}
\hbox to#1{\SW\cleaders\hbox{\bott}\hfill\SE}
}}
```

2. All the dimensions of \box2 are set to 0 by \wd2=0pt \ht2=0pt \dp2=0pt.

3. The output routine now ships out \copy2 (which does not move the reference point), followed by \box255.

```
\output={
  \shipout\vbox{
    \copy2 \vskip.25in
    \moveright.25in\box255}
  \advancepageno}
```

16.10 Mailing Labels

This example introduces the concepts of logical and physical pages. The material placed in \box255 constitutes a *logical page*. The material actually shipped out by the OTR is a *physical page*. Usually one physical page is shipped for each logical page generated. In this example, since the mailing labels are small, several mailing labels are combined and shipped together as one physical page.

The data for the labels is assumed to reside on an external file, named labels, which contains information with a format:

\name the name\\
\address several lines of address\\

The mailing labels are 3.5in wide and 1.5in tall each. There are four labels arranged vertically on a 6in tall page, without any gaps in between. The data file is \input and macros \name, \address are invoked automatically. Three approaches are described:

Approach 1. Macros \name and \address typeset the name and address in the desired format. The value of \vsize is set to the size of four labels. The OTR is thus invoked with a logical page consisting of four labels, and it simply ships it out as one physical page.

```
\nopagenumbers \vsize=6in \hsize=3.5in
\def\name#1\\{\nointerlineskip
  \vbox to.25in{#1 \vfil}}
\def\address#1\\{\nointerlineskip
 \vbox to1.25in{\kern.1in#1 \vfil}\penalty0}
\output={\shipout\box255}
\obeylines\parindent=0pt
\input labels
\bye
```

The \obeylines guarantees that each line in file labels will become a typeset line on the final page. The \nointerlineskip suppresses the normal interline glue that would otherwise be inserted between the boxes in the MVL. The \penalty0 is necessary to supply a valid page breakpoint. The reader will recall that a page can only be broken at a glue, a kern, or a penalty [110]. Without the \penalty0, the page builder would have no place to break the page and would place the entire document, as one page, in \box255 at the end of the job.

Approach 2. Macro \name typesets the name in a \vbox. Macro \address typesets the address in another \vbox. \vsize is set to 1.5in, the size of one label. The
OTR is thus invoked for each label and receives, in \box255, a logical page consisting
of one label. It collects four such pages and ships them out as one physical page.
This approach distinguishes between a logical and a physical page. Note also that,
three out of four times, this OTR goes through a dead cycle.

```
\nopagenumbers \vsize=1.5in \hsize=3.5in
\def\name#1\\{\vbox to.25in{#1 \vfil}}
\def\address#1\\{\nointerlineskip\vbox to1.25in{\kern.1in#1 \vfil}}

\newcount\four \newbox\physpage
\output={\global\setbox\physpage=\vbox{\unvbox\physpage\box255}
  \global\advance\four by1
  \ifnum\four=4 \shipout\box\physpage\global\four=0\fi
      }

\obeylines\parindent=0pt
\input labels
\bye
```

The serious reader should try to understand what happens at the end of the
document. Suppose we have six labels. The first four will be printed on the first
page, and the last two will be accumulated, in \box\physpage, for the second
page. When \bye is found, TeX finds out that \deadcycles ≠ 0. It goes into
its "endgame" [264], where it prepares an empty page and invokes the OTR. This
is repeated twice, accumulating two empty logical pages in \box\physpage, and
then the OTR executes a \shipout, which clears \deadcycles, thereby stopping
the "endgame."

Approach 3. Macro \name typesets the name in a \vbox. Macro \address appends the address to the same box, sets its height to 1.5in, and invokes the OTR.
The OTR again collects four such boxes and ships them out as one physical page.
This approach is interesting since it makes minimum use of the MVL, the page
builder, and \box255. It does not use \vsize, which, consequently, can be set to
any value.

```
\nopagenumbers \vsize=0in \hsize=3.5in
\def\name#1\\{\strut\setbox0=\vbox{#1}}
\def\address#1\\{\setbox0=\vbox to1.5in{%
 \unvbox0\kern.1in #1 \vfil}\eject}
\newcount\four \newbox\physpage \newbox\toss
\output={\setbox\toss=\box255
  \global\setbox\physpage=\vbox{\unvbox\physpage\box0}
  \global\advance\four by1
  \ifnum\four=4
    \shipout\box\physpage \global\four=0
  \fi}
```

```
\obeylines\parindent=0pt
\input labels
\bye
```

We want macro \address to invoke the OTR. However, the only part of TeX that invokes the OTR is the page builder. It does that when it calculates a page break with infinite cost, or when it moves a penalty $\leq -10,000$ to the current page. Macro \address thus says \eject, which is essentially a \penalty-10000. The penalty is placed in the recent contributions, and the page builder is exercised. The page builder tries to move the penalty to the current page, but since the current page is empty, the penalty is discarded [112]. The result is that the OTR is never invoked, and the job terminates without shipping out any pages.

To avoid this situation, macro \name typesets a strut. The strut is moved to the current page and, since the current page is no longer empty, the page builder agrees to move the \penalty-10000 to the current page. This causes the page builder to break the page, place the current page (just the strut) in \box255, and invoke the OTR. The OTR does not need \box255 and simply empties it.

It is possible, of course, to use any text instead of the strut.

16.11 \vsplit

It is important, when working with OTRs, to fully understand the \vsplit operation. Its syntax is '\vsplitbox number to dimen', and the result is a box. Most often it appears in an assignment such as \setbox1=\vsplit0 to2.6in. This sets \box1 to a height of 2.6in, moves material from the top of \box0 to \box1, and keeps the remainder in \box0.

To allow better control of box shape, TeX places the special glue, \splittopskip [124], at the top of the remaining \box0. This glue is similar to h and has a plain format value of 10pt [348]. It can, of course, be easily changed.

The most important thing to keep in mind is that a box can only be split *between* lines of text. If we perform a \vsplit to an "inconvenient" size, \box1 will come out underfull.

Example: \vsize=375pt. This creates, in \box255, a page of 31 lines (since $12(31-1)+10 = 370$). The assignment \setbox0=\vsplit255 to 184pt will set \box0 to a height of 184pt with 15 lines of text. Since 15 lines of text occupy 178pt, the remaining 6pt should be filled with some flexible glue. If there is not enough flexible glue, \box0 will come out underfull. If there are, for example, two paragraphs in \box0, then its total stretchability is 2pt. TeX will stretch it by 6pt and will report an underfull box with a glue set ratio of 3 (300%). The remaining 16 lines in \box255 will now occupy $12(16-1)+10 = 190$pt. Any flexible glues in \box255 will return to their natural size.

Here is an OTR which that the page, ships out the top part, and returns the rest to the MVL (actually, to the recent contributions):

```
output={\setbox0=\vsplit255 to1in
 \shipout\box0 \unvbox255}
```

16.12 Splitting a Box

Imagine a box with n lines of text and with no flexible vertical glues. How can it be split in two equal parts? If n is even, this is easy. However, if n is odd, it is impossible. If H is the height of the box, then $H = b(n-1) + h$. Our approach is to split the box such that the top part will have $m = \lceil n/2 \rceil$ lines, and its height will thus be $H' = b(m-1) + h$. To calculate H', we start with:

$$m = \lceil n/2 \rceil = \begin{cases} n/2, & n \text{ even}; \\ \lfloor n/2 \rfloor + 1, & n \text{ odd}. \end{cases}$$

Since TeX always performs an integer-by-integer division, the case where n is even is simple. It satisfies $m = n/2$ or $m - 1 = \frac{(n-2)}{2}$ and, therefore,

$$
\begin{aligned}
H' = b(m-1) + h &= \frac{b(n-2)}{2} + h \\
&= \frac{b(n-1) + h + h - b}{2} \\
&= \frac{H + h - b}{2}.
\end{aligned}
$$

We therefore split to $H + h - b$. For even n, this is ideal. For odd n, the top part will contain one less line than the bottom. Since we want the top part, in such a case, to be larger, we loop, increasing the split size slightly, until the top part becomes larger than the bottom one.

```
\halfsize=\ht0
\advance\halfsize by\topskip
\advance\halfsize by-\baselineskip
\divide\halfsize by 2
\splittopskip=\topskip
{\vbadness=10000
  \loop
    \global\setbox3=\copy0
    \global\setbox1=\vsplit3 to\halfsize
  \ifdim\ht3>\halfsize
    \global\advance\halfsize by1pt
  \repeat}
```

This method is used on [417] to implement double-column pages (section 16.13). It can also serve to split a box into k equal parts [397]. The principle is to first split the box to size $\lceil n/k \rceil$, then to split the rest recursively.

Next, consider the case where the original box contains flexible vertical glues. This considerably simplifies the problem, since the box can now be split to parts of almost any size, and the glues will be flexed to fill up all the individual parts, eliminating any over/underfull boxes.

16.13 Double-column Pages

Multicolumn pages are common in newspapers and technical publications. Note that, when using narrow columns, the tolerance usually has to be increased, leading to low-quality results (see the experiments in [Ch. 6]).

Three approaches are described. The first one treats the two columns as separate logical pages, which are combined, in the OTR, into one physical page. This approach is described on [257], and is shown here for the sake of completeness. See also [ex. 23.4] for a generalization of this case to three columns.

The second approach [417] creates one long and narrow column and breaks it into two equal parts, which are typeset side by side. The problem of splitting a box into two equal parts has been discussed earlier.

The third approach uses the technique of the second approach to make it possible to switch between one and two columns at will.

Approach 1. The quantity `\lr` is a new control sequence, which is defined and redefined by the OTR depending on the current column (left or right). When the OTR is invoked for the first time (or the third, fifth... times), it saves `\box255` (containing the left column) in another box. These are dead cycles. When it is invoked for the second time (or the fourth, sixth... times), it ships out both columns.

```
\hsize=3.2in
\newdimen\fullhsize \fullhsize=6.5in
\let\lr=L \newbox\leftcolumn
\output={
  \if L\lr
    \global\setbox\leftcolumn=\box255
    \global\let\lr=R
  \else
    \shipout\hbox to\fullhsize{\box\leftcolumn\hfil\box255}
    \global\let\lr=L
    \advancepageno
  \fi}
```

Again, it is important to understand what happens with the last page. If the last page ends somewhere in the left column, the `\bye` invokes the OTR, which saves the left column and returns without shipping out anything (a dead cycle). Since `\deadcycles` is now nonzero, TeX places an empty page in `\box255` and invokes the OTR again. This time it ships a full page, so `\deadcycles` is reset. The last page comes out with unbalanced columns.

It is easy to add a headline and footline to this example:

```
\vsize=2in \hsize=3.2in
\newdimen\fullhsize \fullhsize=6.5in
\let\lr=L \newbox\leftcolumn
\def\fulline{\hbox to\fullhsize}
\output={
  \if L\lr
```

```
    \global\setbox\leftcolumn=\box255
    \global\let\lr=R
  \else
    \shipout\vbox{
      \vbox to3pc{%
        \fulline{\the\headline}\vss}
        \fulline{\box\leftcolumn\hfil\box255}
        \vbox to3pc{\vss\fulline{\the\footline}}}
      \global\let\lr=L
      \advancepageno
  \fi}


\headline{\hfil\bf A Header\hfil}
\footline{\hfil\tenrm\folio\hfil}
```

Approach 2. The OTR is invoked with a long and narrow column. It splits it into two equal parts (see sections 3.27 and 16.11 for \vsplit) and ships out a page consisting of the two parts, laid side by side. The original value of \hsize is saved in \ohsize. \hsize is then set to the width of a single column. \vsize should be set to twice its original value.

```
1  \newdimen\ohsize \ohsize=\hsize
2  \hsize=0.5\hsize \advance\hsize -.1in
3  \newdimen\halfsize
4  \output={
5    \setbox0=\vbox{\unvbox255}
6    \halfsize=\ht0
7    \advance\halfsize by\topskip
8    \advance\halfsize by-\baselineskip
9    \divide\halfsize by 2
10   \splittopskip=\topskip
11   {\vbadness=10000
12     \loop
13       \global\setbox3=\copy0
14       \global\setbox1=\vsplit3 to\halfsize
15     \ifdim\ht3>\halfsize
16       \global\advance\halfsize by1pt
17     \repeat}
18   \shipout\hbox to\ohsize{\box1 \hfil\box3}
19   \advancepageno}
```

On line 5, \box255 is unveiled. This is necessary since \ht255 equals \vsize, but we want to start with a box of height $b(n-1) + h$. Also, for the last page, \box255 may have a large \vfil at the bottom, which should be removed before the split. The right value for the split is calculated on lines 6–9. If the number of lines n is even, the split produces two equal parts. For odd n, the loop on lines

12–17 keeps incrementing the left column until it becomes one line larger than the right one. Both halves are shipped out, on line 18, side by side.

    If the document contains just text, without equations or figures, the following can be used to calculate the best value of \vsize:

```
\newcount\col
\message{Enter number of lines per page: }
\read-1to\ent \col=\ent
\advance\col by-1 \multiply\col by12
\advance\col by10
\vsize=\col pt
```

    It is easy to add a header and footer to the final page. The conscientious reader should do it as an exercise.

Approach 3. Switching between single and double columns. The principles of this method are described here, along with macros taken from the manmac [417]. They are the macros used to typeset the index of *The TEXbook*, which is why they use values such as 14pc and 89pc. The macros are easy to modify for different formats. The reader should also consult (Ref. 38) for two corrections of the macros.

```
\newdimen\pagewidth  \pagewidth=\hsize
\output{\shipout\box255}

\newbox\partialpage
\def\begindoublecolumns{\begingroup
  \output={\global\setbox\partialpage=
    \vbox{\unvbox255\bigskip}}
  \eject
  \output={\doublecolumnout}
  \hsize=14pc \vsize=89pc}
\def\enddoublecolumns{%
  \output={\balancecolumns}\eject
  \endgroup \pagegoal=\vsize}

\def\doublecolumnout
  {\splittopskip=\topskip
  \splitmaxdepth=\maxdepth
  \dimen0=44pc
  \advance\dimen0 by-\ht\partialpage
  \setbox0=\vsplit255 to\dimen0
  \setbox2=\vsplit255 to\dimen0
  \onepageout\pagesofar
  \unvbox255 \penalty\outputpenalty}
\def\pagesofar{\unvbox\partialpage
  \wd0=\hsize \wd2=\hsize
  \hbox to\pagewidth{\box0\hfil\box2}}
\def\balancecolumns
```

```
{\setbox0=\vbox{\unvbox255}
 \dimen0=\ht0
 \advance\dimen0 by\topskip
 \advance\dimen0 by-\baselineskip
 \divide\dimen0 by2 \splittopskip=\topskip
 {\vbadness=10000
  \loop
     \global\setbox3=\copy0
     \global\setbox1=\vsplit3 to\dimen0
   \ifdim\ht3>\dimen0
     \global\advance\dimen0 by1pt
   \repeat}
 \setbox0=\vbox to\dimen0{\unvbox1}
 \setbox2=\vbox to\dimen0{\unvbox3}
 \pagesofar}
```

Macro \begindoublecolumns starts by defining an OTR that saves the page so far (single column) in \box\partialpage. The \eject invokes this OTR. The OTR is then redefined to do double columns by splitting a long, narrow \box255. \enddoublecolumns redefines the OTR to split the page-so-far in two and typeset the two halves, side by side, below the single column in \box\partialpage.

16.14 \raggedbottom

Normally, all pages of a document should have the same vertical size. There are cases, however, where pages of different vertical sizes can improve the appearance of the document. A good example is a document with many short sections. It may be better to leave the current page short and start a new section at the top of the next page.

How can a ragged bottom be implemented? It seems that it's only necessary to place some flexible glue at the bottom of \box255. However, \box255 is constructed by the page builder, not by the user, so it is impossible to insert anything at its bottom. Another idea is to open \box255, in the OTR, and append the flexible glue, something like \shipout\vbox{\unvbox255\vfil}. This, of course, won't work since the \vfil has no reason to stretch to any particular value.

The solution is to use \topskip, since the page builder inserts it into \box255 when the box is being constructed. The trouble is that \topskip is inserted at the top, not the bottom. Simply setting \topskip=10pt plus 5\baselineskip will create pages with a ragged top. However, a flexible \topskip is a step in the right direction, since it creates, in \box255, different amounts of text for different pages. The height of \box255 is still \vsize, but it has a glob of glue at its top.

To transfer the size variation to the bottom, the OTR can simply say '\shipout\vbox to\vsize{\unvbox255\vfil}.' This will reset the glues in the box being shipped. The \vfil will be stretched as much as necessary to bring the height of the box to \vsize. The \topskip, having finite stretchability, will not stretch at all. Note that the box being shipped out has zero depth because its bottom component is glue. The height of the box equals the height plus depth of

\box255. Normally, this poses no problem, but if \box255 has room for exactly n lines, the box shipped out may be overfull.

To understand this point, consider the case where font cmr10 is used and \vsize=142pt. This creates room in \box255 for exactly 11 lines (because $142 = 10 + 11 \times 12$). The height of \box255 is 142pt, and its depth is that of the bottom line. If that depth is positive, then when we open \box255 during the shipout, the resulting \vbox to\vsize will be overfull. The solution is to say '\shipout\vbox to\vsize{\dimen0=\dp255\unvbox255\kern-\dimen0\vfil}'.

If we want a footline below the main body of the text, we can place it below the \vfil by saying

```
\shipout\vbox{
 \vbox to\vsize{\dimen0=\dp255\unvbox255\kern-\dimen0\vfil}
 \baselineskip=24pt
 \line{\the\footline}
}
```

Note that the box being shipped out is made up of two boxes, and its height is greater than \vsize.

The reason for the \baselineskip=24pt is that we want some space between the body of the text and the footline. A \smallskip or a \medskip are inappropriate since they are flexible and will place the footline at different heights on consecutive pages. A \kern24pt is not a good choice either since it will place the 24pt between the bottom of the text and the top of the footline. What we really want is uniform spacing between the bottom of the text and the baseline of the footline, and this is achieved by the \baselineskip=24pt.

This should be compared to macros \pagecontents and \makefootline [256], which are part of the plain OTR.

16.15 Facing Figures

When two figures are textually related, the user may want them typeset on facing pages. The first figure should be typeset at the top of the next even-numbered page and the second one, at the top of the following page.

A macro \facefig#1#2 is defined, with two parameters, the heights of the two figures. It saves the two values in \dimen registers \figa and \figb. The OTR checks the two registers. If the current page number is even and \figa > 0, room is reserved at the top of the current page for the first figure by placing an empty box on the MVL, followed by \box255. If the current page number is odd, \figa = 0 and \figb > 0, the OTR reserves room for the second figure in a similar way. In either case, the OTR goes through a dead cycle.

```
\newdimen\figa \newdimen\figb \newif\ifdead
\def\facefig#1#2{\figa=#1 \figb=#2}
\output={
  \deadfalse
  \ifodd\pageno
    \ifdim\figa=0pt \ifdim\figb>0pt \message{b}
      \vbox to\figb{}\unvbox255
```

```
      \penalty\outputpenalty
      \global\figb=0pt \deadcycles=0 \deadtrue
    \fi\fi
  \else
    \ifdim\figa>0pt \message{a}
      \vbox to\figa{}\unvbox255
      \penalty\outputpenalty
      \global\figa=0pt \deadcycles=0 \deadtrue
    \fi
  \fi
  \ifdead\else
    \shipout\box255
    \advancepageno
  \fi}
```

The simple macros above only use one pair of registers to save the heights of
the figures. In practice, the user may expand \facefig before the OTR has handled
the previous pair of figures. This will place new values in our registers before the
old ones have been used. Our macros should, therefore, be extended so that any
number of pairs of heights can be saved. The following macros save such pairs, as
\kern values, in a \vbox.

```
\newbox\save
\newdimen\figa \newdimen\figb
\newif\ifdone \donetrue \newif\ifdead
\def\facefig#1#2{%
  \setbox\save=\vbox
    {\kern#1 \kern#2 \unvbox\save}}
\output={
  \deadfalse
  \ifvoid\save\else
    \ifdone
    \global\setbox\save=
      \vbox{\unvbox\save
        \global\figb=\lastkern \unkern
        \global\figa=\lastkern \unkern}
    \global\donefalse
  \fi\fi
  \ifdone\else
    \ifodd\pageno
      \ifdim\figa=0pt \ifdim\figb>0pt
        \message{b=\the\figb;}
        \vbox to\figb{}
        \unvbox255
        \penalty\outputpenalty
        \global\figb=0pt \deadcycles=0
        \deadtrue \global\donetrue
```

```
      \fi\fi
    \else
      \ifdim\figa>0pt \message{a=\the\figa;}
        \vbox to\figa{}
        \unvbox255
        \penalty\outputpenalty
        \global\figa=0pt
        \deadcycles=0 \deadtrue
      \fi
  \fi\fi
  \ifdead\else
    \shipout\box255
    \advancepageno
  \fi}
```

Another boolean variable, \ifdone, is declared. It is set to 'false' when two values are extracted from \box\save, and to 'true' when the two figures have been typeset. As long as it is 'false', we are in the process of typesetting two facing figures, and no new values are extracted.

▶ **Exercise 16.1:** The following extensions are left as an exercise:

1. Macro \facefig should check to make sure none of its parameters exceeds \vsize.

2. \facefig should also accept the captions of the two figures, as additional parameters, and save them. The OTR should later retrieve and typeset the captions below (or above) the reserved areas.

Note! Our macros do not use insertions and are therefore incompatible with \midinsert and its relatives. Using both \midinsert and \facefig, the figures would be inserted in an unpredictable order.

16.16 Selective Page Shipout

The following code, part of the manmac format, can be used to produce only a subset of pages. The numbers of the desired pages should be placed on separate lines in a file called pages.tex.

The first line saves \shipout, which is a primitive, in macro \Shipout. Later, \shipout is redefined as either \Shipout or \Tosspage.

```
\let\Shipout=\shipout
\newread\pages \newcount\nextpage
\openin\pages=pages
\def\getnextpage{%
  \ifeof\pages\else
    {\endlinechar=-1\read\pages to\next
     \ifx\next\empty % we should have eof now
     \else\global\nextpage=\next\fi}%
  \fi}
\ifeof\pages\else\message
 {OK, I'll ship only the requested pages!}
```

```
\getnextpage\fi

\def\shipout{%
  \ifeof\pages\let\next=\Shipout
  \else\ifnum\pageno=\nextpage
    \getnextpage
    \let\next=\Shipout
  \else\let\next=\Tosspage\fi\fi
  \next}

\newbox\garbage
\def\Tosspage{\deadcycles=0\setbox\garbage=}
```

16.17 End of the Document

How can the OTR find out if the page it has been given is the last one? The easiest way is to detect the \vfill at the bottom of that page. This can be done by:

```
\def\vfill{\vskip 1sp plus 1fill}
\output={
  \setbox0=\vbox to\vsize{
    \unvcopy255
    \ifdim\lastskip>0pt \message{last page}\fi}
  \shipout\box255
  \advancepageno}
```

(The \lastskip command is explained on page 329.) This usually works but may fail in cases where the last page is full, or almost full. An example is \vsize=1in, which leaves room for about six lines of text on the page. Let's assume that the document has text for six lines, and the last line contains a deep character, such as a ']', whose depth is 2.5pt. Setting \tracingpages=1 generates the following in the log file:

```
%% goal height=72.26999, max depth=4.0
% t=10.0 g=72.26999 b=10000 p=0 c=100000#
% t=22.0 plus 1.0 g=72.26999 b=10000 p=150 c=100000#
% t=34.0 plus 1.0 g=72.26999 b=10000 p=100 c=100000#
% t=46.0 plus 1.0 g=72.26999 b=10000 p=100 c=100000#
% t=58.0 plus 1.0 g=72.26999 b=10000 p=150 c=100000#
% t=70.0 plus 1.0 g=72.26999 b=1168 p=0 c=1168#
% t=72.5 plus 1.0 plus 1.0fill g=72.26999 b=* p=-20000 c=*
```

The cost of breaking the page after the first six lines is 1168, very low. The page builder, however, waits for a point with infinite cost before it decides on a page break. It continues reading the source and finds the \bye. The definition of \bye is [357] \par\vfill\penalty-20000\end. The page builder adds the \vfill to the current page, causing the depth of the current page to become zero. The depth of the bottom line (2.5pt) is now added to the height of the current page, with the result that it is too high (72.5pt). The \vfill is therefore removed, and the last page is shipped out without any fill at the bottom.

splat: Canonical name for an output routine that outputs whatever the local interpretation of "splat" is.

— *Eric Raymond (ed.), The New Hacker's Dictionary*

17. OTR Techniques: I

The following techniques are discussed in this chapter and are applied to practical situations:

- Breaking up `\box255` in the OTR into individual lines by means of the `\last`*xx* commands.
- Identifying individual lines or paragraphs to the OTR by means of `\rightskip`, `\parshape`, or the depth of `\box255`.
- Attaching very small amounts of `\kern` to certain lines of text, to identify those lines to the OTR as special.
- Placing large negative penalties at certain points in the document. This has the effect of invoking the OTR at those points. The OTR does not have to ship out anything.
- Attaching very small vboxes below certain lines, to identify them to the OTR as special lines that require special treatment.
- Using marks. This is a common OTR technique.
- Setting `\vsize` to a very small value. `\box255` consists, in such a case, of just one line of text, which is then easy to examine.
- Using a 2-pass technique where, in the first pass, certain information is written on a file, to be read by the second pass. Certain complex problems may even call for a multipass job.

17.1 Technique: Special Penalties

Penalties are used in TEX to control line breaks and page breaks, depending on the current mode. Penalties generated in horizontal mode are used by the paragraph break algorithm [§831, §859]. To communicate with the OTR by penalties, they therefore have to be generated in vertical mode. A penalty of 10,000 or more is considered infinite and prevents a page break. Similarly, a penalty of −10,000 or less always causes a break. The idea is to say \penalty-10001 at any point that requires the OTR's attention (but TEX must be in vertical mode at that time) in order to invoke the OTR at this point. It is easy to define a macro \def\immed{\vadjust{\penalty-10001}} to do just this. The OTR should check \outputpenalty, and if it equals −10,001, do something special. It can then ship out \box255 or return it to the current page.

This is a good method for communicating with the OTR, and it has only one feature that makes it less than ideal; the special penalty value of −10,001 does not invoke the OTR *immediately*. Instead, it is initially placed in the recent contributions, together with the rest of the paragraph, and has to wait until the page builder is exercised. The problem is that when the page builder is exercised and the OTR invoked, TEX has already read text past the special penalty.

In an experiment such as ..\dimen0=1pt...\immed...\dimen0=2pt..\par the OTR would find \dimen0 to have a value of 2pt.

▶ **Exercise 17.1:** Write an OTR that displays the value of \dimen0, and perform the experiment above.

The reason for this behavior is the way \vadjust is executed. TEX first breaks the entire paragraph into lines that are placed in the recent contributions. Only then does it place the \vadjust material at the proper point between two lines [259]. As a result, the OTR is invoked too late.

To solve this problem, a way should be found to exercise the page builder immediately. The page builder is exercised (see [117]) at the start and at the end of a paragraph; so, if the user wants to invoke the OTR at the end of a paragraph, a \penalty-10001 is the ideal technique. The page builder is also exercised before and after a display formula, which suggests a way to exercise it inside a paragraph. The user should place, in the paragraph, a \penalty-10001, preceded by an empty display formula, at the point where the OTR should be invoked.

An empty formula is easy to create by $$ $$. Furthermore, the large flexible glues surrounding a display are easily eliminated by

```
\abovedisplayskip=1sp
\belowdisplayskip=1sp
\abovedisplayshortskip=1sp
\belowdisplayshortskip=1sp
```

To eliminate any extra interline spaces around the display, we place the command \openup-\baselineskip in it. Finally, a \postdisplaypenalty=-10001 places the special penalty right below the display formula, to make sure that the OTR is invoked.

The result is made into a new definition of macro \immed:

```
\def\immed{$$\postdisplaypenalty=-10001
 \openup-\baselineskip$$}
```

The expansion \immed terminates the current line (same as \hfil\break), places an empty, invisible display formula following the line, and *immediately* invokes the OTR with \outputpenalty=-10001. The paragraph is not terminated.

To see the point where the formula is placed, \immed can be temporarily changed to

```
\def\immed{$$\postdisplaypenalty=-10001
 \openup-\baselineskip+$$}
```

In a test such as ..\dimen0=1pt...\immed...\dimen0=2pt..\par, the OTR would find \dimen0 to have a value of 1pt.

This method is, again, not ideal, since it terminates the current line.

17.2 The \lastxx Commands

The OTR can examine the contents of \box255 and also break it up into its components, by means of the four \lastxx commands [§424, §996]: \lastbox, \lastskip, \lastkern, and \lastpenalty [271]. To use those commands, the OTR should first open \box255, by means of an \unvbox. If the last item in \box255 is a glue, its value will be reflected in \lastskip. Two things can be done at this point: (1) \skip0=\lastskip; (2) \unskip. The first saves the glue value for future use, and the second removes it [280]. This is similar for \lastkern and \lastpenalty. If the last item is a box, the command \setbox0=\lastbox will both set \box0 *and* remove the last box. To quote from [278] "it (\lastbox) refers to (and removes) the last item of the current list, provided that the last item is an hbox or vbox."

17.3 Technique: Breaking up a Page

➮   The OTR may use the \lastxx commands in a loop, to identify successive components of \box255. In such a loop it is, of course, important to check at each iteration and find out what the next item is, before copying and removing it. If the next item is not a glue, \lastskip will have a value of 0pt. Similarly, \lastkern will be 0pt, \lastpenalty will be 0, and \lastbox will be void. A macro \breakup can thus be defined, consisting of a \loop...\repeat to remove successive elements off \box255.

```
\newif\ifAnyleft \newcount\pen
\def\breakup{%
  \loop \Anyleftfalse
    \ifdim\lastskip=0pt\else \Anylefttrue
     \skip0=\lastskip \unskip \fi
    \ifdim\lastkern=0pt\else \Anylefttrue
     \dimen0=\lastkern \unkern \fi
    \ifnum\lastpenalty=0 \else\Anylefttrue
     \pen=\lastpenalty \unpenalty \fi
    \setbox0=\lastbox
    \ifvoid0 \else \Anylefttrue \fi
```

```
\ifAnyleft \repeat}
```

Note the use of variable `\Anyleft` to check if there is anything left in the box after each repetition of the loop. The loop repeats until none of the four items is found. The OTR simply says `\unvcopy255 \breakup`.

An alternative definition of `\breakup`, using nested `\if`s, is

```
\newif\ifAnyleft
\def\breakup{%
  \loop \Anyleftfalse
    \ifdim\lastskip=0pt \ifdim\lastkern=0pt \ifnum\lastpenalty=0
      \setbox0=\lastbox \ifvoid0       % end of breakup loop
    \else \Anylefttrue \fi             % box encountered
    \else \Anylefttrue \unpenalty \fi  % penalty encountered
    \else \Anylefttrue \unkern \fi     % kern encountered
    \else \Anylefttrue \unskip \fi     % glue encountered
  \ifAnyleft \repeat}
```

Before discussing specific applications of the breakup technique, let's look at its main problems.

1. We have to test `\lastskip` for 0pt. Unfortunately, TeX does not have `\ifskip` or `\ifglue` tests. We thus have to use `\ifdim`, which tests a dimension, not a glue. The test `\ifdim\lastskip`... first converts the glue to a dimension. The problem is that such a conversion discards the stretch and shrink components of the glue [118]. Thus if the next glue item has the form 0pt plus.. minus.., our macro will consider it zero.

The solution: change the values of certain common vertical glues that have this form to 1sp plus... minus... . We thus declare

```
\parskip=1sp plus1pt
\def\vfil{\vskip1sp plus1fil}
\def\vfill{\vskip1sp plus1fill}
\abovedisplayshortskip=1sp plus3pt
```

2. A similar problem exists with penalties. A math display formula is followed by a `\postdisplaypenalty` [189], whose default value is zero. As a result, any construct using the math display mode, such as `\verbatim` or `$$\vbox{\halign{...}}$$`, suffers from the same problem. The solution is to set `\postdisplaypenalty=1`.

There is also an `\interlinepenalty` parameter, which goes between the lines of a paragraph. It is usually zero but can be changed to a large value [406] to discourage a page break inside a paragraph. We set it to 1.

The above definitions are all consolidated into a new macro `\zeroToSp`, which should be used in conjunction with any page breakup.

```
\def\zeroToSp{\parskip=1sp plus1pt
  \def\vfil{\vskip1sp plus1fil}
  \def\vfill{\vskip1sp plus1fill}
  \abovedisplayshortskip=1sp plus3pt
  \postdisplaypenalty=1
```

```
\interlinepenalty=1}
```

3. When breaking up a box using the \last*xx* commands, it is easy to identify the four types: box, glue, kern, and penalty. There seems no way, however, to identify the other three components of vertical lists, namely *rules*, *marks*, and *whatsits*. When our breakup loop gets to one of them, it stops, assuming that this is the end of \box255. A whatsit (a \special or a \write) can usually be specified in horizontal mode, which will bury it inside an \hbox and out of harm. A mark, on the other hand, tends to migrate outside horizontal lists [400] and into the top level of \box255. It therefore causes an incomplete breakup, and its use should be avoided when this technique is employed.

A similar problem is presented by a rule. An \hrule at the top level of a \vbox is considered a box [110]. However, the \lastbox operation cannot identify it as such, which results in an incomplete breakup.

A solution: Place the \hrule in its own \vbox, so it does not appear at the top level of the larger \vbox.

Partial relief: Such a case, where the breakup stops prematurely, can be detected by setting a new box (\brk) to the remains of \box255 after the breakup. When the breakup stops, \ht\brk should be zero. An OTR can thus be written which breaks up a copy of \box255 and checks to see if anything is left.

```
\newbox\brk
\output={
  \setbox\brk=\vbox{\unvcopy255 \breakup}
  \ifdim\ht\brk>0pt
    \message{Incomplete breakup}\fi
  \shipout\box255 \advancepageno}
```

▶ **Exercise 17.2:** Implement the above OTR and use it to typeset several pages, some containing rules or marks.

Here are a few simple applications of the breakup technique.

17.4 Duplicating a Page

Macro \breakup can be modified to place broken up components from \box255 in \box1 in the *original order*, creating a copy of the current page.

```
\zeroToSp
\newif\ifAnyleft \newcount\pen
\def\duplicate{%
 \loop \Anyleftfalse
  \ifdim\lastskip=0pt \ifdim\lastkern=0pt \ifnum\lastpenalty=0
    \global\setbox0=\lastbox \ifvoid0 % end of breakup loop
  \else \Anylefttrue                  % box present
    \global\setbox1=\vbox{\box0 \unvbox1} \fi
  \else \Anylefttrue                  % penalty present
    \pen=\lastpenalty
    \global\setbox1=\vbox{\penalty\pen\unvbox1} \unpenalty\fi
  \else \Anylefttrue                  % kern present
```

```
    \dimen0=\lastkern
    \global\setbox1=\vbox{\kern\dimen0 \unvbox1} \unkern\fi
 \else \Anylefttrue                    % skip present
    \skip0=\lastskip
    \global\setbox1=\vbox{\vskip\skip0 \unvbox1} \unskip\fi
 \ifAnyleft \repeat}
```

A test such as

```
\newbox\brk
\output={
  \setbox\brk=\vbox{\unvcopy255 \duplicate}
  \ifdim\ht\brk>0pt
    \message{Incomplete breakup}\fi
  \shipout\box255 \shipout\box1
  \advancepageno}
```

is particularly interesting. It typesets pairs of pages, with the same page numbers. Two *physical pages* are printed for each *logical page* generated. The two pages in a pair are duplicates of each other, but are they identical?

It turns out that they are not. The main difference between \box255 and \box1 is their heights. The heights are different because of the flexible glues on the page. Normally, \box255 contains some flexible vertical glues. Those glues are flexed to adjust \ht255 to equal \vsize. When \box255 is opened, however, *the glues return to their natural size.*

This can easily be seen by an experiment such as the following:

```
\newbox\brk
\output={
  \setbox\brk=\vbox{\unvcopy255 \duplicate}
  \ifdim\ht\brk>0pt
    \message{Incomplete breakup}\fi
  \message{[\the\ht255:\the\ht1]}
  \shipout\box255 \shipout\box1
  \advancepageno}

\parskip=6pt plus6pt minus6pt
\input source
\bye
```

The \parskip glue is given a lot of flexibility, and the heights are shown in the log file. Such a test also shows that the last pair of pages may differ a lot in their heights. This is because the last page of a document is normally only partly full and has a \vfill glue at the bottom. When \box255 is opened, the \vfill returns to its natural size, which is 0pt.

How can we make sure that the two pages in a pair have the same heights? The simplest approach is to flex \box1 in the OTR, just before it is shipped out, by saying \setbox1=\vbox to\vsize{\unvbox1}. Now the two pages in a pair have exactly the same height and the same glue set ratio; they are identical. Our OTR thus becomes

```
\newbox\brk
\output={
  \setbox\brk=\vbox{\unvcopy255 \breakup}
  \ifdim\ht\brk>0pt
    \message{Incomplete breakup}\fi
  \setbox1=\vbox to\vsize{\unvbox1}
  \shipout\box255 \shipout\box1
  \advancepageno}
```

Two `\showbox` commands can temporarily be placed in the OTR to dump `\box1` and `\box255` onto the log file and to verify that they have identical components. It is important to (temporarily) increase the value of `\showboxbreadth`. Also, to make the dumps more manageable, `\vsize` should be set to a small value, such as 1in.

17.5 Reversing a Page

It is now trivial to modify the definition of `\duplicate` so that it breaks up items from `\box255` and places them in `\box1` *in reverse order*. This is, perhaps, a useless operation, but since our aim is to gain an understanding of OTRs, let's ask ourselves how `\box255` and `\box1` differ.

1. They are the reverse of each other, which means that each glob of `\base-lineskip` glue that used to be below a line of text is now above it. The interline spacing in `\box1` is thus all wrong. This is not very noticeable when the entire page is typeset with the same font. Mixing different font sizes, however, results in a funny-looking reversed page. Also, the `\parskip` glues are misplaced, but since they are normally zero, this is not noticeable. Changing `\parskip` to some nonzero value results in large spaces following the first line of each paragraph (which are last lines on the reversed page).

2. They have different vertical dimensions. The height of `\box255` is `\vsize`, and its depth is usually the depth of the last line of text. `\box1`, on the other hand, ends with `\topskip`, which is glue and thus has no depth, so `\dp1=0pt`. Also, `\box1` starts with the bottom line of `\box255`. To guarantee that `\ht1+\dp1` equals `\ht255+\dp255`, we should force `\ht1` to be the sum `\ht255+\dp255`.

▶ **Exercise 17.3:** Write a macro `\reversepage` to reverse `\box255` into `\box1`.

17.6 Counting the Lines

The `\breakup` macro can now easily be modified to count the number of lines of text in `\box255`. We assume that `\box255` does not contain rules, marks, or whatsits, and we break it up, counting the number of `\hboxes` found. Items that we don't want to count should be placed in a `\vbox`. The macros are

```
\zeroToSp
\newif\ifAnyleft \newcount\lineCount
\def\countlines{\global\lineCount=0
```

```
  \loop \Anyleftfalse
    \ifdim\lastskip=0pt \ifdim\lastkern=0pt \ifnum\lastpenalty=0
      \setbox0=\lastbox \ifvoid0
    \else \Anylefttrue \ifhbox0 \global\advance\lineCount by1 \fi \fi
    \else \Anylefttrue \unpenalty \fi
    \else \Anylefttrue \unkern \fi
    \else \Anylefttrue \unskip \fi
  \ifAnyleft \repeat}
```

```
\newbox\brk
\output={\setbox\brk=\vbox{\unvcopy255 \countlines}
        \ifdim\ht\brk>0pt \message{Incomplete breakup} \fi
        \message{\the\lineCount}
        \shipout\box255 \advancepageno }
```

17.7 Breaking up a Line of Text

Can we use the same technique to break up individual lines of \box255? It
seems easy to define a macro \hbreakup that would use \lastxx commands to
break up a line of text. Unfortunately, this does not work, because a line of text
contains individual characters, which the \lastbox command cannot recognize as
boxes. It is interesting to note that a character of text is, in general, considered a
box [63], but evidently there are differences between a general box and a character
box. One such difference is that a character box cannot appear in a vertical list
[110]. Another difference is the one mentioned above, concerning \lastbox, and
this difference is easy to verify with a test such as

```
\setbox0=\hbox{ABC}
\unhbox0 \setbox1=\lastbox
\showbox1
\bye
```

which shows \box1 to be void and typesets 'ABC'. In contrast, the test

```
\setbox0=\hbox{AB\hbox{C}}
\unhbox0 \setbox1=\lastbox
\showbox1
\bye
```

shows \box1 to consist of an hbox with the 'C' and typesets only 'AB'.
This is an unfortunate situation. The ability to break up a line of text would
have meant a full and complete communication with the OTR. The user could have
hidden, say, a strut with a special depth in the line, and the OTR could have easily
found it and done something with, or added something to, the line at that point.
The strut could even have been left in the line.

17.8 Technique: Using \rightskip

Even though \lastbox cannot be used to break up a line of text, \lastskip can be used to detect glue at the right end of such a line. This suggests a way to identify certain lines to the OTR. How can a glob of glue be placed at the end of a line? It turns out that TEX places the \rightskip glue at the end of every line of text when the paragraph is broken into lines. The plain format value of \rightskip is 0pt, so setting \rightskip=1sp will not be visually noticeable and can be used to communicate with the OTR. Unfortunately, "TEX uses the same \rightskip value in all lines of a paragraph" [393]. This method can thus be used to identify certain paragraphs, but not individual lines, to the OTR.

An application demonstrating this technique is shown in section 17.26. It has to do with "special boxes" in a textbook. Following are two examples that are not developed in detail, since they are easier to do in other ways:

1. Suppose that a vertical rule should be typeset on the left margin of certain paragraphs. The OTR can do this by placing a rule, the size of a strut, on the left of each line that ends with \rightskip=1sp. However, this is easier to do by typesetting the paragraph in a \vbox and placing a rule on the left of the box.

2. If only one or two lines of the paragraph appear on (the bottom of) the page, we want to move them to the next page and to \vfill up the present one. This can be done by the OTR checking the \rightskip glue of the bottom line or two. However, it may be easier to do with \filbreak [111].

17.9 Technique: Using \parshape

If we want the OTR to do something special with, say, the second line of a paragraph, we can identify this line by making it 1sp longer or shorter than the other lines. This can easily be done with \parshape. Again, there are no practical applications as yet for this technique.

17.10 The Depth of \box255

The following quote, from [400], is relevant to this technique: "Perhaps the dirtiest trick of all is to communicate with the OTR via the *depth* of \box255." After mastering the techniques described here, the reader will agree that this is no longer the dirtiest trick, but is a special case of the breakup technique. Examples of applications of this technique are the following:

■ In certain religious texts, if a chapter ends on a certain page and less than half a page remains, the next chapter should start on the following page; otherwise, it should start on the same page.

■ Business contracts usually consist of clauses. In certain legal situations it is desirable to break a page between clauses. If the page must be broken inside a clause, a special footer should be typeset, saying Continued.... This can be done by ending each clause with \endclause, a macro defined as {\unskip\vrule height0pt width0pt depth3.5002pt}. The \unskip backspaces over any possible space preceding the special strut, thus making sure that the strut will end up on the same line as the preceding word.

The OTR simply tests

```
\ifdim\dp255=3.5002pt \else
\footline={\hfil\sevenrm Continued...}
\fi
```

■ Certain lines should not appear at the bottom of the page. A business contract
is again a good example. If a certain line contains the most important words or
money sums in the contract, it is better that it not appear at the bottom of the
page, where it is less visible. Again, a special strut can be used to identify the line,
and if the OTR detects such a line, it should alert the user, who can then correct
the situation by rewording the document, or by moving things around.

Beware, certain businessmen do just this!

17.11 Technique: Communication by Kern

Small amounts (a few sp worth) of `\kern` can be placed between the lines of
text and will be detected by the OTR when breaking up `\box255`. The problem is
that a kern is discardable, so we have to make sure that our special kern is not
discarded. The general rule is that a page can be broken at a kern only if the kern
is immediately followed by glue. Therefore, we will have our special kern followed
by another kern. In fact, we will place two consecutive, identical pieces of special,
small kern after the text line that we want to identify to the OTR. This is done
by `\vadjust{\kern1sp\kern1sp}`, which places the kerns immediately below the
current line, that is, they are placed between the line and the `\baselineskip` that
normally follows it. If the line should be followed by a penalty, the order is the line
of text, the pair of kerns, the penalty, the `\baselineskip`. The places where a page
can be broken have been mentioned earlier.

17.12 Practical Examples of OTRs

The techniques described earlier, plus a few others, are now applied to practical
problems.

17.13 Example: Start a Chapter

Proposed by Robert Batzinger

The problem: If a chapter ends on a certain page and less than half a page
remains, skip the rest of the page; otherwise, start the new chapter on the same
page.

Solution: Macro `\chapter` is expanded at the start of each chapter. It appends
a special line to the end of the preceding chapter (only if there is a preceding chapter)
and invokes the OTR by an `\eject`. The special line consists of just a small `\hbox`
with a rule of depth 1sp, and width and height zero.

Each time the OTR is invoked, it checks to see if the following two conditions
are satisfied: (1) `\dp255=1sp`; (2) `\ht255<0.5\vsize`. If yes, the OTR returns
`\box255` to the MVL (it is an end-of-chapter and more than half a page remains);
otherwise, `\box255` is shipped out (either less than half a page remains or is not
end-of-chapter).

Actually, the details are a bit more involved. If `\dp255=1sp`, then `\box255`
contains text, followed by a `\vfill`, and by the special box. The last two items
have to be removed before `\ht255` can be tested. To do this—

1. `\box255` is opened, the special box at the bottom is removed by a `\lastbox`
(see later), and the `\vfill` is skipped over by an `\unskip`. The result then goes

back in \box255. The new \box255 now has just the original text, and its height
can be measured.

2. If \ht255<0.5\vsize, \box255 is opened, and a message (unv) goes in the
log file. Otherwise, a new box is shipped, consisting of \box255, a \vfill, and
a footline. The size of the new box is \vsize+12pt, and it has to be explicitly
specified.

A listing of the macros follows. As usual, they are kept simple.

```
\newif\ifFirstCh \FirstChtrue \newif\ifRet \newcount\chnum

\newdimen\Hvsize \Hvsize=\vsize \divide\Hvsize 2
\newdimen\Nvsize \Nvsize=\vsize \advance\Nvsize 12pt

\def\chapter#1\par{\advance\chnum 1
                \ifFirstCh \FirstChfalse
                \else \vfill\nointerlineskip
                        \hbox{\vrule width0in height0pt depth1sp}
                        \eject \fi
                \bigskip\noindent{\bf\the\chnum.\ #1}
                \medskip}

\footline={...}

% if \dp255=1sp: unvbox255, lastbox (the line with dp=1sp),
%    skip over the \vfill by \unskip, and return to MVL.

\output={\Retfalse
        \ifdim\dp255=1sp
          \setbox255=\vbox{\unvbox255 \setbox0=\lastbox \unskip}
          \ifdim\ht255<\Hvsize \Rettrue \fi \fi
        \ifRet \unvbox255 \message{unv}
        \else
          \shipout\vbox to\Nvsize{\box255\vfill\line{\the\footline}}
          \advancepageno \message{ship} \fi
        }
```

17.14 Example: A Religious Hymn

One way of communicating with the OTR, proposed in [App. D], is by the use
of special penalty values. Any penalty value $\leq -10,000$ will cause the OTR to be
invoked. Values $< -10,000$ can therefore be used to tell the OTR to do something
special.

Note that the OTR is not invoked when TeX first sees the penalty. It is only
invoked when the page builder detects the penalty while moving items from the
recent contributions to the current page [§1005].

The OTR should check the value of parameter \outputpenalty. If that value
$< -10,000$, it should do something special and then return \box255 to the MVL

Proposed by
Robert
Batzinger

without shipping out anything (a dead cycle). If, however, \outputpenalty equals
$-10,000$, the OTR should do a normal \shipout.

The example shown here has to do with typesetting a religious hymn. A hymn
consists of one chorus and a number of stanzas. The chorus is usually printed after
the first stanza and is sung after each stanza. The problem is that a long hymn
may occupy more than one page. In such a case, the chorus should be printed at
the top of each successive page.

The solution is to write macros that will typeset a copy of the chorus if we are
still within the same hymn but have moved to a new page. The original text of the
chorus is saved in a \toks register, so it can be used as often as necessary.

Macro \hymn is expanded at the start of each hymn. It invokes the OTR with
penalty $-10,001$ and the OTRsimply saves the current page number in the count
register \oldpage. Note that the OTR does not ship out anything and returns
\box255 to the MVL.

Macro \stanza is expanded at the start of each stanza. It invokes the OTR with
\penalty-10002. The OTR then tests \ifnum\pageno>\oldpage (we have moved
a page or two since the last printing of the chorus) and sets the boolean variable
\prtCorus to true. The OTR then returns \box255 to the MVL. Macro \stanza
tests \prtCorus and, if it is true, invokes macro \setchorus to typeset the chorus.

Here are the macros used:

```
\newif\ifprtCorus \newcount\oldpage
\hsize=3.5in \vsize=2.2in

\def\hymn#1#2{\bigbreak\bigskip
             \noindent{\bf #1. #2}\nobreak\medskip
             \nobreak \penalty-10001}

\def\stanza#1\endstanza{\medbreak
                \vbox{\noindent#1}\medskip\penalty-10002
                \ifprtCorus \setchorus\fi }

\def\chorus#1\endchorus{\toks2={#1}\setchorus}

\def\setchorus{\medskip
            \moveright.5in\vbox{\noindent
                            \hbox to 0pt{\hss\bf Chorus:\ }%
                            \the\toks2\medskip}
                \global\prtCorusfalse }

\output={\ifnum\outputpenalty=-10001
            \global\oldpage=\pageno
            \global\prtCorusfalse
            \unvbox255
         \else \ifnum\outputpenalty=-10002
           \ifnum\pageno>\oldpage
```

```
    \global\prtCorustrue \global\oldpage=\pageno \fi
    \unvbox255
\else
    \shipout\vbox{\box255\smallskip \line{\the\footline}}
    \advancepageno
\fi \fi }
```

Note that this is just a demonstration of a principle. The macros presented here are simple and will not always work. One case where they fail is when a hymn starts at the end of a page and the chorus is typeset on the following page. The chorus will, in such a case, be typeset twice on that page. There may be other problems, but the idea in this book is to keep the macros simple and easy to read.

▶ **Exercise 17.4:** Generalize the above macros so that they do not typeset the chorus on an odd-numbered (right-hand) page if it was typeset on the preceding even-numbered (left-hand) page. This way the chorus would be typeset only once on a pair of facing pages.

17.15 Example: Line Numbering

When writing a draft of a book, a thesis, or a report that should be reviewed by someone else, it is useful to number the lines on each page (see Figure 17.1). This way the reviewer can easily refer to, say, *line 48, page 84*. The numbers should be placed in the margin so that they can be suppressed in the final version without any changes in the layout of the document.

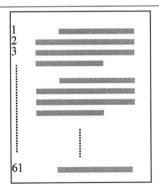

Figure 17.1. Line Numbers on the Margin

The method used here counts the number of lines of text by counting the boxes that make up the page. Macro \countlines below assumes that each box on the page is a line of text and should be numbered. Alternatively, if certain items on the page should not be numbered, they can be placed in vboxes and \countlines revised to count only hboxes.

\zeroToSp

```
\newif\ifAnyleft \newcount\lineCount
\def\countlines{%
  \global\lineCount=0
  \loop \Anyleftfalse
    \ifdim\lastskip=0pt \ifdim\lastkern=0pt \ifnum\lastpenalty=0
      \setbox0=\lastbox \ifvoid0
    \else \Anylefttrue \global\advance\lineCount by 1 \fi
    \else \Anylefttrue \unpenalty \fi
    \else \Anylefttrue \unkern \fi
    \else \Anylefttrue \unskip \fi
  \ifAnyleft \repeat}
```

Note that in an \halign, each line becomes an \hbox and is, therefore, counted
separately. Also note that a blank line preceding a display equation becomes an
empty paragraph and is therefore counted.

The OTR breaks up a copy of the page, removing the lines of text one by one.
At the same time, a new box, \box1, is built, from the bottom up, with the line
numbers in the margin. For each line removed from the page, its height and depth
are measured, and a line with the same size, containing the appropriate number, is
added to the top of \box1. Each glue or kern removed from the bottom of the page
is added to the top of \box1. At the end, the height and depth of \box1 are set to
zero and it (\box1) is typeset, superimposed on the original page.

```
\newcount\pen
\def\breakup{%
  \loop \Anyleftfalse
    \ifdim\lastskip=0pt \ifdim\lastkern=0pt \ifnum\lastpenalty=0
      \global\setbox0=\lastbox \ifvoid0 % end of breakup loop
    \else \Anylefttrue \appendline \fi
    \else \Anylefttrue \pen=\lastpenalty
          \global\setbox1=\vbox{\penalty\pen \unvbox1} \unpenalty \fi
    \else \Anylefttrue \dimen0=\lastkern
          \global\setbox1=\vbox{\kern\dimen0 \unvbox1} \unkern \fi
    \else \Anylefttrue \skip0=\lastskip
          \global\setbox1=\vbox{\vskip\skip0 \unvbox1} \unskip \fi
  \ifAnyleft \repeat}

\def\appendline{%
 \setbox2=\hbox{\vrule height\ht0 depth\dp0 width0pt
  \sevenrm\the\lineCount}
 \global\advance\lineCount-1
 \global\setbox1=\vbox{\box2 \unvbox1}}

\newbox\brk
\output={\global\lineCount=0
        \setbox\brk=\vbox{\unvcopy255 \countlines}
        \global\setbox1=\vbox{}
```

```
\setbox\brk=\vbox{\unvcopy255 \breakup}
\ifdim\ht\brk>0pt \message{Incomplete breakup} \fi
\ht1=0pt \dp1=0pt
\shipout\vbox{\moveleft20pt\box1 \box255}
\advancepageno}
```

This example illustrates both the power of the breakup technique and its main problem. The problem is the flexible glues in \box255. They are flexed, by the page builder [§668, §1017], to adjust \ht255 to \vsize. However, when \box255 is opened, for the breakup, the flexible glues return to their natural size.

A partial solution is to reduce, or even eliminate, the flexibility of those glues (mainly \parskip). This, however, handicaps the page builder in its most important task, namely, finding a good point to break a page.

▶ **Exercise 17.5:** Implement an alternative approach to the line-numbering problem. The new approach should build, in \box1, a duplicate of \box255 with the line numbers inserted on the left.

17.16 Example: Footnote Numbering

Proposed by Lothar Meyer-Lerbs

Another practical problem is to number the footnotes in a document by the line number on the page. This problem is solved here several times, using different approaches. Each approach illustrates different OTR techniques and also involves certain difficulties.

The following quote, from *The Chicago Manual of Style* (see also [125]), is relevant. "Since it is impossible to foresee how footnotes will happen to come out in the make-up, it is impracticable to number them from 1 up on each page. The best way is to number them consecutively throughout an article or by chapters in a book." The problem tackled here is much more complicated than the one proposed in the quote and demonstrates the power of OTRs in TeX.

17.17 A Simple but Wrong Approach

The first approach is simple and intuitive. Macro \Nfootnote uses a penalty of $-10,001$ to invoke the OTR prematurely. The macro is expanded from H mode, and it has to place the penalty at the top level of \box255, between lines of text. This is done with \vadjust. The OTR breaks up \copy255 and counts the number of lines in the page so far. It then returns \box255 to the MVL. Macro \Nfootnote again takes over and typesets the footnote with the number calculated by the OTR.

The macros are very simple:

```
\def\Nfootnote#1{%
  \vadjust{\penalty-10001}%
  \footnote{$^{\the\lineCount}$}{#1}}

\zeroToSp
\newbox\brk \newif\ifAnyleft \newcount\lineCount

\def\breakup{%
  \global\lineCount=0
```

```
\loop \Anyleftfalse
  \ifdim\lastskip=0pt \ifdim\lastkern=0pt \ifnum\lastpenalty=0
    \global\setbox0=\lastbox \ifvoid0
  \else \Anylefttrue \ifhbox0 \global\advance\lineCount1 \fi \fi
  \else \Anylefttrue \unpenalty \fi
  \else \Anylefttrue \unkern \fi
  \else \Anylefttrue \unskip \fi
\ifAnyleft \repeat}


\output={\ifnum\outputpenalty=-10001
        \setbox\brk=\vbox{\unvcopy255 \breakup}
        \ifdim\ht\brk>0pt \message{Incomplete breakup} \fi
        \unvbox255
      \else \plainoutput \fi
      }
```

but they don't work! The serious reader should, by now, know the reason. The
\vadjust with the special penalty does not invoke the OTR *immediately*. Instead,
the penalty is placed following the current line. Thus, in the second part of \Nfoot-
note, when it expands \footnote, the OTR has not yet been invoked.

17.18 A 2-pass Method

The idea in the second approach is to modify the OTR so that it writes
\the\lineCount on a file. This leads to a 2-pass job, shown below.

```
\newcount\lineCount \newbox\brk \newbox\sav \newcount\pass


\newread\aux \immediate\openin\aux=\jobname.lin
\ifeof\aux \immediate\openout\aux=\jobname.lin
 \pass=1 \message{pass 1}
\else \pass=2 \message{pass 2} \fi


\newif\ifAnyleft
\zeroToSp


\def\countlines{%
  \global\lineCount=0
  \loop \Anyleftfalse
    \ifdim\lastskip=0pt \ifdim\lastkern=0pt \ifnum\lastpenalty=0
      \global\setbox0=\lastbox \ifvoid0
    \else \Anylefttrue \global\advance\lineCount by1 \fi
    \else \Anylefttrue \unpenalty \fi
    \else \Anylefttrue \unkern \fi
    \else \Anylefttrue \unskip \fi
  \ifAnyleft \repeat}


\output={\ifnum\outputpenalty=-10001
```

```
        \ifnum\pass=1
          \setbox\brk=\vbox{\unvcopy255 \countlines}
          \ifdim\ht\brk>0pt \message{Incomplete breakup} \fi
          \immediate\write\aux{\the\lineCount} \fi
        \unvbox255 % return to MVL
      \else \plainoutput \fi
      }
% shipout with footnotes

\def\Nfootnote#1{%
  \ifnum\pass=1 \vadjust{\penalty-10001}\footnote*{#1}%
  \else \read\aux to\tmp \footnote{$^{\tmp}$}{#1}\fi}
```

In the first pass, macro \Nfootnote creates the special penalty and also expands \footnote*{...} to typeset the footnote, so it takes the right amount of space on the page. In the second pass, the macro reads the correct number off the file and invokes \footnote with that number. This is still simple and usually works.

It may fail, however, in cases where a footnote appears close to the bottom of the page. Imagine a footnote on line 60 of page 4. Because of the penalty-10001 following this line, TEX will invoke the OTR with a 60-line page. The OTR will write the line count, 60, on the file and then return \box255 (with the 60 lines) to the current page, removing the special penalty. Since the current page is now large, TEX immediately starts looking for a good page break. It may decide, since the special penalty isn't there any more, to break the page after line 59. Line 60 thus becomes line 1 of the next page, but the number 60 has already been written on the file.

17.19 Another 2-pass Solution

The third approach is similar except that instead of being written on a file, the line numbers are saved—by the OTR—in memory. This makes sense since there usually aren't many footnotes on any single page. In the second pass, macro \Nfootnote uses this information to expand \footnote with the correct line numbers. This approach suffers from the same problem as the previous one, but it is shown here because it illustrates how to save the line numbers, each as an \hbox, in a large \vbox. Extracting them later is easily done with a \lastbox.

Note that the 2-pass structure is different from the previous one. Previously, each pass was a separate TEX job, and the line numbers were saved on a file between the jobs. In the present method, however, the line numbers are saved in a box (\sav), which is stored in memory and thus disappears at the end of the job. The two passes must, therefore, be done in the same job. This is faster but requires the source text to be \input from a separate file.

```
\newcount\lineCount \newbox\brk \newbox\sav \newif\ifAnyleft
\zeroToSp

\def\breakup{%
  \global\lineCount=0
```

```
    \loop \Anyleftfalse
      \ifdim\lastskip=0pt \ifdim\lastkern=0pt \ifnum\lastpenalty=0
        \global\setbox0=\lastbox \ifvoid0
      \else \Anylefttrue \ifhbox0 \global\advance\lineCount1 \fi \fi
      \else \Anylefttrue \unpenalty \fi
      \else \Anylefttrue \unkern \fi
      \else \Anylefttrue \unskip \fi
    \ifAnyleft \repeat}


% pass 1
\output={\ifnum\outputpenalty=-10001
  \setbox\brk=\vbox{\unvcopy255 \breakup}
  \ifdim\ht\brk>0pt \message{Incomplete breakup} \fi
  \global\setbox\sav=\vbox{\hbox{\sevenrm\the\lineCount}\unvbox\sav}
  \unvbox255
  \else \plainoutput \fi}
% The above line should later be changed to:
% \setbox0=\box255 \deadcycles=0,
% since we don't really want to shipout pages in pass 1.


\def\Nfootnote#1{%
  \vadjust{\penalty-10001}%
  \footnote*{#1}}


\input source \vfill\eject \pageno=1


% pass 2
\output={\ifnum\outputpenalty=-10001
          \unvbox255
        \else \plainoutput \fi
        }


\def\Nfootnote#1{%
  \setbox\sav=\vbox{\unvbox\sav \global\setbox0=\lastbox}%
  \footnote{\raise4pt\copy0}{#1}}


\input source
```

17.20 A Complex, 3-pass Approach

Approach 4: A 3-pass job. The first pass determines the line numbers (throughout the document) of lines with footnotes. Those numbers are saved in a \vbox called \Asav. The second pass counts the number of lines on each page. Those numbers are also saved, in another box, \Bsav. The third pass uses the numbers from the two boxes to determine the correct line numbers and to typeset the footnotes. This is complex and can perhaps be done in a simpler way. Nevertheless, it has the advantage of demonstrating several useful OTR techniques.

Before describing the three passes in detail, here is a simple numeric example: Let's assume that we have three pages, with 50, 30, and 40 lines, respectively. There are footnotes on lines 3, 15, 15, and 44 of the first page, and lines 25 and 34 of the third page. Pass 1 will save the numbers 3, 15, 15, 44, 105 and 114 in \Asav (note that 15 occurs twice). In pass 2, the line counts 50, 30, and 40 of the three pages are saved in \Bsav. Pass 3 starts by extracting the 50 from \Bsav. The first four times macro \Nfootnote is expanded, it extracts the numbers 3, 15, 15, and 44 from \Asav. Those numbers ≤ 50, so they are used for numbering the first four footnotes. The fifth expansion extracts 105 from \Asav. This > 50, so the next number, 30, is extracted from \Bsav and added to the 50. The current footnote number, 105, still > 80, so the next number, 40, is extracted from \Bsav and added to the 80. The current footnote number, 105, now ≤ 120, so 80 is subtracted and the result, 25, is used. The last number, 114 again ≤ 120, so again 80 is subtracted, yielding 34.

The steps in each pass are as follows:

Pass 1. Macro \Nfootnote computes a running number for each footnote and creates a \mark with that number. The footnote itself is not typeset, but \Nfootnote typesets an asterisk to occupy space on the line, approximately equal to that taken by the final footnote number. \vsize is set to a small value, so the OTR receives a \box255 with just one line [400], which makes it easy to number the lines throughout the document. Each time the OTR is invoked, it checks \firstmark and \botmark and compares them to \topmark. This way it knows if there are any footnotes on the line. If there are any, the line number is saved in box \Asav, once for each footnote on the line.

```
1  \newcount\temp \newcount\footno \newcount\lineno \newbox\Asav
2
3  \def\Nfootnote#1{\advance\footno 1 \mark{\the\footno}*}
4  % typeset an *
5  \output={\global\advance\lineno 1
6  \temp=\botmark \advance\temp -\firstmark
7  \advance\temp 1
8  \if\firstmark\botmark \if\topmark\firstmark \temp=0 \fi \fi
9  % \temp is now the number of footnotes on the current
10   \ifnum\temp>0 \loop                    % page (one line)
11   \global\setbox\Asav=\vbox{\vskip\lineno sp \null\unvbox\Asav}
12    \advance\temp-1
13   \ifnum\temp>0 \repeat
```

```
14  \fi
15  \setbox0=\box255 % get rid of \box255
16  \deadcycles=0}
17
18  %      *** Executable commands ***
19  \message{Pass 1;} \vsize=10pt % small value
20  \footno=0 \lineno=0 \setbox\Asav=\vbox{}
21  \input source \eject
```

Lines 1–19 are macro definitions, and declarations of registers. Lines 22–24 are the actual commands executed in pass 1. \vsize is set, on line 22, to the small value 10pt. As a result, the page in \box255 will consist of just one line of text.

The OTR calculates \temp, on lines 6–8, as \botmark − \firstmark + 1. \temp is now the number of footnotes on the current page (which consists of just one line of text). However, if \botmark = \firstmark = \topmark, there are no footnotes on the current line, and \temp is set, on line 9, to 0.

If \temp ≠ 0, the loop, on lines 12–15, saves register \lineno on top of \box\Asav as glue (in units of sp).

Pass 2. Macro \Nfootnote typesets each footnote with an asterisk. No marks are used. \vsize is set to its normal value, and the OTR breaks up a copy of each page, counts the number of lines, and saves that number, as the top glue item, in box \Bsav.

```
1   \newif\ifAnyleft \newbox\Bsav \newbox\brk
2
3   \def\Nfootnote#1{\footnote*{#1}}% typeset the footnote so
4   % it occupies the right space on the page
5
6   \def\countlines{%
7   \global\lineno=0
8   \loop \Anyleftfalse
9    \ifdim\lastskip=0pt \ifdim\lastkern=0pt \ifnum\lastpenalty=0
10    \global\setbox0=\lastbox \ifvoid0
11    \else \Anylefttrue
12     \ifhbox0 \global\advance\lineno1 \fi \fi % count hboxes
13    \else \Anylefttrue \unpenalty \fi       % on the page
14    \else \Anylefttrue \unkern \fi
15    \else \Anylefttrue \unskip \fi
16   \ifAnyleft \repeat}
17
18  \output={\setbox\brk=\vbox{\unvcopy255 \countlines}%
19   \ifdim\ht\brk>0pt \message{Incomplete breakup}%
20    \showboxbreadth=1000 \showbox\brk \fi
21   \global\setbox\Bsav=\vbox{\vskip\lineno sp \null\unvbox\Bsav}%
22   \plainoutput}
23
24  % *** Executable commands ***
```

```
25 \zeroToSp
26 \message{Pass 2;} \setbox\Bsav=\vbox{}
27 \vsize=2in %  or any desired value
28 \input source \vfill\eject \pageno=1
```

This is a simple pass. It is again divided into declarations and macro definitions (on lines 1–23) and executable commands (on lines 27–29).

Note the \plainoutput on line 22. This causes pages to be shipped out in pass 2, in addition to the final pages created by pass 3. We end up with two sets of pages that should be identical, except for the footnote numbers. Because of the problem mentioned later, the pages may not be identical, and it is therefore important to compare the two sets before they are printed. When the results are finally printed, the pages created by pass 2 should, of course, be suppressed.

Pass 3. Count register \lineshiped is set to zero. Count register \totalines is set to the first value of \Bsav (50 in our example). \vsize remains at its normal value. The OTR ships out pages in the normal way. Each time \Nfootnote is invoked, it (1) extracts the next item from \Asav into \lineno; (2) if \lineno $\leq$ \totalines, the footnote is created with \lineno $-$ \lineshiped; (3) otherwise, \lineshiped is set to \totalines and the next number is extracted from \Bsav and added to \totalines. Step (2) is repeated.

```
1  \newcount\totalines \newcount\lineshiped
2
3  \def\compare{%
4    \ifnum\lineno>\totalines
5      \global\lineshiped=\totalines
6      \global\setbox\Bsav=\vbox{\unvbox\Bsav \setbox0=\lastbox
7      \global\temp=\lastskip \unskip}%
8      \global\advance\totalines by \temp
9    \expandafter\compare % expand recursively for each
10    \fi}                 % page w/o footnotes
11
12 \def\Nfootnote#1{%
13   \setbox\Asav=\vbox{\unvbox\Asav \setbox0=\lastbox
14   \global\lineno=\lastskip \unskip}%
15   % extract bottom glue into \lineno
16   \compare
17   \advance\lineno -\lineshiped
18   \footnote{$^\the\lineno$}{#1}}
19
20 \output={\plainoutput}
21
22 %  *** Executable commands ***
23 \message{Pass 3;} \lineshiped=0
24 \setbox\Bsav=\vbox{\unvbox\Bsav \setbox0=\lastbox
25   \global\totalines=\lastskip \unskip}
26 \input source
```

27 \bye

Macro \compare, lines 3–10, expands itself recursively to implement the following, pseudocode, loop:

```
while \lineno>\totalines
  \lineshiped:=\totalines
  extract \temp from \Bsav
  \totalines:=\totalines+\temp
end while;
```

The \expandafter on line 9 makes sure that the \fi, on line 10, is gobbled up by TeX before \compare is recursively expanded. Without the \expandafter, the \fi would be saved in a stack and popped out at the end of the recursion. In case of a deep recursion, that could overflow the stack.

The macros are deliberately kept simple and readable and, as a result, are not completely general and don't work in all cases. One such case is where there are no footnotes on the first page; there may be other cases. However, in general, this approach seems to work and seems to have just one small problem. Passes 1 and 2 typeset an asterisk '*' in the body of the text where each footnote should be. This is done to occupy space on the line, space that, in pass 3, is taken by the footnote number. Passes 1 and 2 thus end up with the same line breaks, *but pass 3 may not*. The problem is that footnote numbers, in our case, are one or two digits, and thus may be slightly wider or narrower than the '*'. This may, in rare cases, cause different line breaks in pass 3, leading to wrong footnote numbers.

▸ **Exercise 17.6:** Why is it true that footnote numbers, in our case, can be one or two digits, but not three?

Saving numbers in a \vbox. An interesting point is that our line numbers are saved as *glue* in a \vbox. This is done by \vskip\lineno sp \null. The sp is necessary since, otherwise, the value of \lineno would be converted to scaled points. The \null is an empty \hbox to separate the individual pieces of glue in the large \vbox. This technique can only be used if the total number of footnotes in the document is not too large. For a large number of footnotes, there may not be enough room in memory for our boxes, and a file should be used (in our case, two files).

The actual saving of the count register \lineno in box \Asav is done by:

```
\global\setbox\Asav=
  \vbox{\vskip\lineno sp \null\unvbox\Asav}
```

Extracting the bottom glue item from \Asav is done by:

```
\setbox\Asav=
  \vbox{\unvbox\Asav \setbox0=\lastbox
        \global\lineno=\lastskip \unskip}
```

17.21 Tables Broken across Pages

Proposed
by Mary
McClure Another practical problem: In a document with a lot of tables, many times a table is split over two pages. In such a case, the OTR should typeset "Continued..." at the bottom of the page.

Two approaches are shown, one using marks and the other, special boxes, to communicate with the OTR.

The first approach: A \mark{Continued...} is inserted at the start of each \halign (following the preamble), and a \mark{} is inserted just before the end of the table.

The output routine simply typesets \botmark at the bottom of the page, using the right font. The following macros are used:

```
\output={\shipout\vbox{\box255
  \smallskip\line{\sevenrm\hfil\botmark}
  \smallskip\line{\the\footline}}
  \advancepageno}
\def\beginCont{\mark{Continued...}}
\def\endCont{\mark{}}
```

and a typical table looks like this:

```
\halign{...preamble...\cr \beginCont
...1st line...\cr
...
...last line...\endCont\cr}
```

Note that the first mark becomes part of the first table entry (column 1 row 1). The last mark, similarly, becomes part of the last table entry (last column, bottom row). This means that sometimes the mark may be locked inside an internal box. For instance, if the preamble says $#$, then the mark will be buried in the math box. Generally this creates no problem, but if the mark is buried too deeply in \box255, it may not be discovered [259] during \shipout.

A partial remedy is to use \noalign{\beginCont} or \noalign{\endCont}, depending on which mark is missing during \shipout. This way, the mark precedes (or follows) the entire table. The table, in such a case, should end up with ..last line...\cr\endCont}. These constructs should only be used in an emergency, since they also may fail. A typical example is a table that starts at the top of a page. Its \mark{Continued...} may, in such a case, be the last thing in the preceding page.

An interesting feature of this method is the even page height. Each page shipped out contains a line with the \botmark, and this line occupies the same amount of space on the page, regardless of the size of the mark. Thus if the line preceding the mark has a depth of 1.94444pt, and the mark contains the text Continued... (which has a height of 4.78334pt), the \baselineskip glue is set at 5.27222pt. This separates the baselines by $1.94444 + 5.27222 + 4.78334 = 12$pt. However, if the mark is empty, and the line preceding it has a depth of 0.8333pt, the \baselineskip glue right above the mark is set at 11.1667pt, again separating the baselines by $0.8333 + 11.1667 + 0 = 12$pt.

17.22 Communication by Special Vboxes

The second approach uses a \vbox with a special depth to communicate with the OTR. This looks promising, especially since the \vboxes on both sides of a table can be attached to it by means of a \nobreak (which is essentially a \penalty10000). The implementation is similar to the preceding case.

```
\def\beginCont{\noalign{\vbox{
\hrule width0pt height0pt depth1sp}
\nobreak}}
\def\endCont{\noalign{\nobreak
\vbox{\hrule width0pt height0pt depth2sp}}}
```

Note that the \nobreak in \beginCont *follows* the special \vbox, while that in \endCont *precedes* it.

```
\zeroToSp
\newif\ifAnyleft

\def\breakup{%
  \loop \Anyleftfalse
    \ifdim\lastskip=0pt \ifdim\lastkern=0pt \ifnum\lastpenalty=0
      \setbox0=\lastbox \ifvoid0
    \else \Anylefttrue
        \ifvbox0
          \ifdim\dp0=1sp \Anyleftfalse \global\toks0={Continued...}
          \else\ifdim\dp0=2sp \Anyleftfalse \global\toks0={}\fi \fi
        \fi \fi
    \else\Anylefttrue \unpenalty \fi
    \else \Anylefttrue \unkern \fi
    \else \Anylefttrue \unskip \fi
  \ifAnyleft \repeat}

\newbox\brk
\output={\setbox\brk=\vbox{\unvcopy255 \breakup}
        \ifdim\ht\brk>0pt \message{Incomplete breakup} \fi
        \shipout\vbox{\box255\smallskip
                      \line{\sevenrm\hfil\the\toks0}
                      \smallskip\line{\the\footline}}
        \advancepageno
        }
```

This works! Note, however, that macro \breakup stops when it finds the first special \vbox. In such a case, there is no point in finishing the breakup of \box255. This method therefore generates many "Incomplete breakup" messages, and the user should make sure that the text and the tables should not contain any of the things that normally stop the breakup.

▶ **Exercise 17.7:** A variation of the same problem. Each table is preceded by a header. If the table is broken across pages, the header should be typeset at the top of the second page.

▶ **Exercise 17.8:** Add a parameter to macro \beginCont above. The macro should now create a \vbox whose depth is the value of the parameter, in scaled points. Modify macro \breakup such that it will save different messages in \toks0 depending on the depth of the special boxes found.

17.23 Verse Numbers in the Left Margin

Proposed by
Robert
Batzinger

The problem: In the Bible, each chapter is divided into verses. If a verse starts on a certain line, we want the verse number typeset in the left margin of the line. Also, if two or more verses start on the same line, a range of verse numbers, such as 23–24 should be typeset in the left margin.

Solution: Each verse starts with an expansion of macro \verse. The macro computes the verse number and typesets it in the body of the text. In addition, it uses a \vadjust to generate a special \vbox and to attach it, with a \penalty10000, right below the line of text in \box255. The special box has a height and width of zero and a depth equal to the verse number in scaled points. A line of text can thus be followed by any number of such boxes, and no page break can occur in that area. The verse numbers are stored in the \count registers \fVerse (final verse) and \sVerse (start verse).

The OTR expands macro \breakup, which breaks up \box255 and transfers its components to \box1. On identifying a special \vbox, macro \breakup expands \verseline, which (1) converts the depth of the special box into a count; (2) checks for another special box and converts its depth into another count; (3) removes the line of text above the special boxes, attaches the verse number(s) (via \Label) as an \llap, and adds the result to \box1.

After the breakup is complete, the OTR ships out \box1.

```
\newcount\sVerse \newcount\fVerse \newif\iftwo
\def\verseline{%
  \fVerse=\dp0 \unpenalty
  \global\setbox0=\lastbox
  \ifvoid0 \message{error1;}\fi
  \twofalse
  \ifvbox0 \ifdim\dp0<500sp \ifdim\dp0>0sp \twotrue \fi \fi \fi
  \iftwo
    \sVerse=\dp0 \unpenalty
    \def\Label{\hbox to.4in{\hfil\the\sVerse--\the\fVerse\hfil}}
    \global\setbox0=\lastbox
    \ifvoid0 \message{error2;}\fi
  \else\def\Label{\hbox to.4in{\hfil\the\fVerse\hfil}}\fi
  \global\setbox1=\vbox{\line{%
    \llap{\sevenrm\Label\kern6pt}\box0}\unvbox1}
  }
```

```
\newif\ifAnyleft \newcount\pen \newif\ifverseBox

\def\breakup{%
  \loop \Anyleftfalse
    \ifdim\lastskip=0pt \ifdim\lastkern=0pt \ifnum\lastpenalty=0
      \global\setbox0=\lastbox \ifvoid0 % end of breakup loop
    \else \Anylefttrue \verseBoxfalse
      \ifvbox0 \ifdim\dp0<500sp \ifdim\dp0>0sp
        \verseline \verseBoxtrue \fi\fi\fi
        \ifverseBox \else \global\setbox1=\vbox{\box0\unvbox1}\fi \fi
    \else \Anylefttrue \pen=\lastpenalty
          \global\setbox1=\vbox{\penalty\pen\unvbox1} \unpenalty \fi
    \else \Anylefttrue \dimen0=\lastkern
          \global\setbox1=\vbox{\kern\dimen0\unvbox1} \unkern \fi
    \else \Anylefttrue \skip0=\lastskip
          \global\setbox1=\vbox{\vskip\skip0\unvbox1} \unskip \fi
  \ifAnyleft \repeat}

\newbox\brk
\output={\setbox\brk=\vbox{\unvbox255 \breakup}
        \ifdim\ht\brk>0pt \message{Incomplete breakup} \fi
        \setbox1=\vbox to\vsize{\unvbox1}
        \shipout\box1 \advancepageno}

\newcount\versno \versno=0
\def\verse{%
  \advance\versno by 1
  \hskip1em{\bf\the\versno: }\nobreak
  \vadjust{\nobreak\vbox{%
    \hrule width0pt height0pt depth\the\versno sp}}}

\zeroToSp
\input source
\bye
```

17.24 Problems with This Approach

1. To keep our macros simple, they are limited to at most two verses per line. However, it is easy to generalize \verseline to handle up to three verses. It is also possible, although probably not necessary, to generalize it to handle any number of verses per line.

▸ **Exercise 17.9:** Disregarding the statement above, generalize \verseline to handle any number of verses per line. This requires recursive calls to identify and remove any number of consecutive special \vboxes below a text line.

2. The verse numbers are typeset in the left margin, centered in an \hbox to .4in. This is wide enough for three-digit verse numbers. For larger numbers, it

may be necessary to enlarge that box. If no centering is required, it is enough to
say `\def\Label{\the\sVerse--\the\fVerse}`.

3. The verse numbers always start from 1. It is possible to let the user specify
a start number by:

```
\message{enter start verse number:}
\read16to\ent
\versno=\ent
```

instead of `\versno=0`.

4. The macros recognize a special box if its depth is positive and is less than
500sp. In case of many verses, the 500 should be changed to a larger value. The
following quote (from [400]) is reassuring: "A distance of 1000sp is invisible to the
naked eye."

17.25 Verse Numbers, Another Method

Here is an alternative method that does not transfer components from `\box255`
to `\box1`. It breaks up a copy of `\box255`, and each time a special box (or several consecutive special boxes) is discovered, the macro measures the height of the
remaining copy and uses the height to build, in `\box1`, the range of verse numbers in the margin. When the breakup is completed, `\box1` looks like a skeleton
with just the verse number ranges. The OTR then superimposes the two boxes by
`\shipout\hbox{\llap{\box1}\box255}`. (A similar method is used on [391–392].)

Here are the steps in detail:

```
\newcount\versno \versno=0
\def\verse{\advance\versno by 1
  \hskip1em{\bf\the\versno: }\nobreak
  \vadjust{\nobreak\vbox{%
    \hrule width0pt height0pt
          depth\the\versno sp}}}
```

Macro `\verse` creates a special `\vbox` whose depth equals the verse number
(in scaled points) and attaches it, with a `\nobreak`, below the line where the verse
starts.

The output routine copies `\box255` into `\box2` and expands `\0break` to break
up `\box2` and create the necessary information in `\box1`. It then invokes `\reverse-
box` to break up `\box1` and build, again in `\box2`, the correct skeleton. Both steps
are described below. The final step is to ship out a superposition of `\box2` and
`\box255`.

```
\output={\setbox2=\copy255 \0break
  \ifdim\ht\brk>0pt
    \message{Incomplete breakup}\fi
  \reversebox \setbox2=
    \vbox to\vsize{\unvbox2\vfil}
  \wd2=0pt
  \shipout\hbox{\llap{\box2}\box255}
  \advancepageno}
```

Macro \Obreak expands \breakup to break up \box2, and if another verse (another special box) is found, the height of the remaining \box2 is placed, as \kern, in \box1, and \Obreak expands itself recursively. The process repeats until no more verses are found on the page.

```
\newbox\brk
\def\Obreak{%
  \setbox\brk=\vbox{\unvbox2 \breakup}
  \ifanotherverse
    \global\anotherversefalse
    \global\setbox1=
      \vbox{\unvbox1\kern\ht\brk}
    \setbox2=\box\brk
    \expandafter\Obreak
  \fi}
```

Macro \breakup loops and breaks up items from \box2 until it reaches the end, or until it finds an item that is a \vbox with a depth in the range 0–500sp. If it finds such an item, it expands \verseline.

```
\newif\ifAnyleft
\def\breakup{%
  \loop \Anyleftfalse
    \ifdim\lastskip=0pt \ifdim\lastkern=0pt \ifnum\lastpenalty=0
      \global\setbox0=\lastbox \ifvoid0 % end of breakup loop
    \else \Anylefttrue
      \ifvbox0 \ifdim\dp0<500sp\ifdim\dp0>0sp
      \verseline \Anyleftfalse \fi\fi\fi
      \fi
    \else \Anylefttrue \unpenalty \fi
    \else \Anylefttrue \unkern \fi
    \else \Anylefttrue \unskip \fi
  \ifAnyleft \repeat}
```

Macro \verseline removes the \penalty10000 that precedes the special box, and it checks to see if there is another special box right above it. If there is one, the box and the penalty above it are removed, and the boolean variable \iftwo is set to true.

Next, macro \Label is defined, as an \hbox to.4in{\hfil *one or two verse numbers* \hfil} and is inserted, as an \llap, into the margin of \box1.

```
\newcount\sVerse \newcount\fVerse \newif\iftwo \newif\ifanotherverse

\def\verseline{%
  \fVerse=\dp0 \unpenalty
  \global\setbox0=\lastbox
  \ifvoid0 \message{error1;}\fi
  \twofalse
  \ifvbox0 \ifdim\dp0<500sp \ifdim\dp0>0sp \twotrue \fi \fi \fi
```

```
\iftwo
  \sVerse=\dp0 \unpenalty
  \def\Label{\hbox to.4in{\hfil\the\sVerse--\the\fVerse\hfil}}
  \global\setbox0=\lastbox
  \ifvoid0 \message{error2;}\fi
\else \def\Label{\hbox to.4in{\hfil\the\fVerse\hfil}} \fi
\global\setbox1=\vbox{\unvbox1 \line{%
  \llap{\sevenrm\Label\kern6pt}\hfil}}
\global\anotherversetrue}
```

At the end of the process, the output routine expands \reversebox to break up items from \box1, process them, and place them in \box2 in the correct order. To understand this process, let's imagine a page with three verses at a distance of 2in, 3in, and 6in, respectively, from the top of the page (Figure 17.2). The breakup process starts at the bottom of the page, measures the height A of verse 3, then B, and finally C, creating \box1 as in Figure 17.3.

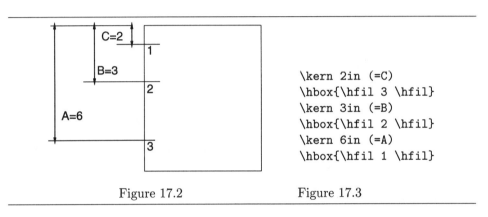

Figure 17.2   Figure 17.3

However, we want a box that looks like Figures 17.4 and17.5, where the \kerns are measured from one verse to the next, not always from the top. We also have to make sure that the lines of text do not take any vertical space, so we add a negative \kern after each line, to skip back to the top of the line.

```
\def\reversebox{\setbox2=\vbox{}
\ifvoid1
\else
 \dimen1=0pt \unvbox1
 \loop
  \dimen0=\lastkern \unkern
  \dimen2=\dimen0
  \advance\dimen0 by-\dimen1
  \dimen1=\dimen2
  \setbox0=\lastbox
  \dimen2=\ht0 \advance\dimen2 by\dp0
  \global\setbox2=\vbox{\unvbox2
```

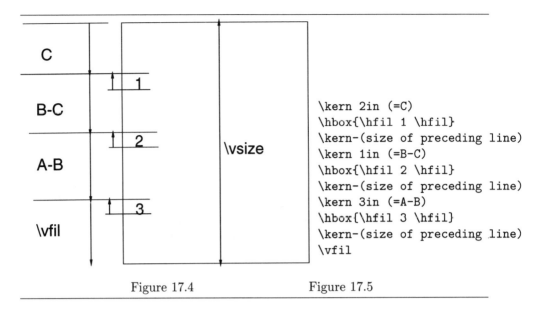

```
\kern 2in (=C)
\hbox{\hfil 1 \hfil}
\kern-(size of preceding line)
\kern 1in (=B-C)
\hbox{\hfil 2 \hfil}
\kern-(size of preceding line)
\kern 3in (=A-B)
\hbox{\hfil 3 \hfil}
\kern-(size of preceding line)
\vfil
```

Figure 17.4   Figure 17.5

```
  \kern\dimen0 \box0 \kern-\dimen2}
 \ifdim\lastkern>0pt\repeat
\fi}
```

Macro \reversebox contains a loop that breaks up \box1, calculates the quantities C, B-C, A-B, and places them in \box2 with the lines of text, each followed by a negative \kern. When finished, the OTR appends a \vfil to end up with a height of \vsize. This way, \box2 has the same height as \box255, and they can be superimposed and shipped out together.

To run the whole thing, just say

```
\zeroToSp
\anotherversefalse
\input source
\bye
```

This is, perhaps, not the most elegant solution, nor is it compact. Each macro, however, has its own well-defined task, making it easier to read and understand the whole thing.

17.26 A "Special Box" OTR

The problem: In many modern science texts, the main flow of text is interrupted by "special boxes." They can be used to develop certain topics in detail, to present a historical background of other topics, or to present the author's opinion or reminiscences. To distinguish such a box from the rest of the text, it may be surrounded by rules on all sides.

The intuitive approach is to place the special text in a \vbox and build the rules as in [ex. 21.3]. This, of course, won't work, since the "special box" may have to straddle two pages, but a \vbox is indivisible.

The approach used here identifies the start and end of the special text by making its lines narrower. Macro \startspbox draws the top \hrule of the special box and expands \narrower. Macro \endspbox terminates the effect of \narrower and draws the bottom \hrule. Note that the hrules are placed in boxes, since otherwise they would cause an incomplete breakup.

The OTR breaks up the page and duplicates it. Each narrow line (a line for which \rightskip> 0) is surrounded with two short rules. To make the rules on successive lines touch, the normal interline glue is suppressed when a narrow line is found.

```
\def\Hrule{\line{\vrule width\hsize height.4pt}}
\def\startspbox{\medskip\Hrule\nobreak\smallskip\begingroup\narrower}
\def\endspbox{\smallskip \nobreak \endgroup \Hrule \medskip}

\zeroToSp

\newif\ifsurround
\def\Strut{\vrule height8.5pt depth3.5pt}
\def\checkline{%
  \setbox2=\hbox{\unhcopy0
                \ifdim\lastskip>0pt \global\surroundtrue
                \else\global\surroundfalse\fi}}

\newif\ifAnyleft \newcount\pen
\def\specialbox{%
  \loop \Anyleftfalse
    \ifdim\lastskip=0pt \ifdim\lastkern=0pt \ifnum\lastpenalty=0
      \global\setbox0=\lastbox \ifvoid0 % end of breakup loop
    \else \Anylefttrue
      \ifhbox0\checkline \ifsurround\setbox0=%
       \hbox{\Strut\box0\Strut}\fi \fi
      \global\setbox1=\vbox{\box0 \unvbox1} \fi
    \else \Anylefttrue \pen=\lastpenalty
         \global\setbox1=\vbox{\penalty\pen\unvbox1} \unpenalty \fi
    \else \Anylefttrue \dimen0=\lastkern
         \global\setbox1=\vbox{\kern\dimen0 \unvbox1} \unkern \fi
    \else \Anylefttrue \skip0=\lastskip
         \ifsurround\skip0=0pt \fi %suppress normal interline glue
         \global\setbox1=\vbox{\vskip\skip0 \unvbox1} \unskip \fi
  \ifAnyleft \repeat}

\newbox\brk
\output={\setbox\brk=\vbox{\unvbox255 \specialbox}
        \ifdim\ht\brk>0pt \message{Incomplete breakup} \fi
        \shipout\box1 \advancepageno}
```

17.27 Example: Revision Bars

(See also section 18.8 for another approach to this problem.) Certain documents—
such as the bylaws of an organization, or the users' manual of a computer system—
may go through many revisions. Sometimes it is desirable to emphasize (or flag)
the revised parts by placing a vertical bar in the left margin of the revised lines. If
the revision is short, affecting only one line, there is no need for a special OTR and
a \vadjust like the one below can be used (see also [ex. 14.28]).

```
\def\rev{\vadjust{\moveleft6pt\vbox to0pt{
\kern-12pt\hrule height10pt width1pt\vss}}}
```

However, if the revision may affect more than one line, the problem becomes much
more complex and the OTR should be involved.

Due to Amy Hendrickson **A simple method.** We start with a relatively simple approach, which is sketched
below but is not implemented.

1. Macro \beginvbars saves the page-so-far in a box \partialpage.

2. Macro \endvbars places a bar on the left of \box255, appends it to \par-
tialpage, and returns the whole thing to the MVL, so that a good page break can
be found.

There are some problems with this approach:

1. The revision may start in mid-paragraph. In such a case, the first part of
the paragraph goes in box \partialpage and eventually has to be seamlessly glued
to the rest of the paragraph. A similar case occurs when the revision ends within a
paragraph.

2. When \box255 is appended to \partialpage, its \topskip glue should be
replaced by the normal interline skip.

A better solution. The approach shown here is different. The start and end of
each revision are flagged with small, special boxes placed between the text lines.
The OTR breaks up \box255 looking for the special boxes. The distance of each
special box from the top of the page is measured. The distances are then used to
prepare vertical rules in a separate box (\box3), which is eventually typeset on the
left of \box255.

Macro \startrev uses \vadjust to place a special \vbox with a height of 1sp
below the line where the revision starts. Macro \endrev places a similar box, with
a height of 2sp, below the last line of the revised text. Note that if the revised text
is short, the two special boxes may end up being placed, one above the other, below
the same line.

```
\def\startrev{\vadjust{%
 \nointerlineskip\nobreak\vbox to1sp{}}}
\def\endrev{\vadjust{%
 \nointerlineskip\nobreak\vbox to2sp{}}}
```

Macro \Obreak expands \breakup with a copy of \box255. The breakup loop
stops when a special box with height=2sp is found. \Obreak then measures the
height of the remaining page, stores that height in \box1 as \kern, stores a flag
indicating that a vertical bar should end at that point, and restarts the loop. When

a box with height=1sp is found, \Obreak does a similar thing, except that it places
a different flag, indicating that the bar should start at that point. The flags are
special hboxes with a width of either 1sp or 2sp.

```
\zeroToSp

\output={\global\setbox1=\vbox{}
        \setbox2=\copy255 \Obreak
        \ifdim\ht\brk>0pt \message{Incomplete breakup} \fi
        \arrangebox
        \setbox3=\vbox to\vsize{\unvbox3\vfil} \wd3=6pt
        \shipout\hbox{\llap{\box3}\box255}
        \advancepageno}

\newbox\brk
\newif\ifstartbar \startbarfalse \newif\ifendbar \endbarfalse
\def\Obreak{%
  \setbox\brk=\vbox{\unvbox2 \breakup}
  \ifstartbar
    \global\startbarfalse
    \global\setbox1=\vbox{\unvbox1\kern\ht\brk\hbox to1sp{}}
    \setbox2=\box\brk
    \expandafter\Obreak
    \fi
  \ifendbar
    \global\endbarfalse
    \global\setbox1=\vbox{\unvbox1\kern\ht\brk\hbox to2sp{}}
    \setbox2=\box\brk
    \expandafter\Obreak
    \fi
  }

\newif\ifAnyleft
\def\breakup{%
  \loop \Anyleftfalse
    \ifdim\lastskip=0pt \ifdim\lastkern=0pt \ifnum\lastpenalty=0
      \global\setbox0=\lastbox \ifvoid0
    \else \Anylefttrue
      \ifvbox0
        \ifdim\ht0=1sp \global\startbartrue \Anyleftfalse \fi
        \ifdim\ht0=2sp \global\endbartrue \Anyleftfalse \fi
      \fi \fi
    \else \Anylefttrue \unpenalty \fi
    \else \Anylefttrue \unkern \fi
    \else \Anylefttrue \unskip \fi
  \ifAnyleft \repeat}
```

At the end of the breakup loop, the OTR expands macro \arrangebox, which reads the kerns and flags from \box1 and uses them to generate the actual vertical bars in \box3. It uses the following algorithm:

```
PrevKern:=0;
read Kern,Flag from \box1
if Flag=start place \kern of size
 Kern-PrevKern in \box3
 PrevKern:=Kern, PrevFlag:=Flag;
if Flag=end    place a rule of size
 Kern-PrevKern in \box3
 PrevKern:=Kern, PrevFlag:=Flag;
if \box1 is empty and
 PrevFlag=start
 place a rule of size
   \vsize-PrevKern in \box3,
End;
```

And here is a listing:

```
\newif\ifcontin
\def\arrangebox{
\setbox3=\vbox{} \dimen1=0pt
\loop
\ifdim\ht1>0pt
 \setbox1=\vbox{\unvbox1
  \global\setbox0=\lastbox
  \global\skip0=\lastkern \unkern}
 \contintrue \dimen0=\wd0
 \dimen2=\skip0 \advance\dimen2 by-\dimen1
  \dimen1=\skip0
 \ifdim\dimen0=1sp
  \setbox3=\vbox{\unvbox3 \kern\dimen2}
 \fi
 \ifdim\dimen0=2sp
  \setbox3=\vbox{\unvbox3
   \hrule height\dimen2 width1pt}
 \fi
\else
 \continfalse
 \ifdim\dimen0=1sp \dimen2=\vsize
  \advance\dimen2 by-\dimen1
   \setbox3=\vbox{\unvbox3
    \hrule height\dimen2 width1pt}
 \fi
\fi
\ifcontin \repeat}
```

As usual, the macros can be improved. The reader may notice that the size and placement of the bars are not ideal and can be improved. This is especially true for cases where only one line of text is revised.

▸ **Exercise 17.10:** Generalize the macros so that they can typeset a revision number, in \sevenrm, to the left of each bar. The number should be specified by the user, as a parameter of \startrev.

17.28 Summary

The examples and techniques described here, even though incomplete and simplified, demonstrate how very powerful TEX is, compared to other typesetting systems.

The main concepts behind TEX, namely, boxes, glue, penalties, and macros, are different from those used by other systems and are more difficult to master. At the same time, they are more powerful, and the person willing to invest the time and effort necessary to learn TEX is rewarded with high-quality results.

black magic: n. A technique that works,
though nobody really understands why.

— *Eric Raymond (ed.), The New Hacker's Dictionary*

18. OTR Techniques: II

Certain typesetting problems can only be handled by the OTR. Many times, such a problem is solved by *communicating with* the OTR. Chapter 17 discusses communication from V mode, where certain clues (such as a small piece of glue or kern, or a box with small dimensions) are placed between lines of text. The OTR searches \box255 for clues and, on finding them, modifies the document in the desired way *in the vicinity of the clue*.

Searching \box255 is done by breaking it up into its components and checking each to see if it is a clue. A component can be a line of text, interline glue, vertical kern, or anything else that can go into a vertical list. The breakup is done by means of the \last$xx$ commands.

The problem with this technique is that the clues can only be placed *between* lines of text, and not *inside* a line. We thus say that it is possible to communicate with the OTR from V mode, but not from H mode. The reason for this is that a line of text is a box made up of characters of text, and a character is not the same as a box. Specifically, the \lastbox command does not recognize a character of text as a box. The following tests are recommended for readers who are uncertain:

```
\setbox0=\hbox{ABC}%              \setbox0=\hbox{AB\hbox{C}}%
\unhbox0 \setbox1=\lastbox       \unhbox0 \setbox1=\lastbox
\showbox1                        \showbox1
\bye                             \bye
```

The first test shows `\box1` to be void and typesets 'ABC'. In contrast, the second test shows `\box1` to consist of an `\hbox` with the 'C' and typesets only 'AB'. Reference 12, p. 217, contains a more detailed discussion of this point.

Communicating with the OTR from H mode is, however, very desirable, since many OTR problems can easily be solved this way. Three methods to do this have consequently been developed. They are described here, each followed by an application to a practical problem. Note that each method has its own limitations, and none is completely general.

The main idea in methods 1 and 2 is to enclose each character of text, as it is being read from the input file, in a box. Now there are no longer any characters, just many small boxes. When the OTR is invoked, each line of text is a box containing other boxes (and glue, kern, and penalties) but no characters. The `\lastbox` command can now be used to break up the line of text into its components and search for clues. (To simplify the discussion, we assume text without any math, rules, marks or whatsits.)

Before discussing the details of the first two methods, their disadvantages should be mentioned. Since we no longer have any characters, just boxes, we lose hyphenation, kerning, and ligatures. As a result, we normally have to use `\raggedright`, so these methods can only be used in cases where a ragged right margin and lower typesetting quality are acceptable.

How can we coerce TeX to place each character, as it is being input, in a box? Here are the principles of the first two methods:

Method 1. Declare every character of text active and define it to be itself in a box. Thus we say '`\catcode`\a=13' followed by '`\def a{\hbox{a}}`' and repeat for all characters. (The simple definition above cannot be used, because it is infinitely recursive. See section 18.1 for how it is really done.) The main disadvantage of this method is that no macros can be embedded in the text. Something like `\abc` will be interpreted as the control sequence '`\a`' followed by the nonletters 'b' and 'c'.

Method 2. Use `\everypar` (and also redefine `\par`) to collect an entire paragraph of text in `\toks0`. Scan `\toks0` token by token and place each nonspace token in a small `\hbox`. Then typeset the paragraph. This method does not have the disadvantage of the previous one, since there are no active characters.

Method 3 is completely different. It does not place characters in boxes and does not use `\lastbox` to break up a line of text. Instead it writes `\box255` on a file, item by item, then reads it back, looking for the clues. This method is slow and tedious, but it does not have the disadvantages of the previous two.

18.1 Method 1

We need to declare all the letters, digits, and punctuations active (actually, I only did this for the lowercase letters, for the uppercase 'L', for the three digits '123', and for '.,;'). Each character should now be defined as a box containing its own character code. Turning 'a', for example, into an active character is done by '`\catcode`\a=13 \def a{\hbox{\char`\a}}`'. When we get to the 'b', however, the command '`\catcode`\b=13 \def b{\hbox{\char`\b}}`' fails because 'a' is no longer a letter, so instead of `\catcode`, TeX sees `\c` followed by a nonletter. The solution is to use '`\let`' to redefine the control sequences `\catcode`, `\def`, `\hbox` and \char. Also the number 13 may cause a problem later, after the digit '1' is declared active. The result is declarations such as:

```
\let\?=\catcode \let\!=\active
\let\*=\def \let\+=\char \let\==\hbox
\let\<=\leavevmode \let\\=\bye
```

following which the active characters can be defined by commands such as '\?'\a\! *a{\={\+'\a}}'. After this is done, any character of text input by TeX is expanded into a box containing that character. Note that TeX does not see any text anymore, just a lot of small boxes. This means that there will be nothing to start a paragraph (we will have to place a \leavevmode explicitly at the beginning of every paragraph). The following example is a simple application of this technique.

18.2 About the Examples

All three examples use the following text, which was artificially divided into two paragraphs.

```
in oLden times, when wishing stiLL heLped one, there Lived a
king whose daughters were aLL beautifuL; and the youngest was so
beautifuL that the sun itseLf, which has seen so much, was
astonished whenever it shone in her face. cLose by the kings
castLe Lay a great dark forest, and under an oLd Lime tree

in the forest was a weLL, and when the day was very warm, the
kings chiLd went out into the forest and sat down by the side of
the cooL fountain;  and when she was bored she took a goLden
baLL, and threw it up on high and caught it; and this baLL was
her favorite pLaything.
```

18.3 Example 1: Widening Letters

This example uses method 1. Before delving into the details of the example, here is the code used to activate characters and to conduct the test:

```
\hsize=3in\tolerance=7500\raggedright\zeroToSp

\let\?=\catcode \let \!=\active \let\*=\def \let\+=\char \let\==\hbox
\let\<=\leavevmode \def\\{\vfill\eject\end}

\?'\a\! \*a{\={\+'\a}}\?'\b\! \*b{\={\+'\b}}
\?'\c\! \*c{\={\+'\c}}\?'\d\! \*d{\={\+'\d}}
\?'\e\! \*e{\={\+'\e}}\?'\f\! \*f{\={\+'\f}}
\?'\g\! \*g{\={\+'\g}}\?'\h\! \*h{\={\+'\h}}
\?'\i\! \*i{\={\+'\i}}\?'\j\! \*j{\={\+'\j}}
\?'\k\! \*k{\={\+'\k}}\?'\l\! \*l{\={\+'\l}}
\?'\m\! \*m{\={\+'\m}}\?'\n\! \*n{\={\+'\n}}
\?'\o\! \*o{\={\+'\o}}\?'\p\! \*p{\={\+'\p}}
\?'\q\! \*q{\={\+'\q}}\?'\r\! \*r{\={\+'\r}}
```

```
\?'\s\! \*s{\={\+'\s}}\?'\t\! \*t{\={\+'\t}}
\?'\u\! \*u{\={\+'\u}}\?'\v\! \*v{\={\+'\v}}
\?'\w\! \*w{\={\+'\w}}\?'\x\! \*x{\={\+'\x}}
\?'\y\! \*y{\={\+'\y}}\?'\z\! \*z{\={\+'\z}}
\?'\L\! \*L{\={\+'\L}}
\?'\1\! \*1{\={\+'\1}}\?'\2\! \*2{\={\+'\2}}
\?'\3\! \*3{\={\+'\3}}
\?'\.\! \*.{\={\+'\.}}\?'\,\! \*,{\={\+'\,}}
\?'\;\! \*;{\={\+'\;}}
```

```
\< in oLden times... Lime tree

\< in the forest... favorite pLaything.
\\
```

The example itself is an interesting OTR problem communicated to the author by Yannis Haralambous and the main reason for developing these OTR methods. If one decides, for some reason, not to hyphenate a certain document, then a ragged right margin is a good choice, which makes the text look best. Certain religious texts, however, don't use hyphenation and also frown on ragged right. They produce a straight right margin by widening certain letters. In the example the 'L' has been selected, since it is easy to design this letter out of two parts that connect with a rule. I did not actually bother to design a special 'L', and I simply extended it with an \hrulefill.

When I started thinking about this problem, it was clear to me that this was an OTR problem, and I tentatively outlined the following steps to the solution:

1. Typeset the text with \raggedright. This makes the interword glue rigid and the \rightskip glue flexible. Each line of text is placed in an '\hbox to \hsize', and \rightskip is stretched as necessary.

2. In the OTR, break \box255 up into individual lines of text. For each line, perform steps 3 through 6.

3. Perform an \unhbox on the line, to return \rightskip to its natural size (zero). Subtract the present width of the line from its original width (\hsize). The difference is the amount by which all the L's on the line will have to be stretched.

4. Break up the line into individual components (mostly characters, glue, and penalties), and count the number of L's in the line.

5. Divide the difference from step 3 by the number of L's from step 4. The result is the amount by which each L will have to be widened.

6. Break up the line again, widening each L. Pack the line in a new \hbox.

7. Rebuild the page from the line boxes generated in step 6, and ship it out.

The only problem was step 4. A line of text cannot normally be broken up into individual characters and examined. However, using method 1, it is possible to break up such a line, since it does not include any characters, and search for clues. A clue, in our case, is a box whose width is the same as that of an 'L' (if other characters happen to have the same width, the width of the 'L' can be changed by 1sp).

> in oLden times, when wishing stiLL heLped one, there Lived a king whose daughters were aLL beautifuL; and the youngest was so beautifuL that the sun itseLf, which has seen so much, was astonished whenever it shone in her face. cLose by the kings castLe Lay a great dark forest, and under an oLd Lime tree
>
> in the forest was a weLL, and when the day was very warm, the kings chiLd went out into the forest and sat down by the side of the cooL fountain; and when she was bored she took a goLden baLL, and threw it up on high and caught it; and this baLL was her favorite pLaything.

1

> in oL___den times, when wishing stiL___L___ heL_ped one, there Lived a king whose daughters were aL_L_ beautifuL_; and the youngest was so beautifuL_ that the sun itseL_f, which has seen so much, was astonished whenever it shone in her face. cLose by the kings castLe Lay a great dark forest, and under an oL_____d L_____ime tree
>
> in the forest was a weL___L___, and when the day was very warm, the kings chiL___d went out into the forest and sat down by the side of the cooL_____ fountain; and when she was bored she took a goL_den baL_L___, and threw it up on high and caught it; and this baL___L___ was her favorite pL_____aything.

2

The seven steps can now be implemented, using the breakup technique of section 17.3.

Step 1. Just say \raggedright.

Steps 2 and 7. The OTR becomes

```
\newbox\brk
\output={\setbox\finPage=\vbox{}%
  \setbox\brk=\vbox{\unvbox255 \breakup}%
  \ifdim\ht\brk>0pt
    \message{Incomplete breakup, \the\ht\brk}\fi
  \shipout\box\finPage \advancepageno}

\newif\ifAnyleft \newcount\pen \newbox\finPage
```

```
\def\breakup{%
  \loop \Anyleftfalse
    \ifdim\lastskip=0pt\else \Anylefttrue
    \skip0=\lastskip \unskip
  \global\setbox\finPage=\vbox{\vskip\skip0 \unvbox\finPage}\fi
    \ifdim\lastkern=0pt\else \Anylefttrue
    \dimen0-\lastkern \unkern
  \global\setbox\finPage=\vbox{\kern\dimen0 \unvbox\finPage}\fi
    \ifnum\lastpenalty=0 \else\Anylefttrue
    \pen=\lastpenalty \unpenalty
  \global\setbox\finPage=\vbox{\penalty\pen \unvbox\finPage}\fi
    \setbox0=\lastbox
    \ifvoid0 \else \Anylefttrue\message{.} \breakupline
  \global\setbox\finPage=\vbox{\box2 \unvbox\finPage}\fi
  \ifAnyleft \repeat}
```

Macro \breakup is essentially the same as in Chapter 17. It places each line of
text in \box0 and expands \breakupline. Note the \global\setbox\finPage=
on certain lines. Those lines rebuild the page, line by line, in \box\finPage (step
7). When the entire process is complete, the OTR ships out \box\finPage.

Steps 3 and 5. Macro \breakupline resets the line of text to its natural width,
calculates the width difference in \diff, expands \countLonline to count the number of L's in the line, divides \diff by that number, and expands \longLline to
actually widen the L's in the line.

```
\newdimen\diff \newcount\Lnum
\def\breakupline{\diff=\hsize
\setbox0=\hbox{\unhbox0} \advance\diff-\wd0
\Lnum=0 \setbox1=\hbox{\unhcopy0 \countLonline}
\ifdim\wd1>0pt \message{Incomplete line breakup}\fi
\ifnum\Lnum=0 \diff=0pt \else \divide\diff by\Lnum\fi
\setbox2=\null \setbox1=\hbox{\unhbox0 \longLline}}
```

Step 4. Macro \countLonline breaks up the line into individual components and
counts the number of L's in the line. It uses a breakup loop similar to the one in
\breakup above. Note how boxes with 'L' are identified by their width.

```
\newif\ifCharleft
\def\countLonline{%
  \Charleftfalse
    \ifdim\lastskip=0pt\else \Charlefttrue
    \skip0=\lastskip \unskip\fi
    \ifdim\lastkern=0pt\else \Charlefttrue
    \dimen0=\lastkern \unkern\fi
    \ifnum\lastpenalty=0 \else\Charlefttrue
    \pen=\lastpenalty \unpenalty\fi
    \setbox1=\lastbox
    \ifvoid1\else \ifdim\wd1=6.25002pt \global\advance\Lnum1 \fi
```

```
\Charlefttrue \fi
  \ifCharleft \countLonline\fi}
```

Step 6. Macro \longLline uses the same technique to break up the line again, extend all the 'L's, and rebuild it in \box2. Macro \extendL packs an 'L' with an \hrulefill in a new \hbox.

```
\newif\ifSomeleft
\def\longLline{%
  \Someleftfalse
    \ifdim\lastskip=0pt\else \Somelefttrue
      \skip0=\lastskip \unskip \global\setbox2=
              \hbox{\hskip\skip0 \unhbox2}\fi
    \ifdim\lastkern=0pt\else \Somelefttrue
      \dimen0=\lastkern \unkern \global\setbox2=
              \hbox{\kern\dimen0 \unhbox2}\fi
    \ifnum\lastpenalty=0 \else\Somelefttrue
      \pen=\lastpenalty \unpenalty \global\setbox2=
                \hbox{\penalty\pen \unhbox2}\fi
    \setbox1=\lastbox
    \ifvoid1\else \ifdim\wd1=6.25002pt \extendL\fi
      \setbox2=\hbox{\box1 \unhbox2}\global\Somelefttrue
    \fi
  \ifSomeleft \longLline\fi}
```

```
\newdimen\Lwidth
\def\extendL{\Lwidth=\wd1 \advance\Lwidth by\diff
  \setbox1=\hbox to\Lwidth{\unhbox1\hrulefill}}
```

The code is somewhat long, but well structured, and most macros use the same breakup technique.

Problems:

1. A line of text without L's is not extended, so it normally comes out shorter.

2. Since there are no letters in our texts, just boxes, there is nothing to signify the start of a paragraph. Each paragraph must therefore start with a \leavevmode command (\< in our case).

3. \box255 may contain only boxes, glue, kern, and penalties. Anything else (such as text, rules, whatsits, or marks) would stop the breakup macros. Note that overfull lines contain rules, so they should be avoided (by increasing the tolerance, increasing the stretch of \rightskip, or rewriting the text).

4. Because of reasons discussed in Chapter 17, glues with a natural size of 0pt stop the breakup macros. Macro \zeroToSp below changes the natural size of several such glues to 1sp. It also changes the plain values of some common penalties from 0 to 1. This macro should be expanded once, at the start of the document.

```
\def\zeroToSp{\parskip=1sp plus1pt
\parfillskip=1sp plus1fil
```

```
\advance\leftskip by1sp
\advance\rightskip by1sp
  \def\vfil{\vskip1sp plus1fil}
  \def\vfill{\vskip1sp plus1fill}
  \abovedisplayshortskip=1sp plus3pt
  \postdisplaypenalty=1
  \interlinepenalty=1}
```

5. To identify boxes with an 'L', we use the width of an 'L' in font cmr10. To guarantee reliable identification, no other character in the font should have the same width.

Possible improvements and applications:

1. If \diff is less than \hfuzz (or some other small parameter), it can be set to zero, since there is no point in widening a letter by a very small amount.

2. The L's on the last line of a paragraph are normally widened a lot. If this is not desirable, the macros can be changed to treat the last line differently.

18.4 Method 2

As mentioned earlier, the principle is to collect the text of an entire paragraph in a toks register, then to scan the register token by token, placing each character token in a small \hbox. We again lose hyphenation, kerning, and ligatures, so we normally have to resort to a ragged right margin. However, we can have control sequences embedded in the text. Care should be taken to identify each control sequence (and its argument) and to expand it, instead of placing it in a box. Here are the macros and the test text:

```
\hsize=4in \tolerance=7500 \raggedright \zeroToSp
\begingroup
\newif\ifargmn
\everypar{\catcode' =12 \toks0=\bgroup}
\def\par{\catcode' =10 \argmnfalse
  \expandafter\Tmp\the\toks0 \end \endgraf}
\def\Tmp#1{\ifx\end#1\def\next{\relax}%
\else
\ifargmn\cs{#1}\argmnfalse
\else
\ifcat\relax\noexpand#1\let\cs=#1\argmntrue%
\else
\ifnum11=\catcode'#1\hbox{#1}%
\else
\ifnum12=\catcode'#1\hbox{#1}%
\else\ifnum'40='#1\ \fi
\fi
 \fi
  \fi
   \fi
\let\next=\Tmp\fi\next}%
```

```
%
in\Mnote{xyz *} oLden times, when \Mnote{abc 2}wishing stiLL
heLped\Mnote{note 3} one, there Lived a king whose
daugh\Mnote{note 4}ters were aLL beautifuL; and the
youngest\Mnote{note 5} was so beautifuL that the sun
itseLf, which has seen so much, was \Mnote{note
6}astonished whenever it shone in her face. cLose by the
kings castLe L\Mnote{note 7}ay a great dark forest, and
under an oLd Lime tree}

in the fore\Mnote{note 8}st was a weLL, and when the day
was\Mnote{note 9} very warm, the kings chiLd went out into
the fores\Mnote{note 10}t and sat down by the side of the
coo\Mnote{note 20}L fountain; \Mnote{note 21}and when she
was bored \Mnote{note 22}she took a goLden baLL,\Mnote{note
12} and threw it up on high and caught it; and this baLL
was her favorite \Mnote{note 13}pLaything.}

\endgroup
\bye
```

The \everypar parameter is modified to place '\toks0=\bgroup' at the start of each paragraph. At the end of a paragraph we need a closing \egroup, which is easy to insert by redefining \par. Unfortunately, the code '\toks0=\bgroup...\egroup' does not work. Using \bgroup is okay, but a right brace (a token of catcode 2) is required instead of the control sequence \egroup. When using this method we unfortunately have to insert a '}' explicitly at the end of every paragraph. This is one of two unsolved problems with this method.

The \par primitive is modified to expand '\the\toks0', to append an \end to it, to expand \Tmp, and to close the paragraph.

Macro \Tmp uses recursion to extract the next token from \toks0 and to test it. Tokens with catcodes 11 and 12 are placed in boxes and appended to the current list (normally the MVL) (except spaces, which are appended as spaces to the MVL). Control sequence tokens are also identified. Each such token is kept in \cs until its argument is identified in the following recursive iteration, where it is expanded. In the current version, any control sequence embedded in the text must have exactly one argument. The changes of \everypar and \par are confined to a group.

Spaces present a special problem. The scanning of tokens skips all spaces. Therefore, the catcode of space had to be changed. It has been changed to 12 (other), and \Tmp identifies spaces by their character code. When a catcode 12 space is identified by \Tmp, a normal (catcode 10) space is appended to \box0.

The second unsolved problem in this method is the end-of-lines. They are converted to spaces, but only after the catcode of a space has been changed. As a result, they appear in \toks0 as normal spaces (catcode 10) and are skipped.

18.5 Example 2: Marginal Notes

Typesetting notes in the margins of a scholarly book is very common. Reference 32 is an interesting example, familiar to many TeX users. Another well-known example is the marginal notes of the mathematician Pierre de Fermat. When trying to prove what is now called Fermat's last theorem (which states that there is no integer $n > 2$ such that $x^n + y^n = a^n$ for rational x, y, and a), he wrote in the margin of the book he was reading (Bachet's *Diophantus*), "I have discovered a truly marvellous demonstration of this general theorem, which this margin is too narrow to contain" (Ref. 43). I like to call this famous note *Fermat's warning*. It warns us not to abuse this useful tool of the author.

When teaching TeX I have always noticed how, when discussing marginal notes, the class suddenly comes to life and starts following the discussion with renewed interest. In the lab that follows, people start writing macros for marginal notes, invariably ignoring Fermat's warning, and overdoing this useful feature.

A single note can easily be placed in the margin of a given line with the help of \vadjust. When writing a text with many marginal notes, however, the writer may end up with two or more notes appearing in the margin of the same line. Because of the limited space in the margin, the notes for the same line of text may have to be rearranged before the page is shipped out, and this is an OTR problem. Rearranging notes may involve placing some on the left, and some on the right margin; it may mean typesetting them in very small type, moving some up or down (if there is room on adjacent lines), or warning the author that there is no room.

In this example, rearranging is done in a simple way. The first note found on a line is typeset on the left. The second one, on the right margin. If more notes are found on the same line, none is typeset, and a warning, with the input line number, is placed in the log file.

The implementation is straightforward. Macro \Mnote places the text of the note in an '\hbox to1sp' inside the paragraph. The macro definition is thus '\def\Mnote#1{\hbox to1sp{#1\hss}}'. Method 2 is used to place every character of text in a box. The OTR breaks up \box255 into its top-level components and identifies the lines of text. Each line is further broken up, and all the clues (boxes of width 1sp) in it located. Depending on how many clues were found, the macros place the notes as described above. The OTR is straightforward:

```
\newbox\brk
\output={\setbox\finPage=\vbox{}
  \setbox\brk=\vbox{\unvcopy255 \breakup}%
  \ifdim\ht\brk>0pt \message{Incomplete breakup, \the\ht\brk}\fi
\shipout\box255 \shipout\box\finPage}
```

Note that it also ships out \box255, for comparison purposes. Macro \breakup rebuilds all the elements of \box255 in \box\finPage, except that each line of text is further broken up by \breakupline (and the notes properly placed in the margins) before being rebuilt and appended to \box\finPage.

```
\newif\ifAnyleft \newcount\pen \newbox\finPage
\def\breakup{%
```

```
\loop \Anyleftfalse
   \ifdim\lastskip=0pt\else \Anylefttrue
    \skip0=\lastskip \unskip
\global\setbox\finPage=\vbox{\vskip\skip0 \unvbox\finPage}\fi
   \ifdim\lastkern=0pt\else \Anylefttrue
    \dimen0=\lastkern \unkern
\global\setbox\finPage=\vbox{\kern\dimen0 \unvbox\finPage}\fi
   \ifnum\lastpenalty=0 \else\Anylefttrue
    \pen=\lastpenalty \unpenalty
\global\setbox\finPage=\vbox{\penalty\pen \unvbox\finPage}\fi
   \setbox0=\lastbox
   \ifvoid0 \else \Anylefttrue\message{.} \breakupline
\global\setbox\finPage=\vbox{\box2 \unvbox\finPage}\fi
  \ifAnyleft \repeat}
```

Macro \breakupline expands \countNotesonline to break up one line of text, and count the number of notes. It then rebuilds the line in \box2 with the notes placed in the margins, and with the special boxes emptied.

```
\newcount\numnotes
\def\breakupline{\numnotes=0
\setbox1=\hbox{\unhbox0 \countNotesonline}%
\ifdim\wd1>0pt \message{Incomplete line breakup}\fi
\ifcase\numnotes
\relax     % \numnotes=0 ==> zero notes for this line
\or        % 1 note
\setbox2=\hbox to\hsize{\llap{\box3\kern3pt}\unhbox2\hfil}%
\else      % 2 or more notes
\setbox2=\hbox to\hsize{\llap{\box4\kern3pt}%
 \unhbox2\hfil\rlap{\kern3pt\box3}}%
\fi}
```

Macro \countNotesonline is a simple application of the breakup technique for one line of text. The first note found in the line is placed in \box3, and the second one in \box4. All subsequent notes are flushed. A small dash is inserted in each special box to show where the note came from.

```
\newif\ifCharleft
\def\countNotesonline{%
  \Charleftfalse
    \ifdim\lastskip=0pt\else \Charlefttrue
     \skip0=\lastskip \unskip
\global\setbox2=\hbox{\hskip\skip0 \unhbox2}\fi
    \ifdim\lastkern=0pt\else \Charlefttrue
     \dimen0=\lastkern \unkern
\global\setbox2=\hbox{\kern\dimen0 \unhbox2}\fi
    \ifnum\lastpenalty=0 \else\Charlefttrue
     \pen=\lastpenalty \unpenalty
```

```
\global\setbox2=\hbox{\penalty\pen \unhbox2}\fi
    \setbox1=\lastbox
    \ifvoid1\else \ifdim\wd1=1sp % a special box
\ifnum\numnotes=0 \global\setbox3=\hbox{\unhbox1}\fi
\ifnum\numnotes=1 \global\setbox4=\hbox{\unhbox1}\fi
\ifnum\numnotes>1
 \global\setbox3=\hbox{!!!}\global\setbox4=\hbox{!!!}%
 \message{Too many notes on line \the\inputlineno}\fi
\global\advance\numnotes 1
\global\setbox2=\hbox{\pop\unhbox2}%
\else % not a special box
\global\setbox2=\hbox{\box1 \unhbox2}\fi
\Charlefttrue \fi
 \ifCharleft \countNotesonline\fi}
```

Finally, macro \pop places a small dash in the special box after it has been emptied. The dash is character "37 of font cmsy (the math symbols). This character is constructed in a box of width 0 and it sticks out on the right. Normally it is followed by a minus sign or a right arrow, to create a "maps to" symbol (Ref. 4., p. 515).

```
\def\pop{\leavevmode\raise4pt\hbox to1sp
 {\hss\smash{\tensy\char"37}\kern1.2pt\hss}}
```

18.6 Tests

The two paragraphs used for the test were shown earlier. The first diagram below shows \box255 before any changes. Note how the text of the notes overlap the text of the paragraphs, since they are saved in boxes *inside the paragraph*.

inxoldzen times, when whishing stiLLheLpedmote; there Lived a king whose daughterswere aLL beautifuL; and the youngestnote-5 was so beautifuL that thesun itseLf, which has seen so much, was astterfished whenever itshone in her face. cLose by the kings castLe Layte-great darkforest, and under an oLd Lime tree

in the forestwea8 a weLL, and when the day wasnotey9varm, the kings chiLd went out into the foresthated18at down by theside of the cooLofred20ain; and22hen she was boredshote33k a goLden baLL, notd-threw it up on high andcaught it; and this baLL was her favorite phteytBing.

The next diagram shows the final result shipped out.

!!!   in oLden times, when wishing stiLLheLped one, there Lived !!!
note-4 a king whose daughterswere aLL beautifuL; and the youngest   note-5
          was so beautifuL that thesun itseLf, which has seen so much,
note-6 was astonished whenever itshone in her face. cLose by the kings
note-7 castLe Lay a great darkforest, and under an oLd Lime tree
note-8   in the forest was a weLL, and when the day was verywarm, note-9
note-10 the kings chiLd went out into the forest and sat down by theside
    !!! of the cooL fountain; and when she was boredshe took a goLden   !!!
note-12 baLL, and threw it up on high andcaught it; and this baLL was
note-13 her favorite pLaything.

In practical use, sophisticated macros can be developed that will set the notes
in small type, will number them consecutively, and will move them vertically, if
necessary. However, as long as they are based on the principles shown here, ragged
right will normally have to be used, which is not always acceptable.

18.7 Method 3

This method is based on a 2-pass job. In the first pass the text is typeset in the
normal way, with characters, not boxes. Clues are inserted in the text, to be found
later by the OTR, in pass 2. Pages can either be shipped out or trashed, but the
OTR writes \box255 on a file, to be read by pass 2. Experienced users know that
a box cannot be written on a file in the normal way, using \write. The novelty of
this method is that a box can be written on the *log file*, using \showbox.

The user has to make sure that the log file is saved after pass 1. Pass 2 reads
the contents of \box255 from the file, searches for the beginning of each line of text
and for clues inside the line. If successful, pass 2 knows what clues are stored in
each text line. Pass 2 then reads the source file, typesets it in the usual way, and
has the OTR modify \box255, before shipping it out, according to the clues read
earlier.

Note that \lastbox is not used. The details of each line of text in \box255
are read from the file. The main advantage of this approach is that none of the
high-quality typesetting features, such as hyphenation, kerning, and ligatures, is
lost.

The main problem with this approach is how to read and analyze the contents
of \box255 from the log file in pass 2 (an example of such a file is shown below for
the benefit of inexperienced readers). This turns out to be easy, and it involves the
following tasks:

1. Certain records contain backslashes that should be ignored. Examples are
'..\tenrm i', '.\glue(\topskip) 3.05556', and '..\glue 3.33333 plus 1.6666
minus 1.11111'. To ignore these, pass 2 uses the following declarations (inside a
group):

```
\let\vbox=\relax \let\glue=\relax \let\topskip=\relax
\let\kern=\relax \let\rightskip=\relax \let\baselineskip=\relax
\let\parfillskip=\relax \let\parskip=\relax \def\shipout\box{\bgroup}
\let\showbox=\egroup \let\discretionary=\relax
```

2. Other records are important and should be identified. Examples are

a. '> \box255=' (this signals the start of the box)

b. '.\hbox(6.94444+1.94444)x216.81, glue set 0.45114' (this signals a new line of text).

b. '..\hbox(0.0+0.0)x0.00002, glue set ...' (this is a box of width 1sp, denoted a clue of type 1).

c. '! OK (see the transcript file).' (this signals the end of the box).

Records of type *a* are identified by \def\box255={\global\clues={(}}. The definition of \box255 is changed (locally) to insert a '(' in the toks register \clues.

Records of type *b* are identified by redefining \hbox.

```
\def\hbox(#1)x#2 {\toks0={}\one#2\end \tmp=\the\toks0 pt
\ifnum\tmp=1sp\appendclue1
\else
\ifnum\tmp=2sp\appendclue2
\else
\ifnum\tmp=\Hsize\appendclue+
\fi\fi\fi}
\def\one#1{\def\arg{#1}\ifx\end#1\let\rep=\relax
\else\ifx\comma\arg\let\rep=\one
\else\toks0=\expandafter{\the\toks0 #1}\let\rep=\one\fi\fi\rep}
\def\appendclue#1{\global\clues=%
 \expandafter{\expandafter#1\the\clues}}
```

Parameter '#2' is the width of the \hbox. In the records that interest us, it is either \hsize or 1sp or 2sp. The examples in *b* above show that the width is followed by a comma and a space, but there are records on the log file (such as the paragraph indentation '..\hbox(0.0+0.0)x20.0') where the width is followed by a space. This is why '#2' in the definition of \hbox is delimited by a space. If the width is followed by a comma, the comma is removed by macro \one. The width is stored in the \dimen register \tmp.

Macro \hsize thus identifies the important records and appends the tokens '+', '1', or '2' to the toks register \clues every time a line of text or a clue of type 1 or type 2 is found.

The end-of-box in the log file is identified when a type *c* record is found. We simply compare each record read to the string '! OK (see the transcript file). '. When finding it, a '(' is appended to \clues, and the loop reading the file is stopped. Note that our macros are supposed to stop reading when the end-of-box is found. They are never supposed to read the end-of-file. If an end-of-file is sensed while reading the log file, an error must have occurred.

All the clues found in the log file for one page (a single \box255) are stored in the toks register \clues, so that later macros can easily find out what clues were found in what text lines. A simple example is the tokens ')21++2+++121+(' where the ')' and '(' stand, respectively, for the end and start of \box255 in the log file, each '+' stands for a line of text, and each '1' or '2', for a clue of type 1 or 2 found in that line. Thus in the above example, a type 2 followed by a type 1 clue were found in the bottom line, another type 2 clue, in line 3 from the bottom, and three more clues, in line 6 (the top line).

Pass 1 normally writes several boxes on the log file, each corresponding to a page. The following appears in the log file between pages and has to be "neutralized."

```
<output> {\showbox 255
                    \shipout \box 255}
```

This is achieved by the weird definitions '\def\shipout\box{\bgroup}' and '\let\showbox=\egroup'. The method is illustrated below by applying it to a practical example.

18.8 Example 3: Revision Bars

(See also section 17.27 for a different approach to this problem.) Certain documents, such as the bylaws of an organization, go through periodic revisions. It is good practice to typeset each new revision with vertical bars to the left of parts that have been revised. This is an OTR problem (note that a revision may be broken across pages), and the solution shown here requires the two passes mentioned earlier. Pass 1 involves these steps:

1. Two macros are defined, to indicate the start and end of each revision.

```
\def\({\leavevmode\raise4pt\hbox to1sp{%
\hss\smash{\tensy\char"37}\kern1.2pt\hss}}
\def\){\leavevmode\raise4pt\hbox to2sp{%
\hss\smash{\tensy\char"37}\kern1.2pt\hss}}
```

The macros also place small dashes in the text, to indicate the boundaries of the revision.

2. The OTR writes \box255 on the log file and can also ship it out, for later comparison. If a shipout is not required, the OTR can say \box255=\null \deadcycles=0 instead of \shipout\box255.

```
\hsize=3in \vsize=2.2in \tolerance=7500
\showboxbreadth=1000 \showboxdepth=10
\output={\showbox255 \shipout\box255 \advancepageno}

\input source
\vfill\eject
```

The log file is saved between the passes. Note that the two passes can be parts of the same TeX job, and the log file can be saved when TeX stops, as usual, for a user's response after the \showbox. Pass 2 starts by opening the log file, if it exists:

```
\newtoks\clues \newdimen\tmp \newif\ifmore \moretrue \newread\logfile
\newdimen\Hsize \Hsize=\hsize \newbox\brk \newbox\bars
\newif\ifendRev  \newif\ifbegRev \newif\ifRev \newif\ifSplitrev

\immediate\openin\logfile=Log
\ifeof\logfile\errmessage{No log file}\fi
```

Note that the file name 'Log' is used here. In general, it is possible to read the name from the keyboard. Now comes the OTR. It is divided into two phases. Phase 1 reads a chunk off the log file, corresponding to one page, and prepares tokens in \clues. Phase 2 starts the breakup of \box255 and ships out \box\bars (stretched to \vsize) and \box255, side by side.

```
\output={%
% Phase 1. Read a chunk off the log file and
% prepare codes in \clues
\begingroup
\def\appendclue#1{\global\clues=%
 \expandafter{\expandafter#1\the\clues}}
\def\OK{! OK (see the transcript file). }
\def\comma{,}
\def\box255={\global\clues={(}}
\def\hbox(#1)x#2 {\toks0={}\one#2\end \tmp=\the\toks0 pt
\ifnum\tmp=1sp\appendclue1
\else
\ifnum\tmp=2sp\appendclue2
\else
\ifnum\tmp=\Hsize\appendclue+
\fi\fi\fi}
\def\one#1{\def\arg{#1}\ifx\end#1\let\rep=\relax
\else\ifx\comma\arg\let\rep=\one
\else\toks0=\expandafter{\the\toks0 #1}\let\rep=\one\fi\fi\rep}
%
\let\vbox=\relax \let\glue=\relax \let\topskip=\relax
\let\kern=\relax \let\rightskip=\relax
\let\baselineskip=\relax \let\parfillskip=\relax
\let\parskip=\relax \def\shipout\box{\bgroup}
\let\showbox=\egroup \let\discretionary=\relax
\setbox0=\vtop{\hsize=\maxdimen
\loop
\read\logfile to\rec
\ifeof\logfile\morefalse\message{end of log file!}
\else
 \ifx\rec\OK\appendclue)\morefalse\fi
 \rec
\fi
\ifmore\repeat}
\endgroup
\nextclue \if)\clue\else\message{Bad clue}\fi
% Phase 2. Breakup \box255 and use the clues
\global\setbox\bars=\vbox{}%
\global\endRevfalse \global\begRevfalse \global\Revfalse
  \setbox\brk=\vbox{\unvcopy255 \breakup}%
```

```
\ifdim\ht\brk>0pt \message{Incomplete breakup, \the\ht\brk}\fi
\shipout\hbox{\vbox to\vsize{\unvbox\bars}\kern4pt\box255}
\advancepageno}
```

3. Macro \breakup breaks up \copy255 into its top-level components. For each component with a dimension, the macro places either a skip or a vrule in \box\bars. It is important to realize that when we say, for instance, \skip0=\lastskip we lose the specific glue set ratio of \box255. This is why the rules are placed in \box\bars using \leaders and not \vrule. This way \box\bars can later be stretched to \vsize, and all the leaders in it will be stretched.

▶ **Exercise 18.1:** Why is it that a rule placed by means of \vrule height\skip0 cannot be stretched later?

```
\newif\ifAnyleft \newcount\pen
\def\breakup{%
  \loop \Anyleftfalse
    \ifdim\lastskip=0pt\else \Anylefttrue
    \skip0=\lastskip \unskip
\global\setbox\bars=\vbox{\ifRev\leaders\vrule\fi
                      \vskip\skip0\unvbox\bars}\fi
    \ifdim\lastkern=0pt\else \Anylefttrue
    \dimen0=\lastkern \unkern
\global\setbox\bars=\vbox{\ifRev\leaders\vrule\fi
                      \kern\dimen0\unvbox\bars}\fi
    \ifnum\lastpenalty=0 \else\Anylefttrue
    \pen=\lastpenalty \unpenalty\fi
    \setbox0=\lastbox
    \ifvoid0 \else \Anylefttrue
\dimen0=\ht0 \advance\dimen0 by\dp0
 \setbox2=\vbox{\unhbox0 \searchclues}%
\ifbegRev
\ifendRev
%TT
\global\Revfalse \global\endRevfalse
\global\setbox\bars=\vbox{\leaders\vrule\vskip\dimen0\unvbox\bars}
\else
%TF
\global\Revfalse
\ifSplitrev \global\Splitrevfalse
\global\setbox\bars=\vbox{\leaders\vrule\vskip\ht\bars}
\global\setbox\bars=\vbox{\leaders\vrule\vskip\dimen0\unvbox\bars}
\else
\global\setbox\bars=\vbox{\vskip\dimen0\unvbox\bars}
\fi\fi
\else
\ifendRev
%FT
```

```
\global\Revtrue
\global\setbox\bars=\vbox{\leaders\vrule\vskip\dimen0\unvbox\bars}
\else
%FF
\global\Revfalse
\global\setbox\bars=\vbox{\vskip\dimen0\unvbox\bars}
\fi\fi\fi
  \ifAnyleft \repeat}
```

When a line of text is found, macro \searchclues is expanded (see below) to update variables \begRev and \endRev. Four cases are possible:

a. Case FF. Both variables are false. This means no revisions have been found yet. A skip, equal in height to the current line of text, is appended to \box\bars. Variable \Rev is set to false, indicating that any future components found in \box255 should become skips in \box\bars.

b. Case FT. \begRev is false and \endRev is true, meaning the current line contains the end of a revision. A rule, the height of the current line, is appended to \box\bars. Also, \Rev is set to true, indicating that any future components found in \box255 should become rules in \box\bars.

c. Case TT. A revision starts on this line. A rule is appended to \box\bars but \Rev is set to false. (Also, \endRev is set to false, so case TF will be in effect from now on.)

d. Case TF. Normally this indicates a line with no revisions, but if \Splitrev is true, we have just found the start of a revision that will end on the next page. In this case, \box\bars (which has only skips in it so far) is filled up with a rule.

Macro \searchclues removes the next token from \clues and, if it is 1 or 2, sets \begRev or \endRev to true, respectively. Note that a revision may start and end on the same line. If the start of a revision is found while \endRev is false, it means that the revision will end on the next page. In such a case, variable \Splitrev is set to true, indicating that the entire \box\bars should be filled with a rule.

```
\def\searchclues{\nextclue
\if+\clue\let\Next=\relax
\else
\if(\clue\let\Next=\relax\message{bad Clue}
\else
\if2\clue \global\endRevtrue \global\begRevfalse
 \let\Next=\searchclues
\else
\if1\clue \global\begRevtrue \let\Next=\searchclues
 \ifendRev\else\global\Splitrevtrue\fi
\else
\message{bad clue}
\fi\fi\fi\fi\Next}

\def\nextclue{\expandafter\extr\the\clues X}
```

```
\def\extr#1#2X{\gdef\clue{#1}\global\clues=\expandafter{#2}}
```

The rest of pass 2 is straightforward.

```
\zeroToSp
\input source
\bye
```

For a multipage document, the OTR performs the same tasks for each page. It first receives \box255 of page 1. It reads the corresponding lines from the log file, looking for clues and storing them in \clues. The OTR then breaks \box255 up, isolating the lines of text from the bottom. It uses the tokens in \clues to modify only the right lines. At the end, \box255 (and \box\bars) is shipped out. The process repeats for each successive page sent to the OTR.

For each page, the OTR reads another chunk off the log file. This is why the two passes must typeset the same text. The best way to handle this is to \input the text in the two passes from the same source file. It is possible to make the macros more robust by checking in pass 2 to see that the chunk read from the log file actually has the same number of text lines as the current \box255.

The source file for our test is, as usual:

```
in oLden times, when wishing stiLL heLped one, there Lived a king
whose daughters were aLL beautifuL; and th\(e youngest was so
beautifuL that the sun itseLf, which has seen so much, was
astonished whenever it shone in her face. cLose by the kings
castLe Lay a great dark forest, and under an oLd Lime tree

in the forest\) was a weLL, and when the day was very warm, the
kings chiLd went out into the forest and sat down\( by the side
of the cooL fountain;  and when she was bored she took a \)goLden
baLL, and threw it up on high and caught it; and this baLL was
her favorite pLaything.
```

Following are the final result and parts of the log file produced by pass 1.

```
Textures 1.5 (preloaded format=plain 92.6.1)  21 OCT 1992 17:54
(test (source)
> \box255=
\vbox(158.99377+0.0)x216.81, glue set 3.04933fill
.\glue(\topskip) 3.05556
.\hbox(6.94444+1.94444)x216.81, glue set 0.45114
..\hbox(0.0+0.0)x20.0
..\tenrm i
..\tenrm n
..\glue 3.33333 plus 1.66666 minus 1.11111
..\tenrm o
..\tenrm L
..\tenrm d
......
```

> in oLden times, when wishing stiLL heLped one, there Lived a king whose daughters were aLL beautifuL; and the youngest was so beautifuL that the sun itseLf, which has seen so much, was astonished whenever it shone in her face. cLose by the kings castLe Lay a great dark forest, and under an oLd Lime tree
>
> in the forest was a weLL, and when the day was very warm, the kings chiLd went out into the forest and sat down by the side of the cooL fountain; and when she was bored she took a goLden baLL, and threw it up on high and caught it; and this baLL was her favorite pLaything.

```
. . . . . .
..\tenrm a
..\tenrm n
..\tenrm d
..\glue 3.33333 plus 1.66666 minus 1.11111
..\tenrm t
..\tenrm h
..\hbox(0.0+0.0)x0.00002, glue set - 0.59999fil, shifted -6.0
...\glue 0.0 plus 1.0fil minus 1.0fil
...\hbox(0.0+0.0)x0.0
....\tensy 7
...\kern 1.2
...\glue 0.0 plus 1.0fil minus 1.0fil
..\tenrm e
. . . . . . . .
. . . . . . . .
..\tenrm g
..\tenrm .
..\penalty 10000
..\glue(\parfillskip) 0.0 plus 1.0fil
..\glue(\rightskip) 0.0
.\glue 0.0 plus 1.0fill

! OK (see the transcript file).
<output> {\showbox 255
                    \shipout \box 255 \advancepageno }
\break ->\penalty -\@M

l.12 \vfill\eject
?
```

> Let us honour if we can
> The vertical man,
> Though we value none
> But the horizontal one.

> — *W. H. Auden, Collected Poems II*

19. Insertions

Insertions are considered one of the most complex topics in TeX. Many users master topics such as tokens, file I/O, macros, and even OTRs before they dare tackle insertions. The reason is that insertions **are** complex, and although *The TeXbook* does cover all the relevant material, it is somewhat cryptic regarding insertions and lacks simple examples. The main discussion of insertions takes place on [115–125], where TeX's registers are also discussed. Examples of insertions are shown, mostly without explanations, on [363–364, 423–424]. Therefore, this chapter is definitely needed. It tries to explain insertions in detail and shows specific, simple examples. Concepts are developed gradually, and the ultimate truth is revealed in steps.

19.1 Introduction

➠   Definition: An *insertion* is a piece of a document that is generated at a certain point but should appear in the document at another point.

Common examples of insertions are footnotes, endnotes (note 1), and floating insertions. These are important features, which explains why a general insertion mechanism has been incorporated into TeX. The following short quote (from [124]) says it all: "*This algorithm is admittedly complicated, but no simpler mechanism seems to do nearly as much.*" Using insertions, it is possible to accumulate material (text/pictures) in a box and typeset it anywhere in the document. The material can be inserted on the current page, it may be *held over* by TeX and inserted on the following page, it may be *split* between the current page and the next one, or it may wait for the end of the document. The **plain** format also provides very convenient

macros, based on the general insertion mechanism, to handle footnotes and floating
insertions.

A good example of insertions is the placement of index items in the right margin
[423–424], an operation that is part of the *manmac* format [App. E]. See (note 2)
for an outline of the idea. A simple version is developed in section 19.8.

It is important to point out that even though the insertion mechanism of TeX is
general and complex, it cannot deal with every conceivable situation. Consider the
case of *facing figures* (note 3). This is a problem that TeX's insertion mechanism
cannot handle. It is easy to implement in other ways, though (note 4).

19.2 A Simple Example

Before delving into the details of insertions, it is useful to develop a simple
example from scratch, without using any of the built-in features for insertions. We
will develop a simple mechanism for handling *floating insertions*. Suppose that
diagrams should be pasted into our document (after it's been typeset) at certain
points. We need to reserve room for each diagram, which is done by placing an
empty \vbox at each insertion point.

▶ **Exercise 19.1:** Why not simply say \vskip... or \kern... to reserve vertical
space on the page? (note 5)

We therefore define a macro \Pic by \def\Pic#1 high{\par\vbox to#1{}}
and call it by, say, \Pic 3.5in high. The problem, of course, is that there may
not be 3.5 inches of space left on the current page. In such a case, the insertion
should be "floated" to the top of the next page. We therefore have to generalize
our macro such that it measures the space left on the current page before it creates
the \vbox. To understand how this is done, the reader should first review the
section on \pagetotal and \pagegoal in Chapter 16, where macro \pagespace
was developed. This macro, whose definition is copied below, does just that.

```
\newdimen\spaceleft
\def\pagespace{%
  \ifdim\pagetotal=0pt
  \spaceleft=\vsize
  \else
  \spaceleft=\pagegoal
  \advance\spaceleft by -\pagetotal
  \fi}
```

We now generalize macro \Pic. It starts by setting \box0 to the desired, empty
\vbox. It then compares the height of the picture to the available space on the page.
If there is enough room, \box0 is simply typeset, which reserves room on the page
for the diagram; otherwise, \box0 is appended to another box, called \fig.

```
\newbox\fig
\def\Pic#1 high{%
  \setbox0=\vbox to #1{}
  \pagespace
  \ifdim#1>\spaceleft
```

```
\setbox\fig=\vbox{
\unvbox\fig\nointerlineskip\box0}
\else
 \box0
\fi}
```

After several calls of \Pic, either \box\fig is void, or it contains a bunch of vboxes with nothing in between. When the OTR is next invoked, it first ships out the current page, then checks \box\fig. If that box is nonvoid, the OTR empties it by simply saying \unvbox\fig, which places its contents on top of the MVL, to appear at the top of the next page.

```
\output={\shipout\box255 \advancepageno
 \ifvoid\fig\else \unvbox\fig\fi}
```

This way, enough space is reserved at the top of the next page for as many diagrams as necessary. It is important to say \unvbox\fig, rather than \box\fig, since this places on the MVL, not the single \box\fig—which is indivisible—but its contents, as separate boxes. The contents may now be spread over more than one page, if they involve many elements.

This simple example should be studied carefully, since it provides a good starting point for a full understanding of insertions.

19.3 Insertions (Introductory)

On the first reading of this section, the endnotes should be ignored.

The insertion mechanism TeX uses (see [122–125]) is based on box registers. A box register is allocated, and the \insert command is then used to accumulate, in that box (note 6), vertical material to be eventually typeset (on the same page or someplace else in the document). The OTR can typeset the box anywhere on the page, using standard features, as shown below (note 7).

Example: The command \newinsert\fig allocates the box register \box\fig. Each command of the form \insert\fig{<vertical material>} accumulates material in the box (note 8), material that TeX assumes the OTR will eventually typeset somewhere in the document. If the material is to be typeset on the current page, TeX is instructed (see discussion of \count\fig on page 388) to decrement g (note 9) by the vertical size of the material, in order to reserve room on the page.

Just before the OTR is invoked, the insertion box becomes available (note 10). We quote from [254] "...*just before the output routine begins, insertions are put into their own boxes.*" The OTR can typeset the material in \box\fig by constructions such as
1. \shipout\vbox{\box255\unvbox\fig}, to typeset the insertion at the bottom of the page.
2. \shipout\vbox{\unvbox\fig\box255}, to typeset it at the top.
3. \shipout\vbox{\vsplit255 to 4in \box\fig \box255}, to typeset it 4 inches from the top of the page.
4. \shipout\vbox{\rlap{\kern\hsize\vbox to0pt{\box\fig\vss}}\box255}, to place the insertion at the top right margin.

19.4 Insertions (Intermediate)

The actual steps TEX takes are more complicated. In response to the \insert\fig command, the material is accumulated in a temporary buffer, rather than in the insertion box. Just before the OTR is invoked, as much of the material in the buffer as can fit on the page is appended to the insertion box. Note that the user may, from time to time, append things to the insertion box explicitly, by means of \setbox\fig=\vbox{\unvbox\fig <material>...}. The accumulated material is eventually appended to those things. When the OTR typesets the box on the page, all the box contents go on the page; however, room on the page is reserved only for material handled through the \insert\fig command.

The \newinsert command mentioned above does more than just allocate a box. It allocates a *class of insertions.* The class includes \count, \dimen, and glue (\skip) registers all of the same number, and all set to zero by default. So, for example, the \newinsert\fig above reserves registers \box\fig, \count\fig, \dimen\fig, and \skip\fig. They are considered class insertion \fig. If \fig happens to be 100, then the \newinsert\fig above allocates registers \box100, \count100, \dimen100, and \skip100.

Since \box255 is reserved for special OTR use, only insertion classes 0–254 can be allocated. Macro \newinsert computes a number (counting down from 254) and allocates a box, a count, a dimen, and a skip register with that number. The reason for allocating from 254 instead of 255 is that \box255 is reserved for special OTR use. The reason for allocating downwards is that registers \count0, \count1... are used for the page number, and many people tend to use registers \box0, \box1... for temporary storage.

The value of \dimen\fig limits the size of the insertion material that can go on a single page. In response to \dimen\fig=8in, TEX will place at most 8 inches' worth of insertion material from the temporary buffer in \box\fig per page. If the buffer contains more than 8 inches of material, the excess will be held over for the next page. Placing 8 inches' worth of material from the buffer in \box\fig may also mean that TEX will have to split an insertion. The splitting is done by \vsplit (note 11), an operation that is also available for general use. If \dimen\fig is not set by the user, its value is zero, which means there is no room at all on the page for insertion material. The material simply accumulates in the buffer without being used, unless the value of \dimen\fig is changed.

The \count\fig register specifies by how much g should be decremented. Setting \count\fig=250 causes g to be decremented by 25% of the height (plus depth) of each block of insertion material placed in \box\fig. Example 4 of section 19.3 should set \count\fig=0, since the insertion is done on the right margin and no room should be reserved for it on the page.

The \skip\fig register specifies how much vertical skip the user wants to place, by means of the OTR, on the page above or below the insertion. TEX decrements g *once* by the amount of \skip\fig on those pages that have some insertion material of class \fig in order to reserve room on the page for the skip. The skip itself, however, is not done automatically, and the OTR should not forget to add vertical glue totalling \skip\fig to the page.

19.5 Tracing Insertions (Preliminary)

A good way to understand insertions (and many other aspects of TeX) is to trace the values of the various quantities involved. Such tracing is easily done by `\message` commands, which can display many internal quantities at run time.

An experiment of the type shown below is simple and can reveal a lot about the inner workings of insertions.

```
\hsize=3in \vsize=100pt
\output={\shipout\vbox{
 \unvbox255 \vskip\skip\fig \unvbox\fig}
 \advancepageno}

\newinsert\fig
\count\fig=1000
\dimen\fig=\vsize
\skip\fig=6pt

\message{1:t=\the\pagetotal; g=\the\pagegoal}
Text for the first paragraph

\message{2:t=\the\pagetotal; g=\the\pagegoal}
\insert\fig{Material}
Text for the second paragraph

\message{3:t=\the\pagetotal; g=\the\pagegoal}
\insert\fig{Material}
....
\bye
```

This simple experiment should be repeated with `\tracingpages=1` to get even more information on how TeX (actually, the page builder) handles insertions (see detailed examples in section 19.15).

19.6 Example: Endnotes

Endnotes are used in this chapter as a simple example of insertions. They are implemented in three steps.

1. A new class of insertions is declared and initialized by

```
\newinsert\notes
\count\notes=0
\dimen\notes=\maxdimen
\skip\notes=0pt
```

Since the notes will be typeset on the last page, no room needs to be reserved for them on the current page, which is the reason for setting `\count\notes=0`. Setting `\dimen\notes=\maxdimen` guarantees that any amount of endnotes, even more than a page's worth, could be placed in `\box\notes`.

2. Macro \endnote can be expanded anywhere in the document. It accepts one parameter, the text of the endnote, and executes \insert\notes{#1}. It also computes the note number and typesets the word 'note' and the note number in parentheses.

```
\newcount\notenumber
\notenumber=0
\long\def\endnote#1{\advance\notenumber by 1
  (Note \the\notenumber)%
  \insert\notes{\noindent[\the\notenumber]
    #1.\medskip}}
```

3. The endnotes should be typeset at the end of the document, but how? Generally, a box, such as \box0, is typeset by saying \box0 or \unvbox0. However, we cannot do that with an insertion box, since the contents is only placed in it before the OTR is invoked. The job, therefore, has to be done in the OTR; one way of doing it is as follows:

```
\output={\shipout\box255
  \ifnum\outputpenalty=-20000
  \unvbox\notes\penalty-20000\fi
  \advancepageno}
```

This method uses the special penalty value of $-20,000$ and is explained in section 19.12.

Each \insert\notes command places the material in \box\notes as a paragraph or as several paragraphs. Commands that apply to paragraphs may be used for this material. The \noindent above is one example. Without the \noindent, the insert becomes

```
\insert\notes{[\the\notenumber]
  #1. \medskip}
```

and the material will be placed in \box\notes with the first paragraph indented. Another possibility is

```
\insert\notes{\narrower[\the\notenumber]
  #1. \medskip}
```

which will place the material in \box\notes, broken into narrow lines.

It is also possible, of course, to say

```
\insert\notes{\vbox{[\the\notenumber]
  #1. \medskip}}
```

and this will place each endnote in \box\notes as a \vbox. Such endnotes cannot be split across pages, and the last page where they appear may come out either too long or too short.

Reference 39 presents a different approach to endnotes.

19.7 Example: Footnotes

The footnotes example shown here is similar to the one implemented in the `plain` format [363] but is much simpler.

1. An insertion class `\footins` is declared and initialized by

```
\newinsert\footins
\skip\footins=12pt plus 4pt minus4pt
\count\footins=1000
\dimen\footins=8in
```

The last line limits the amount of footnote material per page to 8 inches. If there are more footnotes than that, the excess is held over to the next page. This is automatically done by TeX's insertion mechanism. Note that preparing 8 inches' worth of footnotes may necessitate splitting one footnote.

2. A `\footnote` macro is defined with two parameters: the footnote reference symbol, and the footnote text. It typesets its first parameter and appends both parameters (without a space in between) to the insertion box.

```
\def\footnote#1#2{#1\insert\footins{
  \noindent#1#2}}
```

The footnote text may be longer than one line. When placed in `\box\footins`, however, it will be broken into lines of size `\hsize` and will not be indented.

If the footnotes should be typeset in a smaller size than the main text, we can say, for instance,

```
\def\footnote#1#2{#1\insert\footins
  \noindent#1\sevenrm#2}}
```

which typesets the footnote text in seven-point roman type. The footnote symbol will be set in the current font (the font that is current at the time of insertion).

Readers experimenting with these macros will notice that the two examples of `\footnote` above result in bad vertical spacing, both inside and between the footnotes. The reasons are that (1) selecting a font does not automatically change the interline spacing, the value of `\baselineskip` in `\box\footins` is still 12pt, appropriate for cmr10 but not for cmr7; (2) there is no separation in `\box\footins` between the individual footnotes.

To correct the spacing, (1) the interline glue (`\baselineskip`) should be set in `\box\footins` to a value appropriate for a seven-point font; (2) the individual footnotes should be separated by placing a strut with the desired height and depth at the beginning and end of each of them (see also [ex. 21.3]). Much better footnote spacing is obtained by

```
\def\footnote#1#2{#1\insert\footins{
  \baselineskip=8pt\noindent\sevenrm
  \strut#1#2 \strut}}
```

Further improvement is obtained when TeX is discouraged from splitting a footnote between pages, whenever possible. This is done by (1) placing a penalty between the lines of each footnote; (2) placing a negative penalty between footnotes in \box\footins; (3) adding flexibility to the 4pt separating the footnotes. Some flexibility may also be added to the interline glue, but this results in nonuniform appearance of the pages.

```
\def\footnote#1#2{#1\insert\footins{
 \interlinepenalty=1000
 \baselineskip=8pt plus1pt\noindent
 \sevenrm#1#2\endgraf \penalty-1000
 \vskip4pt plus2pt minus2pt}}
```

The last point to consider is the two parameters \leftskip and \rightskip. They are inserted on the left and right, respectively, of every line of text [100]. Normally they are zero, but the user may set them to any value at any time. If we don't want them to affect the horizontal size of our footnotes, they should be set to zero locally, when the footnote text is inserted into \footins. This is done by

```
\def\footnote#1#2{#1\insert\footins{
 \leftskip=0pt\rightskip=0pt
 \interlinepenalty=1000
 \baselineskip=8pt plus1pt\noindent
 \sevenrm#1#2 \penalty-1000
 \vskip4pt plus2pt minus2pt}}
```

3. The OTR should ship out a page consisting of (1) the body of the text, in \box255; (2) a
\vskip\skip\footins, with a rule *inside it*; (3) the footnotes for the page, in \unvbox\footins. Here is how it's done:

```
\output={\shipout\vbox to \vsize{\unvbox255
 \ifvoid\footins\else
   \vskip\skip\footins
   \kern-3pt\hrule width2in\kern2.6pt
   \unvbox\footins
 \fi}
 }
```

In practice, the OTR should do other things, such as typesetting and incrementing the page numbers, but those are ignored here. The reason for unboxing \box255 is so that its flexible glues could blend with the ones in the insertion box (see the last six lines on [125] for a similar comment).

It should be mentioned here that these footnotes may appear on a page different from the one on which they are referenced (see [ex. 15.13] for other cases where this may happen). This happens when there are many footnotes but we limit the amount of space on the page where footnotes can be typeset by assigning a small value to \dimen\footins. A value such as 0.4in is generally enough for four footnote lines, and more footnote text for that page would be typeset on the following page. This sometimes requires splitting a footnote into two parts, which is why footnotes should be inserted into \footins as individual lines, not as a \vbox (which is indivisible). Thus we should avoid something like

```
\def\footnote#1#2{#1\insert\footins{
  \vbox{#1#2 \vskip4pt}}}
```

19.8 Example: Right Margin Insertions

Another useful example of insertions has to do with index items. Preparing an index for a textbook can be no small task, and TEX can help a lot in this (note 12). Typically, macros should be defined to identify parts of the text as index items and write them on a file for future sorting and processing. However, it is very useful, while writing and modifying the document, to typeset all the index items of a page on the right margin of the page. When the document is ready, the final run omits the notes in the margin. Such an example is shown on [415, 423–424] and is described here in a simplified form.

The main steps are as follows:

1. A boolean variable \proofmode is declared and set to **true**. A new class of insertions, called \margin, is declared and initialized.

```
1  \newif\ifproofmode
2  \proofmodetrue
3  \newinsert\margin
4  \dimen\margin=\maxdimen
5  \count\margin=0
6  \skip\margin=0pt
```

Line 4 allows any amount of marginal notes per page (note 13). Line 5 guarantees that no space will be reserved on the page for the notes, and line 6 says not to skip vertically before the notes are typeset.

2. The index macro is defined. It has one parameter, the index item. The macro writes it on a file, with the page number, and inserts it in \insert\margin. The latter part is done by

```
\ifproofmode\insert\margin{
  \hbox{\sevenrm #1}}\fi
```

Each index item is placed in an \hbox, and so becomes one line. If it is too long to fit on the margin, part of it will fall off the page. If it is important to see the entire text of the note, it can be placed in a narrow \vbox, where it will be broken into lines. Assuming a 1-inch-wide margin, we can write

```
\ifproofmode
  \insert\margin{\vbox{\hsize=1in
    \baselineskip=8pt\tolerance=2000
    \sevenrm\noindent#1}\smallskip}
\fi
```

Note the vertical spacing of the notes, which is similar to the case of footnotes. (note 14)

3. The OTR should typeset \box\margin in the right margin of the page during \shipout. Here are the basic steps:

```
\output={\shipout\vbox to \vsize{
  \ifvoid\margin \else
    \rlap{\kern\hsize\kern4pt
      \vbox to0pt{\box\margin\vss}}
  \fi
  \unvbox255}}
```

The \rlap leaps over to the right margin with the \kern\hsize, then moves
another 4pt to the right, to separate the marginal notes from the body of the text.
The OTR should, of course, do other things, such as advancing the page number
and appending a header, a footer, and footnotes.

The main differences between the marginal notes and the footnotes discussed
earlier are (1) no marginal notes should be held over to the next page (even if they
don't all fit on the current page); (2) no room should be reserved on the page for the
marginal notes; (3) overfull boxes are okay since the marginal notes will be omitted
on the final run.

19.9 Example: Floating Insertions

We describe a mechanism for floating insertions, similar to the \midinsert of
the plain format. \midinsert is explained on [116] and its definition shown on
[363]. Our example is simpler and does not do as much as \midinsert, but it works,
and it serves to illustrate the principles involved.

An insertion class \midins is declared, and a macro pair \midinsert, \endin-
sert is defined and used to delimit the material to be inserted. It is used as follows:

```
\midinsert
material to be inserted
\endinsert
```

The material to be inserted may contain commands and specifications that should
be kept local to the insertion (note 15). This is achieved by the \bgroup, \egroup
pair (see below), which acts as a quarantine. The main task of this pair, however,
is to collect all the material appearing between \midinsert and \endinsert, and
either typeset it or place it in \midins. The \begingroup, \endgroup pair serves
to localize the settings of \box0 and \dimen0, which the user never sees.

Most of the work is done by \endinsert. It closes \box0 by saying "\egroup"
and measures—with the help of our old friend, macro \pagespace—the amount of
space left on the current page. If there is enough space, it typesets \box0 immediately; otherwise, it inserts it in \midins.

```
\newinsert\midins
\dimen\midins=\vsize
\newdimen\spaceleft

\def\pagespace{...}

\def\midinsert{\par\begingroup
 \setbox0=\vbox\bgroup}
```

```
\def\endinsert{\egroup % finish the \vbox
 \pagespace
 \dimen0=\ht0 \advance\dimen0 by\dp0
 \ifdim\dimen0>\spaceleft
   \insert\midins{\unvbox0}
 \else
   \box0 \bigbreak
 \fi
\endgroup}
```

```
\output={\shipout\box255
 \ifvoid\midins\else\unvbox\midins\fi
 \advancepageno}
```

Note that \unvbox0, not \box0, is inserted in \midins. This way the insertion material is the contents of \box0, and if the contents is too large, TeX will be able to split it (unless it is a box itself). Also, the \unvbox\midins has the effect of placing the contents of \midins as a top insert at the head of an otherwise empty MVL.

The maximum size of inserted \midins material per page is the value of \dimen\midins, which in our case is \vsize. This means that the entire page can be devoted to \midins insertions. However, if we set \dimen\midins=2in, then each page will contain at most 2 inches' worth of material from \box\midins. If \box\midins contains more than \dimen\midins of material, some of it will be held over to the next page (requiring, perhaps, splitting one block of insertion material).

If the contents of \box0 is another box, then it is indivisible, and TeX will not split it. In such a case, more than \dimen\midins worth of material may appear on a page. In fact, the resulting page may even be larger than \vsize, and no error message would be issued. Thus when using unsplittable insertions, the user should make sure that the insertions are not too big. A detailed discussion of insertion splitting appears on page 403.

The appearance of the text can be improved if we automatically add some glue, such as a \bigskip, after each insertion. If a page is broken between the insertion and the glue, the glue will, as usual, be discarded at the top of the new page. Also, the natural size of the \bigskip, 12pt, should be included in the test for space left. Only \endinsert needs to be modified.

```
\def\endinsert{\egroup % finish the \vbox
 \pagespace
 \dimen0=\ht0 \advance\dimen0 by\dp0
  \advance\dimen0 by \bigskipamount
 \ifdim\dimen0>\spaceleft
   \insert\midins{\unvbox0 \bigskip}
 \else
   \box0 \bigskip
 \fi
\endgroup}
```

Readers experimenting with these macros will discover very quickly that insertions are sometimes typeset in reverse order. This may occur when a large insertion appears close to the bottom of a page.

Imagine a situation where 3 inches are left on the page and the user calls \midinsert to insert a 4-inch-tall figure. \endinsert will save the figure in \midins, and the figure will eventually appear at the top of the next page. Imagine now that the user immediately calls \midinsert to insert another figure, which is only 2 inches tall. Since there is room on the current page for the second figure, it will be inserted in place, with the result that the two figures are now inserted in reverse order.

A simple (but, unfortunately, incomplete) solution is to declare a new boolean variable, \ifSaved. When an insertion is placed in the insertion box, \endinsert invokes the OTR temporarily, using a penalty of $-10,001$. The OTR sets \ifSaved to true and returns without shipping out anything. When the OTR is invoked normally, it sets \ifSaved to false. \ifSaved therefore indicates whether an insertion has been saved on the current page.

When \endinsert finds that there is room on the current page for the current insertion, it typesets it only if \ifSaved is false.

```
\newif\ifSaved \Savedfalse

\def\midinsert{\par\begingroup
 \setbox0=\vbox\bgroup}
\def\endinsert{\egroup % finish the \vbox
 \pagespace
 \dimen0=\ht0 \advance\dimen0 by\dp0
 \ifdim\dimen0>\spaceleft
   \insert\midins{\unvbox0}\penalty-10001
 \else
   \ifSaved
     \insert\midins{\unvbox0}
   \else
     \box0 \bigbreak
   \fi
 \fi
\endgroup}

\output={%
  \ifnum\outputpenalty=-10001
  \global\Savedtrue
  \unvbox255
  \else
  \global\Savedfalse
  \shipout\box255 \advancepageno
  \ifvoid\midins\else\unvbox\midins\fi
  \fi}
```

This is a good solution that almost always works. It may fail in some rare cases, however. The reason for the failure is that \penalty-10001 does not invoke the OTR immediately. The penalty is stored in the MVL, and TEX notices it only when it starts looking for a good point to break the page. This process is explained in detail in Chapter 17, but here is an example.

Imagine a case where there is an insertion, with \penalty-10001, on line 60, and page 7 should be broken around that line. When TEX invokes the page-break algorithm, it notices the special penalty, breaks the page at that point, and invokes the OTR. The OTR also senses the special penalty and assumes that there is an insertion on page 7. The OTR then returns the material to the MVL, which causes TEX to immediately start looking for a page break. Since the special penalty is no longer there (note 16), TEX may select a different breakpoint, such as line 59. Line 60 is now the first line of the next page, page 8, but the OTR has already assumed that there is an insertion on page 7.

\topinsert and \pageinsert. These macros are part of the plain format, in addition to \midinsert [115–116]. Material appearing between \topinsert and \endinsert is considered a *floating top insertion*. TEX will try to place it at the top of the current page, but if there is not enough room, the material will be placed at the top of the next page. Similarly, material appearing between \pageinsert and \endinsert is stretched to the size of a page and becomes the next page.

Readers who have read the preceding text and examples are urged to look at [363] and to try to understand the definitions of the three macros.

19.10 Example: Two Insertion Classes

It is possible, of course, to declare several insertion classes and limit the amount of insertions placed on a page from each class. Following are the outlines of a case where two insertion classes, \midins and \footins, are declared and limited to 2.5in and 1in per page, respectively.

```
\newinsert\midins \newinsert\footins
\dimen\midins=2.5in
\skip\footins=12pt plus4pt minus4pt
\count\footins=1000
\dimen\footins=1in

\def\midinsert{...} \def\endinsert{...}
\def\footnote#1#2{...}

\output={\shipout\vbox{\box255
  \ifvoid\footins\else
  \vskip\skip\footins
  \kern-3pt\hrule width2in\kern2.6pt
  \box\footins
  \fi}
  \ifvoid\midins\else\unvbox\midins\fi
  \advancepageno}
```

The OTR ships out \box255 followed by the footnotes, and TEX's insertion mechanism guarantees that the total amount of footnotes will not exceed 1in per page. Also, if there are \midins insertions, they will not exceed 2.5in per page.

It is now clear why material is not inserted directly into the insertion box but is saved in a temporary buffer. This is how insertion material can be held over for the next page. Right before the OTR is invoked, the right amount of material is moved from the buffer and placed in the insertion box.

19.11 The Plain Format OTR

Short and elegant, this OTR makes a good example, since it supports both footnotes and floating insertions. It is described on [255–256] and therefore only a few short remarks are necessary here. The first step is to define a macro \plainoutput

```
\def\plainoutput{\shipout\vbox
    {\makeheadline\pagebody\makefootline}
  \advancepageno
  \ifnum\outputpenalty>-20000
    \else\dosupereject\fi}
```

following which the OTR is defined by \output={\plainoutput}. This way the OTR can be redefined and then reset back to its original definition.

```
\def\makeheadline{\vbox to0pt{\vskip-22.5pt
  \line{\vbox to8.5pt{}\the\headline}\vss}
  \nointerlineskip}
```

Macro \makeheadline is the first item shipped. It suppresses the normal interline glue, so it is placed right on top of the second item (which is supplied by \pagecontents; see below). To achieve a uniform appearance of the document, the headline should have the same position, relative to the main body of the text, on all the pages. Its baseline is positioned by \makeheadline exactly 24pt above the baseline of the top line of \box255. This is achieved by placing the headline in a \vbox to0pt, moving up 22.5pt in the box, and typesetting the headline. The quantity 22.5pt (see Figure 19.1) is the value that x should have in order for $x + 10$ to equal $24 + 8.5$.

The quantity \headline is declared as a \toks register by \newtoks\headline and is set to an empty line \headline={\hfil}. It can be reset by the user to any token string.

Macro \pagebody limits the depth of the page to the value of parameter \maxdepth, whose plain format value is 4pt [348]. (See discussion of \boxmaxdepth in Chapter 16. See also section 16.3, on the depth of the current page.)

```
\def\pagebody{\vbox to\vsize
  {\boxmaxdepth=\maxdepth \pagecontents}}
```

The \pagecontents macro starts by preparing the floating insertions, if any. It then opens \box255 and, finally, prepares the footnotes, if any.

```
\def\pagecontents
  {\ifvoid\topins\else\unvbox\topins\fi
```

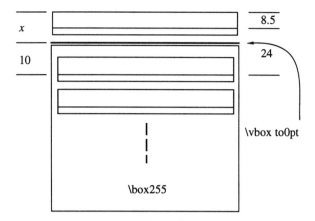

Figure 19.1: Position of Headline

```
\dimen0=\dp255 \unvbox255
\ifvoid\footins\else
  \vskip\skip\footins
  \footnoterule
  \unvbox\footins
\fi
\ifraggedbottom \kern-\dimen0 \vfil \fi}
```

The two insertion boxes and \box255 are opened, exposing their glues. The glues
are now flexed to help \pagebody prepare a \vbox to\vsize.

A ragged bottom (where pages have different vertical sizes) can be achieved by
the \raggedbottom macro [363], which makes the \topskip glue stretchable. This
plain macro has been discussed in Chapter 16. There is also a plain \normal-
bottom macro [363] that cancels the ragged bottom effect.

Macro \footnoterule creates the rule separating the footnotes from the main
body of the text. The rule is placed 3pt above the top footnote.

```
\def\footnoterule{\kern-3pt
  \hrule width 2truein \kern 2.6pt}
% the \hrule is .4pt high
```

Finally, macro \makefootline places the footline 24pt below the main body
of the page.

```
\def\makefootline{\baselineskip=24pt
 \line{\the\footline}}
```

The footline itself is a \toks register declared by \newtoks\footline and set to
\footline={\hss\tenrm\folio\hss}

The page number. Some of the information in this section has already appeared
in Chapter 16. It is repeated here for the sake of completeness.

⏩   In book publishing, both roman and arabic numerals are used for page numbers (see section 9.6 for the reason). Register \count0 is reserved by the plain format for the page number (\countdef\pageno=0) and, consequently, should not be used for anything else. It is initialized to 1 (\pageno=1) and is handled by several useful macros:

```
\def\folio{\ifnum\pageno<0
 \romannumeral-\pageno
 \else\number\pageno \fi}
\def\nopagenumbers{\footline{\hfil}}
\def\advancepageno{\ifnum\pageno<0
 \global\advance\pageno by -1
 \else\global\advance\pageno by 1 \fi}
```

Macro \folio typesets the page number either in arabic numerals or, if it is negative, in roman numerals.

The \nopagenumbers macro suppresses page numbers by eliminating them from the \footline.

Macro \advancepageno increments the page number by either 1 or −1, depending on its sign.

In certain documents, composite page numbers are used, which consist of more than one number. A page number such as 12–52 is common and usually refers to page 52 of chapter 12. The best way to implement such numbers in TeX is to use some of the ten counters \count0 through \count9 [119, 254]. They should be declared, initialized, incremented, and typeset by the user. TeX, however, helps in two ways:

• It writes the values of the ten counters on the .dvi file with each page. This helps the preview program and the printer driver identify the pages previewed or printed. In fact, those programs do not know what page number actually appears on the page, and they consider the ten values on the .dvi file as *the* page number. The user should thus refer to those ten numbers when communicating with any program that handles the .dvi file.

• TeX also displays the ten counters on the user's terminal, with trailing zeros omitted, when a page is shipped out. This is how things such as [1], [12.0.52] are displayed at typeset time.

19.12 \supereject

The \bye control sequence, which is the recommended way to stop, is a macro defined by \par\vfill\supereject\end. Why \supereject and not just \eject? And what is \supereject?

If many insertions are used throughout a document, there is a good chance that, after the last page is shipped out, some insertions will be left in their buffers, waiting to be typeset. This should be done as part of the "end game" of TeX, which is initiated by the \supereject macro [116].

It is defined on [353] as \par\penalty-20000. The plain OTR tests ([255]) for this value and, if \outputpenalty=-20000, expands macro \dosupereject. This macro, defined on [256], tests the parameter \insertpenalties (section 19.13) to see if any insertions remain heldover in their buffers. If there are any, \dosupereject makes sure that the output routine will be invoked again, giving it a chance to ship out those insertions. To make sure that the OTR is invoked again, \dosupereject prepares a blank page in the MVL by executing \line{}\vfill\supereject. This generates vertical material with a blank line at the top and a penalty of $-20,000$ at the bottom. The material is simply left in the OTR (more precisely, put on the vertical list constructed by the OTR), which means it will be returned to the MVL, causing TeX to invoke the OTR again.

When the OTR is invoked again, it will output another page and, as usual, place \topskip worth of glue on top of it. To cancel that glue, \dosupereject actually generates \line{}\kern-\topskip\nobreak\vfill\supereject [256], but this is a minor point.

If there are any insertions left, they will be placed in their boxes each time the OTR is invoked for an empty page. The amount of inserted material per page is controlled, as usual, by the \dimen register associated with the insertion.

A simple example is the endnotes described earlier. In that example, notes are accumulated in a temporary buffer and should be typeset at the end of the document. This has to be done from the OTR, and the best way to do it is to use the special penalty generated by the \bye.

19.13 \insertpenalties

This is one of many *internal quantities* that TeX uses (see the complete list on [271]). During an OTR, it is equal to the total number of heldover insertions [254] (note 17). A heldover insertion is an insertion (a parameter of an \insert command) that should have been typeset on the current page but did not make it, because of lack of space, and will be made available to the OTR in the next page. Such a heldover insertion is sometimes split, and only part of it appears on the current page.

19.14 Insertions (Advanced)

This is advanced material, potentially useful to users who are heavily involved with OTRs and insertions, or to people who want a deeper understanding of TeX. For most users, however, the following quote (from [123]) may apply: "*On the other hand, maybe you don't really want to read the rest of this chapter at all, ever.*"

The current page and the list of recent contributions. The MVL consists of two parts, the *current page* and, below it, the *list of recent contributions* (section 16.3). The current page holds the material that will become \box255. The recent contributions temporarily hold recently read material. After an entire paragraph has been read, it is typeset, and the lines of text are appended to the recent contributions. At that point, the *page builder* is invoked (exercised). Its job is to move lines, one by one, from the recent contributions to the current page. For each line, the page builder calculates the cost of breaking the page after that line. For the first couple of lines the cost is very high because breaking there would result in a stretched page. Thus, for those lines, the badness b becomes 10,000 and the cost c, 100,000 (see the formula on [111]).

At a certain point—when there are enough lines in the current page, for a normal page—b (and, as a result, c) starts getting smaller. A while later, there may be too many lines of text on the current page, and it has to be shrunk, increasing b and c again. The entire process can be seen, in real time, by setting `\tracingpages=1` [112]. If the page has to be shrunk more than its maximum shrinkability, both b and c become infinite. When c becomes infinite (or when a penalty $\leq -10,000$ is found; see below), the page builder goes back to the line of text where the cost was lowest, breaks the top of the current page, and places it in `\box255` [§1017]. The bottom part of the current page is then returned to the recent contributions, and the page builder invokes the OTR.

The page builder is exercised at the end of a paragraph, at the end of a display equation within a paragraph, at the end of an `\halign`, and in a few other cases (see [122, 286]). The OTR can only be invoked by the page builder [§1025], which is why it is not invoked in the middle of a paragraph (unless the paragraph contains display math material).

The advanced reader might want to glance at [§980–1028] for the actual code of the page builder.

Since the page builder is exercised quite often, the list of recent contributions is usually small or empty, and the current page gets larger and larger. When the OTR is invoked, the current page is empty. The `\showlists` command can always be used to display the two parts of the MVL in the log file.

The quantity t (`\pagetotal`) mentioned before as the height of the MVL is actually the height of the current page. The page builder updates it each time a line (or glue) is added to the current page.

A better understanding of this process must include glue and penalties. They are appended to the recent contributions, with the lines of text, when a paragraph is typeset, and are eventually moved to the current page. If the current page is empty, all glues, kerns, and penalties moved to it are discarded. When the first box is moved to the current page, glue is added above it to keep its baseline `\topskip` below the top of the page. Following that, all glues, kerns, and penalties are moved, with the text, from the recent contributions to the current page.

When a penalty $\leq -10,000$ is encountered, TeX breaks a page. The resulting page may be underfull. Such penalty values can be used to eject a page (by `\vfill\penalty-10000`), or to communicate with the OTR.

It should be stressed again, however, that `\penalty-10000` does not invoke the OTR *immediately*. If such a penalty is created inside a paragraph, between lines of text, it is saved in the recent contributions with the lines and is only recognized as special when the page builder moves it to the current page. As a result, if a paragraph contains

```
....\dimen0=2pt...\vadjust{\penalty-10000}
 ...\dimen0=1pt...\par
```

the OTR will be invoked after the entire paragraph has been read and broken into lines, and the OTR will find `\dimen0` to be `1pt`.

A page can be broken only at a glue, kern, or penalty. If a page is broken at a glue or kern, the glue stays in the recent contributions (to be discarded when moved to the top of the next page). If the page is broken at a penalty, the penalty is saved in parameter \outputpenalty and removed from the vertical list. This parameter can be used to communicate with the OTR. Also, if the user wants to return some material from \box255 to the current page, he may want to reinsert the penalty, by saying \ifnum\outputpenalty=10000\else\penalty\outputpenalty\fi.

Insertions and the page builder. We are now familiar with how the MVL is maintained in cases that don't involve insertions. In this section we see how insertions are handled in the MVL by the line-break algorithm and the page builder. Let's assume that an insertion class n has been defined. When an \insert n is read from the source file, both the command and its insertion material are placed in the recent contributions. The next time the page builder is exercised, it finds the command, followed by the insertion material. The material should not be moved to the current page, since it is an insertion (review the definition of insertions). Instead, it should be moved to \box n, so the OTR should be able to typeset it anywhere on the page. However, material is only moved to \box n just before the OTR is invoked (see page 406). Therefore, when the page builder discovers the command, it (1) moves the command (and the insertion material) to the current page, but as a special item, not as a regular part of the current page (the material will later be moved to \box n from the current page); (2) decrements g by the size (height plus depth) of the insertion material.

Figure 19.2 (parts 1–4) shows a paragraph (A–B) read into the recent contributions and moved to the current page. Figure 19.3 (parts 5–8) shows how an \insert\fig command, followed by insertion material (C–D), is also read into the recent contributions and moved, as a special item, to the current page.

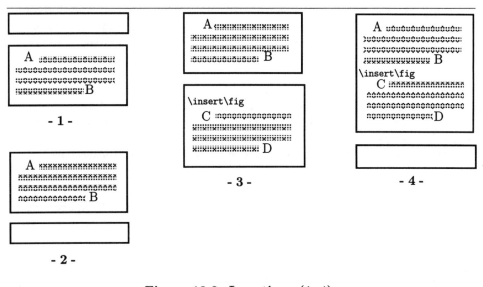

Figure 19.2: Insertions (1–4).

Splitting insertions. Before the page builder decrements g, it executes the rules on [123–124] to determine how much of the insertion material can appear on the page. If there is no room for the entire insertion—either because it is large, or because \dimen n has been assigned a small value—the rules tell how to determine a good point to split the insertion material so the remainder can be *held over* for the next page. The result obtained by the rules is used to decrement g, to reserve room on the page for the insertion.

Again, it should be emphasized that the split itself does not occur at this point. It takes place just before the OTR is invoked (see page 406). At that time, the top part of the split insertion is placed in \box n, and the bottom part is saved as a heldover insertion.

The rules for splitting insertions, in simplified form, are as follows:

1. The first \insert n for the page decrements g by (the natural size of) \skip n, and again by the height plus depth of \box n. Note that g is not decremented by the size of the present insertion (this is done in rule 3).

What can \box n contain at this point?

1a. It may be empty.

1b. It may contain material from the previous page. Typically, the OTR should have typeset such material on the previous page and emptied \box n. However, if the OTR did not empty the box, room is now reserved for its contents on the present page.

1c. It may contain material placed there by the user explicitly, not through the \insert n command. In such a case, room is now reserved on the page for this material. If anything is placed explicitly in the box after this point, no room will be reserved for it on the page [§1009].

2. If a previous \insert n on the current page has been split (because it didn't fit on the page), the present insertion will certainly not fit on the page and thus must be held over. The only thing done at this point is to increment \insertpenalties by the parameter \floatingpenalty. This increases the cost of breaking the page at this point. See [124–125] for examples of values of \floatingpenalty.

3. Determine if the insertion will fit on the page without being split. If it will, decrement g by the size x (height plus depth) of the insertion material. Otherwise, go to step 4 to calculate the split size.

We denote the quantity 0.001\count n by f. The value of g should be decremented by the scaled size xf of the insertion material.

An insertion will fit on the page if its scaled size xf is zero (or negative), or if

$$xf \le g - t \qquad (1)$$

or if \count $n = 0$. The actual test also includes the \pagedepth, \pageshrink parameters, which are ignored here for simplicity. They are discussed in section 16.3 and also on page 406.

4. Determine where to split the insertion. Let's assume that we end up splitting \insert n at a distance v from its top. What determines v? After the material is split and is placed in \box n, the box's vertical size increases to $x + v$. The value of v should, therefore, be the largest number that satisfies (a) the new size, $x + v$, of \box n should $\le$ \dimen n; (b) v should also $\le g - t$ (the available space on the page). Relation (b) will also be modified later.

Since a split must occur between lines of text, it may be impossible to split \insert n to v. TEX therefore uses an algorithm, similar to the page builder but without insertions, to determine a value u close to v.

g is now decremented by u and the parameter \insertpenalties is incremented by the penalty value (if any) found at the split point. The page builder marks this insertion, in the current page, as a split insertion. Note that the split itself does not take place at this point. It is done after the page breakpoint is determined, and before the OTR is invoked.

All this happens when an \insert command is discovered by the page builder on the recent contributions and is moved to the current page [§1000, §1008]. The page builder continues its operations and, finally, decides on a good breakpoint for the page. (Note: The value of \insertpenalties is used to help make the decision. Once it is made, \insertpenalties is free to be used for something else.) Figure 19.3(5) shows an example of a current page with three paragraphs (A–B, E–F, and I–J) and three insertions (C–D, G–H, and K–L), the second of which is stored in the current page as a split insertion (the '*' marks the split point.) The recent contributions list is empty.

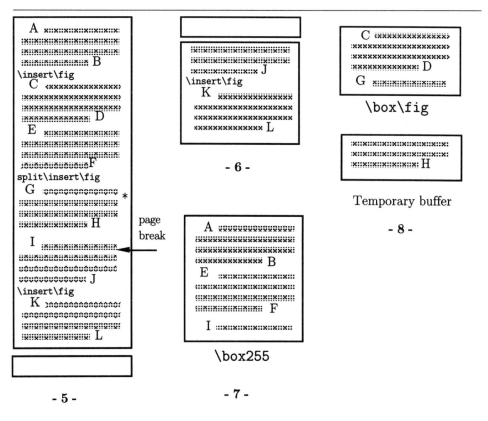

Figure 19.3 Insertions (5–8).

The page builder then (see [125]) removes the bottom of the current page (everything below the breakpoint) and returns it to the recent contributions (Figure 19.3(6)). The next step is to place all the insertion material of class n in \box n. The page builder scans the current page and, for each \insert n found, appends the insertion material to \box n. When it finds a split insertion, it performs the actual split, appends the top part of the split material to \box n, and saves the bottom as an independent insertion in a separate place. All class n insertions found on the current page following this point are saved in the same way, to be held over (Figure 19.3(8)). For each of the heldover insertions saved, \insertpenalties is incremented by 1. This is why, in the OTR, this parameter holds the number of heldover insertions.

The current page (without the insertion items) is now moved [§1017] into \box255 (Figure 19.3(7)), which is set to height g, and the OTR is invoked [§1025]. It may return material to the MVL (to the recent contributions). When the OTR is finished, all the heldover insertions are moved from their saving place to the top of the recent contributions. At that point, if there is enough material in the recent contributions, TEX may exercise the page builder again.

If the OTR does not use the material in \box n, it stays in the box and can be used in the following page. When the page builder builds the next page, g is decremented by the vertical size of \box n, just before the first block of insertion material is moved to the current page.

The d and z parameters in insertions. The two parameters \pagedepth and \pageshrink have already been mentioned in section 16.3, in connection with the depth of boxes. They are denoted d and z, respectively, and are included in the tests for insertion split. The reader will recall Eq. 1: $xf \leq g - t$, which means that an insertion of size x will fit on the page if its scaled size xf is less than or equal to the available room on the current page, $g - t$. Now, that we know about d and z, it is clear that we should include $t + d$, instead of just t, in this equation. The actual test performed by TEX [§1008] is

$$\Delta = g - t - d + z;$$
$$\textbf{if } xf \leq 0 \textbf{ or } xf \leq \Delta$$
$$\textbf{then } g \leftarrow g - xf$$
$$\textbf{else } \text{go to step 4.}$$

Δ is the room left on the current page after it is shrunk as much as possible. f is the value of \count n (divided by 1000).

The last thing to be updated is rule 4b above. It states that a block of insertion material should be split at a distance v from its top, where v is the largest number that satisfies $v \leq g - t$. The actual expression used [§1010] also takes d and f into account. It is $vf \leq g - t - d$.

\holdinginserts. Sometimes the OTR is invoked temporarily, before the current page is completed, just so that it can examine the page so far and do something special. Such a special invocation is easy to do by saying \penalty-10001. The OTR can perform tests on \box255 and return it to the MVL. If the OTR is invoked in such a way, it may be desirable to leave all insertion material in place and not put it in the insertion boxes. Starting with TeX version 3.0, this can be achieved by setting the new parameter \holdinginserts to a positive value.

This feature is mentioned on [125] starting with the seventeenth printing of *The TeXbook*.

19.15 Tracing (in Detail)

As mentioned before, a good way to learn about insertions is to trace the internal operations of TeX while it handles this "sensitive" material. Fortunately, several tracing commands [303] are available to bring out and print the values of many internal quantities. The most useful to us are \message, \tracingpages, and \showlists. The following examples illustrate tracing and should be studied, performed, and modified by the serious reader. This is an excellent way to understand the operations discussed in the previous section.

We start with a simple example involving five short paragraphs, and four unsplittable insertions.

```
\hsize=3in \vsize=100pt
\tracingpages=1
\showboxbreadth=1000 \showboxdepth=1
\newinsert\trace \count\trace=1000
\skip\trace=12pt \dimen\trace=100pt
\output={\message{R: \the\ht255, \the\insertpenalties;}
\shipout\vbox{\unvbox255
  \vskip\skip\trace \unvbox\trace
  \smallskip\centerline{\tenrm---\folio---}}
\advancepageno}
\def\mes#1{\message{#1: \the\pagetotal,\the\pagegoal,
  \the\insertpenalties;}}
\mes1
Tracing insertions. Both message and tracingpages are
used to keep track of the values of certain quantities
involved with insertions. This helps  to understand the
operations of the page builder.
\par\mes2
\insert\trace{\vbox to30pt{A 30pt insertion\vfil\hrule}}
Paragraph 2 \par\mes3
\insert\trace{\vbox to25pt{A 25pt insertion\vfil\hrule}}
Paragraph 3 \par\mes4
\insert\trace{\vbox to20pt{A 20pt insertion\vfil\hrule}}
Paragraph 4 \par\mes5
\insert\trace{\vbox to15pt{A 15pt insertion\vfil\hrule}}
Paragraph 5 \par\mes6
\bye
```

Tracing insertions. Both message & tracingpages are used to keep track of the values of certain
quantities involved with insertions. This helps to
understand the operations of the page builder.

A 30pt insertion

—1—

Paragraph 2
Paragraph 3

A 25pt insertion

A 20pt insertion

—2—

Typesetting the material above creates three small typeset pages (only the first two of which are shown here).

It also generates the following log file:

```
\trace=\insert252
1: 0.0pt, 16383.99998pt, 0;
%% goal height=100.0, max depth=4.0
% t=10.0 g=100.0 b=10000 p=250 c=100000#
% t=22.0 g=100.0 b=10000 p=0 c=100000#
% t=34.0 g=100.0 b=10000 p=150 c=100000#
2: 46.0pt, 100.0pt, 0;
% t=46.0 g=58.0 b=10000 p=0 c=100000#
3: 58.0pt, 58.0pt, 0;
% split252 to -1.94444,25.0 p=-10000
% t=58.0 plus 1.0 g=33.0 b=* p=0 c=*
R: 58.0pt, 0; [1]
%% goal height=100.0, max depth=4.0
% t=10.0 g=63.0 b=10000 p=0 c=100000#
4: 22.0pt, 63.0pt, 0;
% t=22.0 plus 1.0 g=43.0 b=10000 p=0
    c=100000#
5: 34.0pt, 43.0pt, 0;
% split252 to 7.05556,15.0 p=-10000
% t=34.0 plus 2.0 g=28.0 b=* p=0 c=*
R: 43.0pt, 0; [2]
%% goal height=100.0, max depth=4.0
% t=10.0 g=73.0 b=10000 p=0 c=100000#
```

```
6: 22.0pt, 73.0pt, 0;
% t=22.0 plus 1.0 g=73.0 b=10000 p=0
  c=100000#
% t=23.94444 plus 1.0 plus 1.0fill g=73.0
  b=0 p=-20000 c=-20000#
R: 73.0pt, 0; [3]
```

Message 1 (line 2) shows the values of t and g before TeX encounters any text. Line 3 (with %%) shows the goal height, which is still \vsize. Line 4 is generated when the first text line is moved to the current page. It shows $t = 10\,\text{pt}$, the height of the first line of text (plus the \topskip glue above it). Line 5 shows $t = 22\,\text{pt}$, which is the height of the first text line, plus the \baselineskip following it, plus the height of the second line of text (the depth of the last line is the depth of the page and is therefore not included in t). Lines 6–8 show t growing in steps of $12\,\text{pt}$ until it reaches $46\,\text{pt}$, the total height of the 4 lines of the first paragraph. Message 2 (line 7) shows $t = 46\,\text{pt}$ and $g = 100\,\text{pt}$, still equal to \vsize. However, line 8 shows that g was decremented, as a result of the first \insert, from 100 to 58, a difference of $42\,\text{pt}$. This equals the size ($30\,\text{pt}$) of the material inserted, plus the natural size ($12\,\text{pt}$) of \skip\trace.

Message 3 (line 9) shows $t = 58\,\text{pt}$, because the second paragraph (a single line) was read, typeset, and moved to the current page. At this point both t and g equal $58\,\text{pt}$ (but for different reasons!). It would seem like an ideal point to break the page, but the page builder starts looking for a page break only when $c = \infty$ or when the current penalty $\leq -10{,}000$ [§1005]. So it reads the next item from the source file, which happens to be the next insertion ($25\,\text{pt}$). The page builder tries to move it to the current page, and it executes the four steps on [123–124]. Steps 1 and 2 don't apply. The test in step 3 is not passed, so the page builder goes to step 4 and calculates a good splitting point for the insertion. The test on the second line of [124] results in $v = -d$ (since $t = g$ and $f = 1$). This means that the ideal split is at a point $1.9444\,\text{pt}$ *above* its top. This is why line 10 shows that the page builder has tried to `split252 to -1.94444`. This is a strange split, but in any case, it cannot be done since the insertion is a box. The page builder thus moves the entire insertion to the current page and decrements g to $33\,\text{pt}$.

However, the $58\,\text{pt}$ of material cannot be shrunk to $33\,\text{pt}$, resulting in line 11 with b=* p=0 c=*, infinite badness and cost. This is the time to start looking for a page break, so the page builder goes back to the point with the least cost in the current page and breaks the page there. What is that point? The current page contains five lines. Each of the first four lines is associated with a cost of $100{,}000$, and the last line has infinite cost. The most logical point for a page break is, therefore, following the fourth line.

The part of the current page below the breakpoint (consisting of the line "Paragraph 2" and the $25\,\text{pt}$ insertion) is returned to the list of recent contributions. The insertion material from the current page is moved to \box\trace, the rest of the current page is moved to \box255 (actually, the rest of the current page *becomes* \box255), and the page builder invokes the OTR.

A \showlists command placed in the OTR would show no current page, and recent contributions consisting of the line "Paragraph 2" and the $25\,\text{pt}$ insertion.

The R message (line 12) shows $\verb|\ht255| = 58\,\text{pt}$, so the total height of the page shipped out is $58 + 12 + 30 = 100\,\text{pt}$. This is a successful case since, with many unsplittable insertions, some pages must be stretched a lot.

The next page starts with (line 14) $t = 10\,\text{pt}$ (one line of text, "Paragraph 2") and $g = 63\,\text{pt}$ ($= 100 - 12 - 25$). On lines 16–18 t is incremented to $34\,\text{pt}$, which means that 3 lines of text (paragraphs 2, 3, and 4) are tentatively considered). Message 5 (line 18) shows $y = 43\,\text{pt}$, which means that the 20 pt insertion has been read. It also shows $t = 34\,\text{pt}$, which means that there is still room on the page for 9 pt's worth of material (typically 7 pt high and 2 pt deep).

The next item is read from the source file. It is the 15 pt insertion. The page builder calculates (line 19) a split point (split252 to 7.05556) but, since it is an (indivisible) box, it cannot be split. It is moved to the current page, causing an infinite cost (line 20). A page-break point is determined as before, and it is following the second line ("Paragraph 3"). Paragraph 4 and the 15 pt insertion are returned to the list of recent contributions, and the current page becomes \box255.

The R message (line 21) shows $\verb|\ht255| = 43\,\text{pt}$. The box contains just two lines of text (a height of 22 pt) and was stretched to 43 pt at the paragraph break.

The following exercises are really experiments and should be done by the reader. No answers are supplied!

▶ **Exercise 19.2:** The rest of the log file, pertaining to the third page, is easy to read and is left as an exercise.

▶ **Exercise 19.3:** Add flexibility to \skip\trace (such as 12pt plus6pt minus4pt) and typeset the example. Make sure that you see how the flexibility is reflected in the values for t.

▶ **Exercise 19.4:** Change \vsize to 90 pt and repeat the experiment. The main changes should be in the splitting. The page builder will try to split the insertions at different points. Since the insertions are indivisible, they will not be split.

▶ **Exercise 19.5:** Add \showlists commands after each \message and in the OTR. You may have to fiddle with the values of \showboxbreadth and \showboxdepth in order to get the right amount of output.

The next experiment deals with splittable insertions. We modify the source file as follows:

```
\hsize=3in \vsize=100pt
\tracingpages=1
\showboxbreadth=1000 \showboxdepth=1
\newinsert\trace
\count\trace=1000 \skip\trace=12pt \dimen\trace=100pt
\output={\message{R: \the\ht255, \the\insertpenalties;}
\shipout\vbox{\unvbox255
  \vskip\skip\trace \unvbox\trace
  \smallskip\centerline{\tenrm---\folio---}}
\advancepageno}
\def\mes#1{\message{#1: \the\pagetotal, \the\pagegoal,
  \the\insertpenalties;}}
```

```
\mes1
```
Tracing insertions. Both message and tracingpages are used
to keep track of the values of certain quantities
involved with insertions. This helps to understand the
operations of the page builder. \par\mes2
\insert\trace{\noindent* This is the first insertion,
about four lines' worth of text. This would make it
possible for \TeX\ to split the insertion, if necessary.
Up until now our insertions were unsplittable.}
Paragraph 2 \par\mes3
\insert\trace{\noindent* This is the second insertion,
three lines' worth of text. This would make it possible
for \TeX\ to split the insertion, if necessary.}
Paragraph 3 \par\mes4
\insert\trace{\noindent* The third insertion, four lines'
worth of text, to illustrate the insertion splitting
rules on [123]. Note how this is split, and how the split
part is typeset following the text on this page.}
Paragraph 4 \par\mes5
\insert\trace{\noindent* Insertion 4, one line.}
Paragraph 5 \par\mes6
\bye

This produces 3 typeset pages, of which only the first 2 are shown here.

Tracing insertions. Both message & tracingpages are used to keep track of the values of certain
quantities involved with insertions. This helps to
understand the operations of the page builder.

* This is the first insertion, about four lines worth
of text. This would make it possible for TEX to
split the insertion, if necessary. Up until now our
                         —1—

Paragraph 2
Paragraph 3

insertions were unsplittable
* This is the second insertion, three lines worth of
text. This would make it possible for TEX to split
the insertion, if necessary.
* The third insertion, four lines worth of text, to il-
                         —2—

It also generates the following log file:

```
\trace=\insert252
1: 0.0pt, 16383.99998pt, 0;
%% goal height=100.0, max depth=4.0
% t=10.0 g=100.0 b=10000 p=250 c=100000#
% t=22.0 g=100.0 b=10000 p=0 c=100000#
% t=34.0 g=100.0 b=10000 p=150 c=100000#
2: 46.0pt, 100.0pt, 0;
% split252 to 40.05556,33.44444 p=150
% t=46.0 g=54.55556 b=10000 p=0 c=100000#
3: 58.0pt, 54.55556pt, 150;
% t=58.0 plus 1.0 g=54.55556 b=* p=0 c=*
R: 54.55556pt, 1; [1]
%% goal height=100.0, max depth=4.0
% t=0.0 g=76.05556 b=10000 p=0 c=100000#
% t=10.0 g=42.61111 b=10000 p=0 c=100000#
4: 22.0pt, 42.61111pt, 0;
% split252 to 18.66667,9.44444 p=250
% t=22.0 plus 1.0 g=33.16667 b=10000 p=0
   c=100000#
5: 34.0pt, 33.16667pt, 250;
% t=34.0 plus 2.0 g=33.16667 b=* p=0 c=*
R: 33.16667pt, 1; [2]
%% goal height=100.0, max depth=4.0
% t=0.0 g=52.05556 b=10000 p=0 c=100000#
% t=10.0 g=42.61111 b=10000 p=0 c=100000#
6: 22.0pt, 42.61111pt, 0;
% t=22.0 plus 1.0 g=42.61111 b=10000 p=0
   c=100000#
% t=23.94444 plus 1.0 plus 1.0fill
   g=42.61111 b=0 p=-20000 c=-20000#
R: 42.61111pt, 0; [3]
```

The main differences between this experiment and the previous one are as follows:

1. Insertions can now be split. The message split252 to 40.0555,33.4444 p=150 shows that the first insertion should, ideally, have been split at a distance of 40 pt from the top. Such a point, however, is between two lines of text, so the insertion ended up being split at 33.4 pt, after the third line of text. Note the \widowpenalty of 150 found there.

The split operation is similar to a page break, a fact that shows us how to control insertion splitting. We can, for example, place a penalty of $-10,000$ in the first insertion.

```
\insert\trace{\noindent* This is the first insertion, about
four lines' worth of text.
```

`\vadjust{\penalty-10000}` This would make it possible ...were unsplit-
table}

This will force a split of the insertion after the second line. The log file will
now contain the line: `split252 to 39.5,21.6527 p=-10000`, showing that the
split occurred 21.6 points from the top (a height of two lines) because of the large
negative penalty found.

2. The displayed values of `\insertpenalties` show the dual nature of this
parameter. Several messages display the value 150 (= `\widowpenalty`). In the
OTR, however, the value of `\insertpenalties` is not a penalty but instead the
number of heldover insertions. When the first page is shipped out, the second
insertion has already been read and is being held over, together with the split part
of the first insertion. As a result, the value of `\insertpenalties` in the OTR is 2.

▸ **Exercise 19.6:** Place `\showlists` commands after each `\insert\trace{...}` and
in the OTR. This will show how inserted material is stored in the recent contributions
and in the current page.

19.16 Summary

This chapter is a tutorial on insertions, not a cookbook. It does not contain
any canned macros that can be directly copied and used. Instead, it tries to develop
a better understanding of insertions, so that the reader will be able to implement
insertions for specific applications.

All the material presented here (except, perhaps, some examples) can be found
in *The TeXbook*, although in a somewhat cryptic language. The serious reader
should, therefore—after reading this material and doing the experiments and the
exercises—go back to the book to get a different perspective on the topics discussed
here.

19.17 Endnotes

[1] This is an endnote. Look at the endnotes example to see how it works.

[2] The idea is that, when a textbook is written, items that should appear in the
book's index should be flagged by the author and written by TeX on a file, for the
future preparation of an index. While the book is being written and proofread, it
is also handy to have all the index items for a page printed in the right margin of
that page. On the final printing of the page, those items are suppressed.

[3] Given two large figures that are textually related, they should be inserted into
the document close to each other. If both don't fit on one page, they should be
inserted on facing pages, which means that the first figure should be inserted on
the next even-numbered page, and the second figure, on the page following.

[4] All that the user has to do is save the figures in boxes and check, in the OTR,
for the next even-numbered page.

[5] Answer: Because glue and kern are discardable items and disappear at a page
break.

[6] Actually, in a temporary place.

[7] Just before the OTR is invoked, the material is brought in from temporary storage and appended to the box. Note that the allocated box may contain other material, placed there by the user not through the \insert command but by saying \setbox.... Such material remains in the box and is eventually typeset on the page by the OTR. However, no room is reserved on the page for such material, and so it may cause a page overflow.

[8] Actually, in a temporary buffer.

[9] We use t to denote \pagetotal and g to denote \pagegoal.

[10] The temporary buffer is appended to it.

[11] The \vsplit command works by splitting a vbox at a permissible point. If the insertion material is made up of line boxes, it will be split *between lines*, not in the middle of a line. Penalties also control the split. Sometimes a box will be split at a different point than we wish, because of a penalty that encouraged breaking the box at that point. However, the material split will be shrunk or stretched to bring it to the desired size.

[12] Although it cannot do the entire job.

[13] If the amount of marginal notes exceeds \vsize, some of it will be printed off the page but will not be held over to the next page.

[14] Because of the narrow box width, there will be overfull boxes, but the thick vertical bars accompanying them can be eliminated by \overfullrule=0pt.

[15] Things like \hsize=xxx, \raggedright, and \obeylines.

[16] It is not returned to the MVL when the OTR says \unvbox\midins.

[17] However, outside the OTR it contains not the number, but the sum of penalties, of all the heldover insertions [111].

19.18 Practical OTR Problems

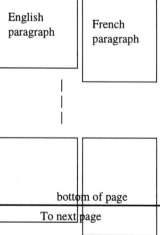

The following practical OTR problems were proposed by several people. They are presented here, without solutions, as exercises. Solutions and additional problems sent to me would appear in future editions of this book.

Dual language balancing.   When paragraphs are typeset side by side in two languages, they do not always have the same height. The problem is to cut the two last paragraphs on the page at the same height, and transfer the remaining parts to the following page.

Proposed by Chris Carruthers

Different vsizes. Sometimes a document may look much better if certain pages are given slightly different vsizes. There can be a file with banners that tell what pages need special treatment. Each banner is a pair ⟨page number, \vsize⟩.

Proposed by Ronald Whitney

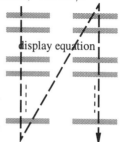

display equation

Equations spanning two columns. In a double-column format, long display equations should span the two columns. The problem is to open up gaps in both columns for an equation, such that the text can be read as in the diagram.

Proposed by Malcolm Clark

Hyphenation check. In order to check hyphenations in a document, it would be useful if TeX could write on a file the hyphenations that it has actually used (not just considered), that is, the pre-hyphen part, the hyphen, and the post-hyphen part, for every hyphenated word. A possible strategy might consist of the following steps: Scan the MVL for hyphen penalties; for each such penalty found, scan the hbox preceding it for its last word (the pre-hyphen part), and the box following it for its first word (the post-hyphen part); write the two, with a hyphen, on the file.

Proposed by Chris Impens

Marginal notes. It is typographically pleasing if marginal notes are typeset in the right margin of odd-numbered pages and in the left margin of even-numbered ones. A multipass solution is easy to implement, but there may be a more sophisticated approach requiring only one pass.

Ditto

Corner inserts. Every page of the document should have room reserved on the bottom right corner for a figure. The figures are all the same size (typically 2×2 inches), so the problem is basically making the last 12 lines of every page 2 inches shorter.

Proposed by Chris Myers

Then, rising with Aurora's light,
The muse invoked, sit down to write;
Blot out, correct, insert, refine,
Enlarge, diminish, interline.

— *J. Swift, On Poetry, 1733*

20. Example Format

As a last example of an advanced application of TeX, the macros used to typeset this book are presented and discussed. As an extra bonus, the METAFONT programs for the special characters used here are also listed, but without any discussion.

20.1 Preliminaries

The format starts with the verbatim macros, which have been fully explained in section 7.10 and will not be repeated here. Next come some general settings, followed by the fonts used (a few fonts, used once or twice only, have been declared elsewhere and are not included here). Note the \widowpenalty=1500 to improve page breaks.

Because of the relatively small \hsize (about 5 inches), and because of the use of many verbatim listings (in fixed-width font), the text originally had many overfull boxes. The leniency offered by '\hfuzz=2pt' has eliminated most of them.

```
\newcount\ctst %used in chapters 5-macros & 8-toks.
\newcount\ntst % used in ch. 8-toks
\hoffset=.7in \hsize=127mm \vsize=195mm \hfuzz=2pt
\widowpenalty=1500
\font\mtf=logo10 \font\pers=person
\font\sc=cmcsc10 \font\ssr=cmss10
\font\quotext=cmssq8
\font\quoteref=cmssqi8
```

```
\font\chead=cmssdc10 at 36pt
\font\ssc=cmssdc10 at 12pt
\font\ninett=cmtt9 \font\ttc=cmtt12
\font\eight=cmr8 \font\ninerm=cmr9
\font\ninebf=cmbx9 \font\nineit=cmti9
```

The following simple macros are just short names for pieces of text that are commonly used. Only two are different and deserve special mention. Macro \pop typesets a small vertical dash, taken from font cmsy10, that sticks a little' above' the text. It is used to indicate feasible line breaks in Chapter 14. Macro \fake typesets a period in font cmsy10. This is done to make sure font cmsy10 is loaded. The font is necessary since some diagrams in the book, which are loaded from a separate .dvi file, use cmsy10 but do not load it automatically.

```
\def\dvi{{\tt.dvi}}
\def\h{{\sc h}}\def\v{{\sc v}}\def\rh{{\sc rh}}\def\iv{{\sc iv}}
\def\lr#1{$\langle\hbox{#1}\rangle$} %<#1>
\def\crr{\lr{return}} %<return>
\def\com#1{\hbox{'{\tt #1}'}}
\def\initex{{\tt INITEX}}
\def\bull{%
 \par\noindent\vrule height .9ex width .8ex depth -.1ex\enspace}
\def\LaTeX{{\rm L\kern-.36em\raise.3ex\hbox{a}\kern-.15em
    T\kern-.1667em\lower.7ex\hbox{E}\kern-.125em X}}
\def\MF{{\mtf META}\-{\mtf FONT}}
\def\otr{{\sc otr}}
\def\fake{\lower1.8pt\hbox{\tensy\char1}\space} % to load font \tensy
\def\pop{\raise4pt\hbox to0pt{%
 \hss\smash{\tensy\char"37}\kern1.2pt\hss}}
\def\toC{{\sc toc}}
\def\PS{{\sc PostScript}}
\chardef\bs='\\
\def\vrt{{\tt\char'\|}}
\def\attention{\ifvmode
\noindent\smash{\lower4pt\llap{\pers\char8\kern6pt}}\indent
\else
\vadjust{\leftline{\smash{\lower1pt\llap{\pers\char8\kern6pt}}}}\fi}
\def\exper{\ifvmode
\noindent\llap{\pers\char11\kern1pc}\indent
\else
\vadjust{\leftline{\smash{\llap{\pers\char11\kern1pc}}}}\fi}
\def\pt{\,{\rm pt}} \def\in{\,{\rm in}}
\def\SP{{\tt\char32 }} %visible space
\def\t{\char'176} % tilde
\def\adots{\mathinner{\mkern2mu\raise1pt\hbox{.}\mkern2mu
 \raise4pt\hbox{.}\mkern2mu\raise7pt\hbox{.}\mkern1mu}}
```

20.2 The Main Macros

The main macros for formatting the book are listed in this section. The most important ones are \chapter, \section, and the macros to create the raw index file.

General cross references. These macros are used to refer to a chapter, section, figure, etc., from any part of the text. They have been discussed in section 9.2 and are shown here for the sake of completeness.

```
\def\label#1{\ifAuxFiles\edef\CrossRef{\write\Aux{%
 \string\expandafter\string\def%
 \string\csname\space#1\string\endcsname%
 {{\expandafter\strip\meaning\lblitem}{\noexpand\folio}}}}%
 \CrossRef\fi}
\def\ifundefined#1{\expandafter\ifx\csname#1\endcsname\relax}
\def\ref#1{\ifundefined{#1}\message{! No ref. to #1;}\else%
 \expandafter\firstparam\csname #1\endcsname\fi}
\def\firstparam#1{\expandafter\paramone #1}\def\paramone#1#2{#1}
\def\pageref#1{\ifundefined{#1}\message{! No ref. to #1;}\else%
 \expandafter\secondparam\csname #1\endcsname\fi}
\def\secondparam#1{\expandafter\paramtwo #1}\def\paramtwo#1#2{#2}
```

Examples and exercises. Several chapters use numbered examples. This is done by macro \exm. Note how this macro (e)defines \lblitem, in case the author wants to refer to an example.

```
\newcount\exno
\def\exm{\advance\exno1\goodbreak\medskip{\bf\the\exno.}\
 \edef\lblitem{\the\exno}}
```

Exercises involve two macros. Macro \exercise is simple; macro \answer illustrates the problem of verbatim copying, discussed in section 7.9. The macro sanitizes the special characters and scoops up the text of the answer, line by line, writing each line on the .ans file, for later typesetting in an appendix. The last line of code prepares the special string \endanswer, where the '\' has catcode 12.

```
\newcount\exernum \newtoks\anStr
\def\exercise{\advance\exernum1 \medbreak\noindent
\llap{\pers\char10\tenrm\enspace}%
 {\bf Exercise \chnum\dotnull\the\exernum:}\enspace%
 \edef\lblitem{\chnum\dotnull\the\exernum}}
\def\answer{\ifAuxFiles
 \immediate\write\exer{\string\answer\chnum\dotnull\the\exernum; }%
\fi\moreans}
\def\moreans{\begingroup\sanitize\makeother\\\makeother\^^M%
 \makeother\ \makeother\|\aux}
{\catcode`\^^M=12 \gdef\aux#1^^M{\def\temp{#1}%
\ifx\temp\enP% end of answer?
```

```
\gdef\next{\relax}%
\ifAuxFiles\immediate\write\exer{\string\endanswer}\fi%
\else \anStr={#1}\ifAuxFiles\immediate\write\exer{\the\anStr}\fi%
\gdef\next{\moreans}%
\fi\endgroup\next}%
}%
{\catcode`\*=0 \makeother\\ *gdef*enP{\endanswer}}%
```

Quotations. Each chapter ends with a quotation, typeset by macro `\quote` below. The only interesting thing is the use of `\raggedleft` (which is discussed on page 454).

```
\def\raggedleft{\spaceskip=.3333em \xspaceskip=.5em
  \parfillskip=0pt \leftskip=0pt plus\hsize}

\long\def\quote#1#2.{\vfill
\line{\hfil\vbox{\hsize=4in\quotext\raggedleft\baselineskip=10pt
\noindent#1\smallskip
\line{\quoteref\hfil---#2}
\par}}}
```

Sections and chapters. Macro `\head` starts a section. It uses `\obeylines` and `\gethead` to scoop an entire line as its argument (à la `\elp`, section 5.20). It typesets the argument (the name of the section), then expands `\SCan` to loop and insert `\noexpand` in front of all control sequences in the argument, so the argument can be written on the `.toc` file without expansion. This requires `\immediate\write`, so the page number is written on a separate file (pagetoc). Both files are later read when the TOC is prepared. The macro also places a `\message` in the log file, so the user can follow the progress of the job.

Macro `\SCan` below scans the tokens in the section name and sets the meaning of all the control sequences found to `\relax`. This way, the name can easily be written on a file verbatim. The macro does not use tail recursion. Each time it expands itself, two `\fi`'s have to be saved in the parameter stack, whose size in many TeX implementations is 60. As a result, section names cannot be very long, and are limited to fewer than 30 characters. The same is true for chapter names.

Note how a special character was used as the `\newlinechar`. This character, similar to a `\bullet`, is available in most of the standard Macintosh fonts and can be typed by pressing 'Option 8' on the keyboard. Also note the use of `\xdef`. This is necessary since macro `\lblitem` is defined (rather edefined) inside a group.

```
\newlinechar=`•
\newcount\headn \headn=0
\def\head{\begingroup\obeylines\gethead}
{\obeylines
\gdef\gethead#1
{\global\advance\headn by 1 \message{•\the\headn. #1}%
\vskip0pt plus.02\vsize\penalty-2500 \vskip0pt plus-.02\vsize
```

```
\ifx\empty\lrhead \mark{\hfil}%
\else\mark{Sec.\ \chnum.\the\headn]\hfil#1}\fi
\medskip\leftline{\bf\llap{\chnum\dotnull\the\headn\ }#1}%
 \xdef\lblitem{\chnum\dotnull\the\headn}%
% **** \xdef used since we are in a group ****
\nobreak\medskip
\ifAuxFiles{%
 \let\Let=\let \let\String=\string \let\The=\the \SCan #1\end
 \immediate\write\toc{\String\se\space\The\headn;#1\String\\}}%
 \edef\SAve{\the\headn}\write\pagetoc\expandafter{\SAve;\folio}%
\fi
\endgroup}}
%
\def\SCan#1#2\end{\def\AuX{#1}%
\ifx\AuX\empty
\else
 \def\AuX{#2}%
 \OneStep{#1}%
 \ifx\AuX\empty
 \else
  \SCan#2\end%
\fi\fi}
%
\def\OneStep#1{\ifcat\relax\noexpand#1\Let#1=\relax\fi}
```

A section may be divided into subsections, but subsection heads are not listed in the TOC. As a result, macro \subhead below is very simple. Macro \gobblepar gobbles up the blank line following \subhead, so a subsection starts on the same line as its name.

```
\def\subhead{\begingroup\obeylines\pickarg}
{\obeylines
\gdef\gobblepar\par{}
\gdef\pickarg#1
{\medskip\noindent\hbox{\bf#1.\ }\endgroup\gobblepar}}
```

▶ **Exercise 20.1:** Why the \par in the definition of \gobblepar?

Macro \chapter typesets the chapter head and writes TOC information on file. Parameter #1 is the chapter name, and #2 is the chapter number. The number is normally a positive integer but can also be a letter (for an appendix) or zero (for an unnumbered chapter, such as the preface or the references). Note the use of the local macro \dotnull.

```
\def\chapter#1 #2\par{\vfill\eject   % last page of previous chapter
\ifodd\pageno\else \advancepageno\fi % Start on an odd page
\tabskip=0pt \exernum=0 \exno=0 \fignum=0 \headn=0 \def\chname{#2}%
\if0#1 \def\chnum{}\def\dotnull{}\def\lrhead{}%
\else\ifcat A#1%
```

```
\def\chnum{#1}\def\dotnull{.}\def\lrhead{[App.\ \chnum}%
\else\def\chnum{#1}\def\dotnull{.}\def\lrhead{[Ch.\ \chnum}%
\fi\fi
\def\lefthead{\chname\hfil\lrhead}\mark{\hfil}\headline={\hfil}%
\topglue1.5in plus.2in minus.2in
\setbox0=\hbox{\chead\chnum\dotnull\ \chname}%
 \edef\lblitem{\chnum}%
\vbox{\copy0 \medskip\hrule height1.5pt width\wd0}%
\vskip2in
\ifAuxFiles%
 \immediate\write\toc{%
   \string\ch\space\chnum\string\\\chname\string\\}%
 \write\pagetoc{Ch;\folio}%
\fi}
```

Indexing. Macros for indexing are discussed in section 20.5. They are powerful and involve several subtle points. Their basic task is to pick up certain items (flagged by a '^') and write them on the .idx file, which is later processed by MakeIndex (and another utility) to create the final index.

The circumflex '^' is defined as the indexing symbol. It is declared active and is defined to be macro \Caret. An index item can be one of the following:

■ ^|abc| where abc may include any special characters. The string abc is typeset verbatim and also written verbatim on the .idx file.

■ ^[abc] where abc is as before. This is a "silent" index item that's only written on the .idx file but is not typeset.

■ ^{xyz} where xyz may contain special characters (including a '^') but not a '\' (since it is sanitized during indexing). The string xyz will be typeset and written on the .idx file.

It is, however, invalid to say ^\abc again because the '\' is sanitized during indexing. One should say '^[\abc]\abc' instead. Also, the argument of a macro cannot have index items. If \xyz is the name of a macro, then the expansion '\xyz{this ^{argument}}' is wrong, since all tokens in the argument get their catcodes assigned when the argument is absorbed, and those catcodes cannot be changed later.

Here are examples of valid index items (see Refs. 41, 42 for the special meaning of the '!', the '@', and the parentheses).

```
^[character!special] ^|\pop| ^[cmsy10] ^[verbatim!listing|(]
^|^| ^[|^|!as an index] ^[null@<null>] ^[|^^M|] ^[|\TeX||)]
^{page break} ^[\dvi\ file] ^{#$%^&}
```

▶ **Exercise 20.2:** How can one index a left (or right) brace?

If the '^' is used outside math mode, it becomes macro \Caret, which expands \indexT, which, in turn, uses \futurelet to peek at the token that follows the '^'. If that token is a '|', macro \inxC is expanded. If it is a '[', macro \inxB is expanded; otherwise, \inxA is expanded. Each of these macros, in turn, expands \finidx, which is responsible for the rest of the job. It works in three steps:

1. It defines macro \idxitem as the index item.
2. It defines macro \INDEX (using \edef).
3. It expands \INDEX.

Macro \INDEX uses (delayed) write to write a record of the form 'indexentry{the index item}{\folio}}' on the .idx file. Note how \strip and \meaning are used to extract the index item, as a string of tokens of catcodes 12 and 10, from macro \idxitem. This has been discussed on pages 174 and 215, but here is a recap: If the replacement text of a macro \abc is 123, then \meaning\abc creates the string 'macro:->123', and applying \strip leaves the string '123' (with catcodes 12, except spaces, which get catcode 10). This way, the index item may include any characters, but they are written on the file as plain text because of their "common" catcodes.

Here is a listing of all the macros involved.

```
\def\Caret{\ifmmode\def\next{^}\else\let\next=\indexT\fi\next}
\catcode`\^=\active \let^=\Caret
\def\indexT{\begingroup\sanitize\makeother\\\obeyspaces%
 \makebgroup\{\makeegroup\}\futurelet\new\inxcase}
\def\inxcase{\begingroup\ifx\new|\aftergroup\tmpC
 \else\ifx\new[\aftergroup\tmpB
 \else\aftergroup\inxA \fi\fi \endgroup}
\def\tmpC{\makeother\}\makeother\{\inxC}% deactivate braces during
\def\tmpB{\makeother\}\makeother\{\inxB}% ^[...] and ^|...|
\def\inxA#1{\finidx{#1}#1}
\def\inxB[#1]{\finidx{#1}}
\def\inxC|#1|{\finidx{|#1|}|#1|}
\def\strip#1>{}
\long\def\finidx#1{\def\idxitem{#1}%
\ifAuxFiles
 \edef\INDEX{\write\inx{\string\indexentry%
 {\expandafter\strip\meaning\idxitem}{\noexpand\folio}}}%
 \INDEX%
\fi \endgroup}
```

Output routine. The OTR is similar to \plainoutput. It temporarily disables the '^' and ships out a page with a header, floating insertions, and footnotes.

```
\newtoks\headline \newtoks\Headline
\output={\let^=\relax
\shipout\vbox{\makeheadline\pagebody}%
\global\headline=\Headline \advancepageno
\ifnum\outputpenalty>-20000 \else\dosupereject\fi}
```

On odd-numbered pages, the headline typesets the section name and number from \firstmark (the marks are set by \head). On even-numbered pages it typesets the chapter name and number from macro \lefthead, itself defined by \chapter.

For a chapter of text, the headline is set to

```
\Headline={\ssr\ifodd\pageno\firstmark\kern1pc\folio%
\else\folio\kern1pc\lefthead\fi}
```

For chapters that don't have sections, such as the Preface, the Answers to Exercises, the References, and the Index, the headline is set to

```
\Headline={\ssr\ifodd\pageno Preface\hfil\folio%
\else\folio\hfil Preface\fi}
```

Macro \chapter says '\headline={\hfil}', to guarantee that the first page of a chapter won't have a headline. The assignment '\global\headline=\Headline' in the OTR is needed to make sure that all other pages in the chapter do have a headline.

Margin notes and endnotes. Margin notes are the task of macro \marnote, which uses \vadjust and is straightforward.

```
\newdimen\vraise
\def\marnote#1#2#3#4;{\vadjust{\setbox0=\vbox{\eight\baselineskip=9pt
\hbox{#1}\hbox{#2}\hbox{#3}\hbox{#4}}\vraise=\ht0 \divide\vraise2
\ht0=0pt\dp0=0pt\llap{\vbox to0pt{\kern-\vraise\box0\vss}\kern4pt}}}
```

Endnotes are used in Chapter 19 to illustrate insertions.

```
\newcount\notenumber \newinsert\notes
\count\notes=0 \dimen\notes=\maxdimen \notenumber=0
\def\endnote#1{\advance\notenumber 1 (Note \the\notenumber)%
\insert\notes{\noindent[\the\notenumber] #1.\medskip}}
```

Pictures. The \special command is used to pick up a picture, either from the picture window in the source file (macro \picture) or from a separate file (macro \pictfile), and paste it in the page (it can also be floated, if necessary). This topic is discussed in Chapter 10.

```
\def\picture#1 by #2 (#3){\vbox to #2{
\hrule width#1 height0pt depth0pt
\vfill \special{picture #3}}}
```

```
\def\pictfile#1 by #2 (#3){\vbox to #2{
\hrule width#1 height0pt depth0pt
\vfill \special{pictfile #3}}}
```

Boxing material. Next, macros for boxing material. Macro \boxit has been discussed on page 75. Macro \Cboxit is more sophisticated and is illustrated on page 188.

```
\def\gap{1}
\long\def\boxit#1#2{\vbox{\hrule \hbox{\vrule \kern#2pt%
\vbox{\kern#2pt #1\kern#2pt}\kern#2pt\vrule}\hrule}}
%
\newcount\numArg
```

```
\def\Cboxit#1;{\setbox0=\null\PickupNext#1,;,}
\def\PickupNext#1,{% Note that #1 may be \null
\if;#1\let\next=\insertrules
\else\let\next=\PickupNext
  \setbox0=\hbox{\unhbox0\vbox{#1}}%
  \advance\numArg1
\fi\next}
\def\insertrules{\setbox1=\null
\dimen0=\ht0 \advance\dimen0 \dp0
\unhbox0
\loop
 \setbox2=\lastbox
 \advance\numArg-1
 \setbox1=\hbox{\boxit{\vbox to\dimen0{\vss\box2\vss}}6%
         \kern-0.4pt\unhbox1}%
 \ifnum\numArg>0
\repeat
\box1}
```

Macro \shade has been described in section 10.2. Here is its definition

```
\newdimen\tmp \newdimen\tmpw \newdimen\tmpk \newdimen\tmph
\newcount\heit \newcount\widf \newcount\strk \newcount\radius
\def\shade#1,#2,#3,#4,#5\\{%
% #1-percent of white in background (typically 0.97)
% #2-size of shaded margins (typically 8pt).
%  Does not include width of stroke
% #3-width of stroke, in pt (integer, typically 1).
%  Zero or empty--> no stroke
% #4-the radius (0 for square corners. Typically 1/10 of the width)
% #5-a box containing the text
\setbox1=\hbox{#5}%
\def\inner{#3}\ifx\inner\empty \strk=0 \else\strk=#3\fi
\def\inner{#4}\ifx\inner\empty \radius=0 \else\radius=#4\fi
\tmp=#2 \tmp=2\tmp \advance\tmp\wd1 \widf=\tmp \divide\widf65536
\tmp=#2 \tmp=2\tmp \advance\tmp\ht1 \heit=\tmp \divide\heit65536
\setshadebox{#1}{\the\heit}{\the\widf}{\the\strk}{\the\radius}%
\tmp=#2
\tmpw=#2 \tmpw=2\tmpw \advance\tmpw by\the\strk pt \advance\tmpw\wd1
\tmpk=\the\strk pt \divide\tmpk by2
\tmph=#2 \tmph=2\tmph \advance\tmph by\the\strk pt \advance\tmph\ht1
\vbox to\tmph{\vfil
 \hbox to\tmpw{%
 \kern\tmpk\lower\tmp\box0\kern-\tmpk\hfil\box1\hfil}\vfil}}
%
% The ps commands are placed, as the argument of a \special,
% at the bottom of \box0 whose width is set to zero.
```

```
\def\setshadebox#1#2#3#4#5{%
% #1-setgray, #2-heit, #3-widf, #4-stroke, #5-radius
\setbox0=\vbox to\tmp{\hsize=0pt\vfil\special{postscript
/fpops{4 {pop} repeat} def
#3 0 moveto 0 0 0 #2 #5 arcto %this calc. the start point & leaves it
in stack
newpath % start the rounded-corner rectangle
moveto % the coordinates are in stack
0 #2 #3 #2 #5 arcto fpops
#3 #2 #3 0 #5 arcto fpops
#3 0 0 0 #5 arcto fpops
0 0 0 #2 #5 arcto fpops closepath
#4 0 ne {gsave #4 setlinewidth stroke grestore} if
% stroke width of zero not recommended
#1 setgray fill}}}
```

\everyjob. Each time this book, or part of it, is typeset, the code below asks the user if the auxiliary files should be used. There are five files: for answers, index items, table-of-contents items (two files), and cross references. This is a typical application of the primitive **\everyjob**. Note the use of macro **\yesno** (page 156).

```
\newread\Aux
\newwrite\Aux
\newwrite\toc
\newwrite\pagetoc
\newwrite\inx
\newwrite\exer
\newif\ifAuxFiles
\everyjob{\footline={\hfil} \everyverbatim{\parindent=0pt}
\immediate\openin\Aux=\jobname.aux
\ifeof\Aux \message{! No file \jobname.aux;}
\else\input\jobname.aux \immediate\closein\Aux
\fi

\message{$\bullet$Create aux files (y/n)? }\read-1 to\AuxFile
\def\yesno{y }
 \ifx\yesno\AuxFile \AuxFilestrue \else \AuxFilesfalse \fi

\ifAuxFiles
 \openout\inx=\jobname.idx
 \immediate\openout\toc=\jobname.toc
 \openout\pagetoc=\jobname.pag
 \immediate\openout\exer=\jobname.ans
 \immediate\openout\Aux=\jobname.aux
\fi} % end of \everyjob
```

The very last thing is, of course, a \dump, to create a format file. Normally \dump can only be executed in INITEX, but Textures, on the Macintosh computer, supports this command.

20.3 The Final Typesetting

The entire book (except the index) was typeset from file 'tatb.tex', whose contents is

```
\everyverbatim{\parindent=0pt}
\input Preface
\input :ch.1:1intro
\input 2advanced
\input 3boxglue
\input :ch.4:4parag
\input 5macros
\input 6if
\input :ch.7:7examples
\input :ch.8:8toks
\input :ch.9:9multi-pass
\input :ch.10:10special
\input 11leaders
\input 12alignment
\input 13math
\input :ch.14:14LinePage
\input 15errors
\input 16otr
\input 17otr
\input :ch.18:18otr
\input :ch.19:19otr
\input 20format
\input References
\input answers
\ifodd\pageno\else\advancepageno\fi
\immediate\write\toc{\string\ch\space\string\\Index\string\\}%
\write\pagetoc{Ch;\folio}%
\bye
```

Notice that chapters with diagrams were placed in special folders (subdirectories) with names such as :ch.1. The final typesetting run took less than five minutes on a Macintosh Quadra 630, and has generated the five auxiliary files discussed earlier.

20.4 The TOC macros

These macros read the two TOC files and create the table of contents (TOC). Since they are only used in one place, they were not part of the format and are included here for the sake of completeness. The bulk of the TOC is written (by macros \chapter and \head) on the \jobname.toc file, with the page numbers written on the \jobname.pag files.

Typical records on 'tatb.toc' are

```
\ch \\Preface \\
\ch 1\\Introduction \\
\se 1;Line Breaking and Page Layout\\
\se 2;Fonts\\
```

Typical records on 'tatb.pag' are

```
Ch;v
Ch;1
1;4
2;5
```

Indicating that one chapter starts on page v, another chapter, on page 1; section 1 starts on page 4, and section 2, on page 5. It is easy to \input file 'tatb.toc' and have macros \ch and \se each read a record off 'tatb.pag'. As a reminder, a similar, simpler method of preparing a TOC is discussed in section 9.3.

```
\Headline={\ssr\ifodd\pageno Contents\hfil\folio%
 \else\folio\hfil Contents\fi}
\newread\pag
\newdimen\Sk \Sk=2.9in
\newdimen\Sm \Sm=.3in
\def\Ch{Ch}
\def\splitit#1;#2 {\def\one{#1}\def\pagenum{#2}}
\def\ch#1\\#2\\{\read\pag to\temp \expandafter\splitit\temp
\ifx\one\Ch
\def\chname{#2}
 \bigbreak\medskip
 \line{\bf\rlap{#1}\kern\Sm\chname\ \hrulefill\ \pagenum}%
 \medskip
\else
 \line{\bf#1\ #2****\hfil****\pagenum}\medskip\nobreak
\fi}

\def\se#1;#2\\{\read\pag to\temp \expandafter\splitit\temp
 \line{\kern\Sm\llap{#1}\kern\Sm\rlap{#2}%
 \kern\Sk\llap{\pagenum}\hfil}}

\immediate\openin\pag=tatb.pag
\input tatb.toc
```

20.5 The Index

Creating the index for this book involves two problems: (1) the index should be typeset in double columns; (2) macros should be developed to typeset items and subitems (`MakeIndex` can even handle subsubitems) and to handle special cases such as **see** and **see also**. The special OTR used for the index is listed below. It was kept simple, for the benefit of the readers, and can be improved in many ways.

```
\newdimen\Hsize \Hsize=\hsize \hsize=.5\hsize \advance\hsize-.05in
\newdimen\Vsize \Vsize=\vsize \vsize=2\vsize
\newdimen\htlast

\headline={\hfil} \Headline={\ssr\ifodd\pageno Index\hfil\folio%
\else\folio\hfil Index\fi}
\output={\setbox0=\vbox{\unvcopy255}
\splittopskip=\topskip \splitmaxdepth=\maxdepth
\htlast=.5\ht0  \setbox0=\vsplit255 to\htlast
\setbox255=\vbox{\hbox to\Hsize{\box0\hfil\box255}}%
\shipout\vbox to203mm{\hbox to\Hsize{\the\headline}%
 \medskip\box255\vss}%
\global\headline=\Headline\advancepageno}
```

To understand the special macros, here is a short sample from the final index file (the `.ind` file), created by `MakeIndex`.

```
\item catcode, vi, 19, 21, \bold{24--25}, 26, 199, 200, 417
  \subitem change, 181
  \subitem of |\relax|, 168
\item |\catcode|, 20, 24, 110, 137--138, 156, 157, 182, 361
\item category code, \see{catcode}{v}
\item |\cdot| ($\cdot$), 256
```

The main macro, `\item`, is followed by an index item and a list of page numbers (some perhaps in boldface). There may also be a `\see` or `\seealso` case, which should be handled by appropriate macros. Note that `\item` does not have parameters. This is because the index item may contain verbatim listings, and those don't work inside a macro argument. The same is true for macros `\subitem` and `\subsubitem`. This makes these macros somewhat confusing to a beginner.

Another point is the parameters of macros `\see` and `\seealso`. When indicating index items in the text, we would like to be able to say, e.g., '`^[abc|see{xyz}]`'. The point is that the braces should be written on the raw index file, so they have to be sensitized before the index item is read. This is why macros `\tmpC` and `\tmpB` have been added. They sensitize the braces before the index item is absorbed by `\inxB` or `\inxC`.

```
\catcode`\^=7 % no index items generated here
\def\item{\par\begingroup\obeylines\hangafter=1 \hangindent=15pt
 \def\par{\endgraf\global\leftskip=0pt\endgroup}}
\def\goodfil{\hfil\penalty-9\hfilneg} %similar to \filbreak [353]
```

```
\def\see#1#2{{\it see }#1}
\def\seealso#1#2{#2; \goodfil{\it see also }#1}
\def\bold#1{{\bf#1}}
\def\indexspace{\par \vskip 10pt plus 5pt minus 3pt\relax}
\def\subitem{\par\leftskip=7.5pt\item}
\def\subsubitem{\par\leftskip=15pt\item}
```

Finally, the current font is changed to 9 points, and the index file is \input.

```
\parindent=0pt \parskip=0pt plus.8pt \tolerance=7000
\pageno=471


\ninerm \baselineskip=11pt plus.1pt minus .1pt
\def\tt{\ninett}\def\bf{\ninebf}\def\it{\nineit}
\vbox to3.5in{\kern1.5in\smash{\chead Index}\medskip\hrule
 height1pt\vss\vskip1in
\smash{\rlap{\vbox{\ninerm
\hbox{Page numbers in boldface indicate the}
\hbox{most definitive source of information}
\hbox{about an item.}}}}%
\vskip1in}% head in first page


\input tatb.ind
\endgroup\let\par=\endgraf
```

▶ **Exercise 20.3:** Why the \endgroup?

20.6 Special Characters

A font has been created in METAFONT, with the special characters needed for this book. The METAFONT programs are listed here for the benefit of curious readers (so they don't have to pester the author with special requests). The file starts with the declarations

```
mode:=laserwriter;
mag=magstep0;
screenchars;


font_identifier "pers"; font_size 10pt#;
wA#=6.5043pt#; wa#:=3.8544pt#; wi#:=1.9272pt#;
% 27, 16 & 8 pixels wide, respectively
thin#=.7227pt#; thick#=1.4454pt#; epsi#=.241pt#;
% 3, 6 & 1 pixels tall
path p,p.l,p.r,outerr;
pair b,center;
picture pict;


mode_setup;
define_pixels(wA,wa,wi,thin,thick,epsi);
```

```
def openit=openwindow currentwindow from origin to(2400,1600)
at(-450,480) enddef;

def hairline=pencircle scaled .5thin; enddef;
def thin_pen=pencircle scaled thin; enddef;
def thick_pen=pencircle scaled thick; enddef;
def thicker_pen=pencircle scaled 2thick; enddef;
```

Chain links. Then come the programs for the eight characters used for a bordered text in Chapters 11 and 16. The characters are ⊐⊏ ⊐⊏ ⊦⊦ ⊦⊦ ⌐⌐ ⌐⌐ ⌐⌐ ⌐⌐. And the programs are:

```
qu#:=.25in#; ei#:=.125in#;

% -------- 8 chars, chain links -------
def top_bot =
s:=.1w;
pickup thick_pen;
bot z1=origin; bot z2=(.25w,0); z3=(.5w,.5h);
p:= z1---z2..{up}(z3+(s,0));
draw p;
pickup thicker_pen;
erase draw p reflectedabout((.5w,0),(.5w,1));
pickup thick_pen;
draw p reflectedabout((.5w,0),(.5w,1));
currentpicture:=currentpicture+currentpicture
reflectedabout((0,.5h),(1,.5h)) reflectedabout((.5w,0),(.5w,1));
enddef;

def rt_lft =
s:=.1h;
pickup thick_pen;
lft z1=origin; lft z2=(0,.25h); z3=(.5w,.5h);
p:= z1---z2..{right}(z3+(0,s));
draw p;
pickup thicker_pen;
erase draw p reflectedabout((0,.5h),(1,.5h));
pickup thick_pen;
draw p reflectedabout((0,.5h),(1,.5h));
currentpicture:=currentpicture+currentpicture
reflectedabout((.5w,0),(.5w,1)) reflectedabout((0,.5h),(1,.5h));
enddef;

def corners =
s:=.1w;
```

```
pickup thick_pen;
rt bot z2=(w,0); top lft z3=(0,h);
draw z2{up}..z3;
enddef;

beginchar(0,qu#,ei#,0); "top";
top_bot;
endchar;

beginchar(1,qu#,ei#,0); "bot";
top_bot;
currentpicture:=
 currentpicture reflectedabout((0,.5h),(1,.5h));
endchar;

beginchar(2,ei#,qu#,0); "rt";
rt_lft;
endchar;

beginchar(3,ei#,qu#,0); "lft";
rt_lft;
currentpicture:=
 currentpicture reflectedabout((.5w,0),(.5w,1));
endchar;

beginchar(4,ei#,ei#,0); "NE";
corners;
endchar;
beginchar(5,ei#,ei#,0); "NW";
corners;
currentpicture:=currentpicture reflectedabout((.5w,0),(.5w,1));
endchar;
beginchar(6,ei#,ei#,0); "SE";
corners;
currentpicture:=currentpicture reflectedabout((0,.5h),(1,.5h));
endchar;
beginchar(7,ei#,ei#,0); "SW";
corners;
currentpicture:=currentpicture
reflectedabout((0,.5h),(1,.5h)) reflectedabout((.5w,0),(.5w,1));
endchar;
```

Attention symbol. The program for the attention symbol ⫸➤ is

```
ht#:=12pt#; wd#:=16pt#; curv:=21; ruledim#:=6pt#;
define_pixels(ht,wd,tail,ruledim);
```

```
beginchar(8,wd#,ht#,0); "attention";
def rect(expr pl,pr) =
fill pl[z6,z3]--pl[z7,z4]--pr[z7,z4]--pr[z6,z3]--cycle;
enddef;
R:=floor h; if not odd R: R:=R+1; fi;
z1=(w,.5R); z2=(.7w,h);
z3=(x2,y1+.5ruledim); z4=(x2,y1-.5ruledim); z5=(x2,0);
z6=(0,y3); z7=(0,y4);
sAngle:=angle(z2-z1)+curv; eAngle:=angle(z1-z5)-curv;
fill z1{dir sAngle}..z2--z3--z4--z5..{dir eAngle}cycle;
fill z6--z7--(x7+1,y7)--(x6+1,y6)--cycle;
rect (.11,.18);
rect (.27,.42);
rect (.5,.73);
rect (.8,1);
endchar;
```

Triangles. The triangles placed at the end of an answer '◄' and the start of an exercise '►' are characters 9 and 10, respectively:

```
def triangl=
x1=x2=w-x3=round .43pt; y3=.5+floor.5h;
z1-z2=(z3-z2) rotated 60;
y1:=.5sqrt3+round(y1-.5sqrt3); y2:=h-y1;
fill z1--z2--z3--cycle;
enddef;

beginchar(9, 5pt#, 6.25pt#, 0); "Left-pointing triangle";
triangl;
currentpicture:=currentpicture reflectedabout((.5w,0),(.5w,1));
endchar;

beginchar(10, 5pt#, 6.25pt#, 0); "Right-pointing triangle";
triangl;
endchar;
```

Experiment symbol. The experiment symbol ⊞ is character 11:

```
beginchar(11,20pt#,8pt#,0); "keyboard";
s:=.1w;
currenttransform:=identity slanted 1/4;
pickup pensquare scaled2;
top z1=(s,h); top z2=(w-s,h); rt z3=(w,h-s);
rt z4=(w,s); bot z5=(x2,0); bot z6=(.5w,0);
draw z1---z2{right}..z3---z4{down}..z5---z6;
z7=(w-s,.5h); z8=(2s,y7);
```

```
p:=z7--z8;
draw p shifted(0,s);
draw subpath(0,.7) of p;
draw subpath(0,.4) of p shifted(0,-s);
z9=(x7,y7-s); z10=(x9,y7+s); p:=z9--z10;
draw p; draw p shifted(-s,0); draw p shifted(-2s,0);
endchar;
currenttransform:=identity;
```

Accents. The special accents in Chapter 3 are characters 20–23 '˘ ˜ ˘ ^'.

```
beginchar(20,wa#,thick#,0); "a crescent";
x1l=0; x2l=.5[x1,x3]; x3l=w;
y1l=y3l=h-epsi; y2l=0;
penpos1(epsi,90); penpos2(4epsi,90); penpos3(epsi,90);
penstroke z1e..z2e..z3e;
endchar;
```

```
beginchar(21,wa#,thick#,0); "a sine";
z1=(0,.5h); z5=(w,y1); x2=.25[x1,x5]; y2r=h; z3=(.5[x1,x5],y1);
x4=.75[x1,x5]; y4l=0;
penpos1(epsi,90); penpos2(2epsi,90);
penpos3(3epsi,90); penpos4(2epsi,90); penpos5(epsi,90);
penstroke z1e..z2e..z3e..z4e..z5e;
endchar;
```

```
beginchar(22,wa#,thick#,0); "a horns";
x2r=0; x5r=w; y2=y5=.5h; x1=x3=.25[x2,x5]; x6=x4=.75[x2,x5];
y1r=y6r=h; y3r=y4r=0; z34=(.5[x3,x4],y3);
penpos1(epsi,90); penpos6(epsi,90);
penpos2(2epsi,180); penpos5(2epsi,0);
penpos3(2.5epsi,-90); penpos4(2.5epsi,-90);
penpos34(3epsi,-90);
penstroke z1e{left}..z2e..z3e---z34e---z4e..z5e..{left}z6e;
endchar;
```

```
beginchar(23,wa#,thick#,0); "a bell";
z1=origin; z5=(.5w,h); z3=.5[z1,z5];
z9=(w,0); z7=.5[z5,z9];
p:= z1..z3..z5..z7..z9--cycle;
fill p;
erase fill p scaled.5 shifted(.25w,0);
endchar;
```

That's all.

Old people like to give good advice, as solace for
no longer being able to provide bad examples.
— *François de la Rochefoucauld*

References

1: Knuth, D. E., *Computers and Typesetting*, Vol A, *The TEXbook*, Addison Wesley, 1986.

2: Knuth, D. E., *Computers and Typesetting*, Vol B, *TEX, The Program*, Addison Wesley, 1986.

3: Knuth, D. E., *Computers and Typesetting*, Vol C, *The METAFONTbook*, Addison Wesley, 1986.

4: Knuth, D. E., *Computers and Typesetting*, Vol E, *Computer Modern Typefaces*, Addison Wesley, 1986.

5: Henderson, D., *Outline fonts with METAFONT*, *TUGboat*, **10**(1) April 1989, 36.

6: Wujastyk, D., *The many faces of TEX*, *TUGboat*, **9**(2) August 1988, 131.

7: Tobin, G. K. M., *The OCLC roman family of fonts*, *TUGboat*, **5**(1) May 1984, 36.

8: Billawala, N., *Metamarks: Preliminary Studies for a Pandora's Box of Shapes*, Rep. STAN-CS-89-1256, Computer Science Department, Stanford University, 1989.

9: Siegel, D. R., *The Euler Project at Stanford*, Department of Computer Science, Stanford University, 1985

10: Haralambous, Y., *Typesetting Old German*, *TUGboat*, **12**(1) March 1991, 129.

11: Fuchs, D., *The Format of TEX's* `.dvi` *Files, Version 1*, *TUGboat*, **2**(2), July 1981, 12.

12: Salomon, D., *Output Routines: Examples and Techniques. Part II*, *TUGboat*, **11**(2), July, 1990, 212–236.

13:   Salomon, D., *Output Routines: Examples and Techniques. Part I*, *TUGboat*, **11**(1), April, 1990, 69–85.

14:   A letter from DEK to Elizabeth Barnhart of TV Guide (June 11, 1987).

15:   Hoenig, A., *Line-Oriented Layout With TₑX*, in M. Clark, ed., *TₑX, Applications, Uses, Methods*, Ellis Horwood, 1990, pp. 159–183

16:   Salomon, D., *Macros for Indexing and Table-of-Contents Preparation*, *TUGboat*, **10**(3), Nov. 1989, 394–399.

17:   Kabelschacht, A., \expandafter *in Conditionals*, *TUGboat*, **8**(2), July 1987, 184.

18:   Schwarz, N., *Introduction to TₑX*, Addison-Wesley, 1989.

19:   Greene, A. M., *BASₓX—An Interpreter Written in TₑX*, *TUGboat*, **11**(3), 1990, 385–392.

20:   Bechtolsheim, S., \csname *and* \string, *TUGboat*, **10**(3), 1989, 203–206.

21:   Hendrickson, A., *Getting TₑXnical*, *TUGboat*, **11**(3), 1990, 359–370.

22:   Hendrickson, A., *Some Diagonal Line Hacks*, *TUGboat*, **6**(2), 1985, 83–86.

23:   Reid, T. J., *Random Numbers*, *TUGboat*, **8**(3), Nov. 1987, 315–319.

24:   Pittman, J. E., *Loopy.TeX*, *TUGboat*, **9**(3), 1988, 289–291.

25:   van der Laan, C. G., *Typesetting Crosswords via TₑX*, in *Proceedings of the 7th European TₑX Conference*, 1992.

26:   Breitenlohner, P., *How to Avoid Writing Long Records*, *TUGboat*, **11**(1), 1990, 62.

27:   Reid, T. J., *Floating Figures at the Right*, *TUGboat*, **8**(3), Nov. 1987, 315–319.

28:   Salomon, D., *Creating Shaded Boxes*, *TUGboat*, **13**(3), Oct. 1992, 327.

29:   Knuth, D. E., *A Course on METAFONT Programming*, *TUGboat*, **5**(2), Nov. 1984, 105–118.

30:   Bechtolsheim, S. v., *TₑX in Practice*, Springer Verlag, 1992.

31:   Snow, W., *TₑX for the Beginner*, Addison-Wesley, 1992.

32:   Graham, R., et. al., *Concrete Mathematics*, Addison Wesley, 1989.

33:   The Times Math postscript fonts, The TₑXplorators Corp., 1992.

34:   Knuth, D. E. & M. F. Plass, *Breaking Paragraphs into Lines*, Software, Practice and Experience, **11**, 1981, 1119–1184.

35:   Lieber, L., *Galois and the Theory of Groups*, Galois Institute, 1956.

36:   Liang. F., *Word hy-phen-a-tion by com-pu-ter*, Stanford CS Rep. 977, 1983.

37:   Plass, M. F., *Optimal Pagination Techniques for Automatic Typesetting Systems*, Stanford CS Rep. 870, 1981.

38:   Platt, C., *Macros for Two-Column Format*, *TUGboat*, **1**(6), March 1985, 29–30.

39:   Durst, L., *Long-Winded Endnotes*, *TUGboat*, **4**(11), Nov. 1990, 581–588.

40:   Hendrickson, A., *A Macro Writing Tool: Generating New definitions*, *TUGboat*, **9**(1), 1988, 64.

41:   Lamport, L., *MakeIndex: An Index Processor for LATₑX*, (available from the many archives carrying LATₑX stuff).

42: Pehong, C., et al, *Index Preparation and Processing*, Soft—Practice & experience **18**(9), 1988, 897–915.

43: Bell, E. T., *Men of Mathematics*, Simon and Schuster, 1937.

You will find it a very good practice
always to verify your references, sir!

— *Dr. Routh*

A. Answers to Exercises

1: Don't worry. Such a thing is rare and will be explicitly mentioned. ◄

2: Sleep on it. Think twice about it. Ask a friend. Ask an expert at work. As a last resort, contact the author. The present address is `dxs@secs.csun.edu`, but an up-to-date address can always be obtained from the TUG office. ◄

1.1: These numbers *are* round when expressed in binary, rather than decimal form. Thus $128 = 2^7$ and $256 = 2^8$. Since computers use binary numbers internally, they find it easier to deal with numbers that are powers of two. ◄

1.2: Perhaps the most common example is a negative kern in a `\noalign` or in a box, where it means movement to the left or up. There are numerous examples in Chapters 12, 16, 17, and 18. ◄

1.3: It is truncated to 10 sp. ◄

1.4: It's easy to verify that $72.2700007227/72.27 = 1.00000001$, so the ratio of the two quantities deviates from 1 by 10^{-8}. ◄

1.5: `\magstep` is a macro whose argument must be a digit between 0 and 5. ◄

2.1: The result of `\number'1` is 49, the character code of '1', being typeset. The result of `\number'12` is the same 49, which is typeset, and is immediately followed by a 2; thus 492. ◄

2.2: The character code of '{' (123) will be typeset, The right brace will cause an error (too many '}'). The % in `\number'%` will be considered a comment, so the result will be the code of whatever character happens to follow (see below on how to get the character code of %). ◄

2.3:

```
\def\text{abcx1}
\begingroup
...
\uccode'x=255
\edef\Text{\expandafter{\uppercase\expandafter{\text}}}
\Text
\endgroup
```

(It is also possible to '`\def\Text`', but '`\edef\Text`' takes less memory.) The main difference is that any control sequences stored in `\text` are expanded when the tokens are moved to `\Text`. With toks registers, this does not happen. ◄

2.4: It is 'ABX'. The parameter is substituted by the argument before the replacement text is executed. ◄

2.5: Both cases result in X. In the first case, when `\lowercase` is executed it scans its argument and finds one character token, namely x (which is already lower case). The `\lowercase` therefore does nothing. The argument ('{\uppercase{\def\abc{x}}\abc}') is then executed. The `\uppercase` again finds only one character token, and converts it to X. The argument of `\uppercase` (which is '{\def\abc{X}}\abc}') is then executed, which yields X. The second case is analyzed similarly. ◄

2.6: The character code of 'e' is 101, so the assignment makes 'e' a comment character. The `\number` is now considered the control sequence `\numb` (which is normally undefined), followed by a comment. The result is the error message
`! Undefined control sequence \numb.` ◄

2.7: When TeX is used to produce graphics, we sometimes want to suppress all spaces (see example on [390]). This can be done by `\catcode'\ =9`. A ⟨return⟩ can be ignored by saying

```
\catcode'\
=9
```

on two separate lines. Also, certain control characters can sometimes be added to a file when it is transmitted between computers, and they should be ignored by TeX. ◄

2.8: Imagine receiving a file where every paragraph is terminated by a ¶. Instead of changing all occurrences of ¶ to `\par`, it is better to declare the ¶ an active character and define it as `\par`. ◄

2.9:

```
\def\abc{'} \abc ' \abc\ '
\def\?{\message{ok}} '\?1' '\? '
\toks0={\abc}
\toks0=\expandafter{\the\toks0 '} \showthe\toks0
```

◄

2.10: This is a result of the general rule that says that a number should normally be terminated by a space. TeX considers the space a terminator and thus does typeset it. To better understand this rule, try the following:

```
\count20=20%
1stop
```

This example sets \count20 to 201 and typesets 'stop'. The reason is that, after reading the '20', TeX reads ahead, attempting to find more digits. It finds the '1' since the '%' does not terminate a number.

Without the space, our three-line example is executed in the following steps:
1. The first line is read, and \abc is defined.
2. The second line is scanned. The '%' is read, which terminates the name \abc. The macro is expanded and the last thing, of course, is the '11'. Therefore, before executing the expansion, TeX reads the next character, hoping to find more digits. Since the next character is the '%', TeX skips to the next line and starts expanding \abc, still hoping to find more digits. The expansion of \abc on the third line, however, does not start with a digit.
3. At this point, TeX realizes that there are no digits following the '11' (from the second line). It therefore executes \abc from the second line, which changes the catcode of '%'.
4. Next, \abc from the third line is executed, which does not change a thing.
5. The '%' of the third line is reread and, since it is now considered a letter, is typeset. (End of answer.) ◄

2.11: See list on [279–281]. ◄

3.1: Crossroads; Split level; Six feet underground; Uptown (Downtown?); GI overseas; Three degrees above zero; Circles under eyes. ◄

3.2: It is ⎍ material ⎍. The total amount of empty space is equally divided among three pieces of \hfil. Two are placed on the left, and one on the right, of the text. The text thus ends up two-thirds of the way from the left edge of the box. ◄

3.3: This is a good demonstration of the flexibility of the interword glue. The normal space between words is considered flexible glue that can be stretched or shrunk, so the results are

| A B C |
| A B C |
| A B C |

with no errors. (However, greater amounts of spread would exceed the maximum flexibility of the glue and cause errors.) ◄

3.4: It is an empty \hbox with a width of 24pt and height and depth of zero (it can be used as a horizontal strut). ◄

3.5: It is a 'Yes', since the box has just been emptied and is now void. ◄

3.6: It is a 'No'; \box0 contains \null, which is an empty \hbox, but \box0 itself is not empty. ◄

3.7: The first component (\box0) and its first component (the 'A') are the only ones shown.

```
> \box1=
\hbox(6.83331+0.0)x34.03894
.\hbox(6.83331+0.0)x25.13893
..\tenrm A
..etc.
.etc.
```

◄

3.8: One way [313] is to typeset a space, which is H mode material. If it is important that the space itself doesn't show, it can be suppressed by an \unskip. Thus the definition \def\leavevmode{\ \unskip} works.

Another way is to create an empty box and open it. This can be done by

```
\newbox\voidbox
\def\leavevmode{\unhbox\voidbox}
```

and is the way \leavevmode is actually defined. ◄

3.9: It is

Text A

The 'T' changes the mode to H, and the box with the 'A' becomes part of that paragraph. ◄

3.10: It is an empty \vbox with a height of 12pt and depth and width of zero (it can be used as a strut). A slightly different construct is \setbox0=\vbox{} \ht0=8.5pt \dp0=3.5pt, following which \box0 (or \copy0) can be used as a strut. ◄

3.11: They should be placed in an \hbox and the second one lowered by the difference of .94444pt.

```
\hbox{
\vbox{\hbox{p}}
\boxmaxdepth=1pt
\lower.94444pt\hbox{%
\vbox{\hbox{p}}}}
```

3.12: It is caused by an end-of-line that has been converted into a space. It can be eliminated by placing a '%' at the end of that line.

```
\hbox{
\vbox{\hbox{p}}%
\boxmaxdepth=1pt
\lower.94444pt\hbox{%
\vbox{\hbox{p}}}}
```

◄

3.13: It is \hsize, since it contains text at top level. ◄

3.14: By a construct such as \vtop{\vbox{...}...}. A \vtop whose top component is a \vbox. The height of the \vtop equals the height of the \vbox. Its depth is that of the \vbox plus the rest of the material. The effect is as if we have added material at the bottom of the \vbox, and the material has contributed to the *depth*, rather than to the height, of the box. Such a construct is used in the \cstok macro [420] and in one of the \boxit examples on page 447. ◄

3.15: In step 1, a \vbox is constructed, with a depth of 3.5pt (the depth of the strut) and a height of 20pt, as required. The total vertical size is thus 23.5pt. In step 2, this is converted to a \vtop by moving the reference point such that the height is that of the top component (zero), and the depth is the rest (23.5pt). ◄

3.16: (0.0+23.5)x0.0, as can easily be verified with a \showbox. ◄

3.17: It is an empty \vtop with depth 3pt and height and width of zero (it can be used as a strut). ◄

3.18: Only top-level components are shown.

```
> \box0=
\vbox(82.74086+0.0)x22.77783
.\kern 20.0
.\hbox(6.83331+0.0)x20.00003 []
.\kern 7.0
.\glue(\baselineskip) 5.16669
.\vbox(6.83331+0.0)x21.68121 []
.\kern 5.0
.\glue(\lineskip) 1.0
.\vbox(28.90755+0.0)x22.77783, glue set 5.03712fil []
.\kern 2.0
```

◄

3.19: This is expanded to .. glues have \hskip 12pt plus5pt plus and minus components .., and the second plus is considered text, not part of the glue, so it takes TeX out of the context of the glue. ◄

3.20: The simplest way to achieve a width of 100pt is to stretch the first glue and leave the second one alone. The glue set ratio will be 0.84723, and the text, as a result, will be right justified. Thus ⬚ ⬚ ⬚ abc. ◄

3.21: It means that the space on the right of 'abc' is twice as large as that on the left. It does not mean—as some may intuitively feel—that the text is centered on the one-third point. ◄

3.22: There are two units of `fil` on the left, and one on the right. The word "Text" is thus pushed two-thirds of the way to the right. ◄

3.23: There are always y units of `fil` on the left, and one on the right, for a total of $y + 1$ units. The `fil` on the left is therefore stretched by $\frac{y}{y+1}$ units, and the equation is $x = \frac{y}{y+1}L$, where $0 \leq y \leq 9$, and L is the width of the box (300pt in our example). ◄

3.24:

```
\newcount\coumt
\def\oval#1#2{\coumt=0
\vbox{\offinterlineskip\def\updown{D}
\loop
\hbox to#2pt{\hfil #1\hskip0pt plus\coumt fil #1\hfil}
\if\updown D\advance\coumt by1
\else \advance\coumt by-1 \fi
\ifnum\coumt=10 \def\updown{U}\fi
\ifnum\coumt>-1\repeat
}}
```

◄

3.25: The glue is set at 31.57407fil, and the result is

⬚ ⬚ ⬚   x

◄

3.26: This is easy:

```
\leftskip=0pt plus1fil
\rightskip=0pt plus-1fil
\parfillskip=0pt plus1fil
```

◄

3.27: Because the first construct has the width of the '=', whereas the second one has the width of the '/'. The two experiments `\setbox0=\hbox{=\llap{/}}` `\showbox0` and `\setbox0=\hbox{\rlap{=}/}` `\showbox0` may help clear up this point. ◄

3.28: It is ⬚AB⬚. The `\dotfill` behaves like flexible glue, and since no width was specified for the `\hbox`, it is set to zero. ◄

3.29: A `1fill` is infinitely greater than `16383fil` or even `16383.99999fil`. However, `16383.999999fil` is considered equal to `16384fil` and produces an error message (`Dimension too large`). Similarly `.1fill` is infinitely greater than `16383.999999fil`, and even `0.00001fill` has that property (see also [72]). However, `0.000001fill` is too small for TeX to handle and is considered equal to 0pt. ◄

3.30: The box should be 4pt wider than its natural width, so $\Delta = -4$pt. The total stretchability is $Y = 10$pt, implying $r = 4/10 = 0.4$. ◄

3.31: $r = 0.4$ implies $r^3 = 0.064$, so the badness = 6. ◄

3.32: Badness $= 100 \Rightarrow r^3 = 1 \Rightarrow r = 1 \Rightarrow \Delta = Y$ or $\Delta = Z$. ◄

3.33: The box is `\vbox to0.9pt{\vskip0pt minus1pt\hrule height2pt}`. It results in a badness of 1,000,000 but is not considered overfull, because its height overrun does not exceed `\vfuzz`. ◄

3.34: The algorithm for interline glue [80, 282] has a special test to suppress that glue when the next component is a rule. ◄

3.35: No. The second one has a height of 5pt, but also a depth. Its depth is unspecified and is thus set equal to that of the '('. ◄

3.36: Place an empty box on both sides. Thus '`\null\hrulc\null`'. ◄

3.37: With `\rlap{Name}\vrule height-2pt depth2.4pt width1in`; but TeX should be in H mode (add `\leavevmode` if necessary). ◄

3.38: The first idea is to place both accentee and accent in hboxes, and place the two boxes in a `\vbox`, with a `\nointerlineskip` in between. Thus we write `\vbox{\hbox{a}\nointerlineskip\hbox{\char'30}}`, which produces ą. This seems satisfactory, except that the accent needs to be moved up a bit (but we will ignore this).

A little thinking shows that this simple method may produce unsatisfactory results if the accentee has a depth. A simple test produces 'gğ', which shows that the depth of the 'g' is now included in the height of the `\vbox`. Our construction has to be modified such that the height of the `\vbox` becomes the height of the accentee, and its depth becomes the depth of the accentee plus the vertical size of the accent. This, fortunately, is easy to achieve by changing the `\vbox` to a `\vtop` `\vtop{\hbox{g}\nointerlineskip\hbox{\char'30}}`. The test above now produces 'gğ'.

What about the horizontal alignment? It seems that the accent is centered about the 'a' and the 'g', but this is only because these characters are about as wide as the accent itself. Trying a wider accentee produces 'Ą'. The two are aligned on the left. To center the accent below the 'A' we say:

```
\vtop{\setbox0=\hbox{A}\copy0\nointerlineskip%
 \hbox to\wd0{\hss\char'30\hss}}
```

Which produces 'Ą'. The accent is centered in a box whose width is that of the accentee. The use of `\hss` instead of `\hfil` makes it possible to use an accent wider than the accentee.

There is one more point to consider. The accentee is typeset as a box, not as a character. Therefore, if it happens to be the first thing in a paragraph, TEX will handle it in vertical mode. To make sure that the accentee is handled in horizontal mode, a \leavevmode should be placed before the box. The final result is a macro \subaccent, which can be used with any accent that has no height.

```
\def\subaccent#1#2{\leavevmode\vtop{\setbox0=\hbox{#2}
\copy0\nointerlineskip\hbox to\wd0{\hss\char#1\hss}}}
```

The tests

```
a\subaccent{'30}a A\subaccent{'30}A g\subaccent{'30}g
q\subaccent{'30}q {\it A}\subaccent{'30}{\it A}
{$\alpha$}\subaccent{'30}{$\alpha$} \subaccent{'30}{ui}
\subaccent{'30}{\hbox{u\kern-.1em i}}
```

produce aą AĄ gg qq *AĄ* αǫ ui ui. Note the pairs of letters in the last two tests. These results should be compared to 'ç Ç g q', produced by the plain macro \c, which places a cedilla under its parameter. \c was specifically designed for a 'C', and is not as general as our \subaccent (end of answer). ◄

3.39: We start with an \hbox with two vphantoms (which act as vertical struts) placed side by side \hbox{\vphantom{\hbox{#1}}\vphantom{\hbox{#2}}}. The height and depth of this box are the required maximums. Next we build a \vbox with two hphantoms, each as wide as one of the parameters, stacked vertically \vbox{\offinterlineskip\hphantom{\hbox{#1}}\hphantom{\hbox{#2}}}. Finally, the two boxes are placed in the desired \hbox.

```
\def\triMax#1#2{
\hbox{\hbox{\vphantom{\hbox{#1}}\vphantom{\hbox{#2}}}%
\vbox{\offinterlineskip\hphantom{\hbox{#1}}\hphantom{\hbox{#2}}}}
}
```

◄

3.40:

```
\vbox{\font\coll=cmr10 at 28pt
\hbox{\phantom{Mc}Co\hphantom{\coll ll}een}
\hbox{McCu\smash{\coll ll}ough}}
```

◄

3.41: Using a \loop, the lines are read from the keyboard (file 16) one by one. Each is placed in \box0 to measure its width, and the ⟨dimen⟩ register \Width is set to the size of the shortest line.

```
\newdimen\Width \Width=\hsize
\vbox{\endlinechar=-1
\loop
 \read16to\Line
```

```
\setbox0=\hbox{\Line}
\ifdim\wd0>0pt              % \wd0=0 means an empty input line
 \ifdim\wd0<\Width \Width=\wd0\fi
 \centerline{\Line}
\repeat
\nointerlineskip\centerline{\vrule width\Width depth.4pt}
   }
```

The \endlinechar=-1 suppresses the end-of-line character that otherwise is appended to every line of input, making it longer (see [ex. 20.18]). ◄

3.42: This is straightforward, and the only subtle point is the '%'.

```
\long\def\boxit#1#2{\hbox{\vrule \vbox{\hrule\kern#2pt\hbox{%
\kern#2pt\vbox{#1}\kern#2pt}\kern#2pt\hrule}\vrule}}
```

◄

3.43: We start with

```
\setbox0=\hbox{#1}
\vbox{\hrule\hbox{\vrule\box0\vrule}\hrule}}
```

But the depth of this \vbox is zero because its bottom component is a rule. We want the bottom component to be the \hbox{\vrule\box0\vrule}. To do this, we create the bottom rule above the \hbox, drop it below the box by means of a \kern, typeset it, and go back by means of a negative \kern. Thus

```
\def\depthit#1{\setbox0=\hbox{#1}%
\vbox{\kern-0.4pt\hrule
\kern\ht0\kern\dp0\kern-0.4pt\hrule\kern-\ht0\kern-\dp0
\hbox{\vrule\box0\vrule}}}
```

The two \kern-0.4pt compensate for the heights of the horizontal rules, so the height of the final box is that of the argument. The width is 0.8pt greater than that of the argument.

Another approach creates the desired result in an \hbox. We start with \vbox{\hrule\hbox{#1}}. The depth of this \vbox is the depth of the \hbox inside it, which is also the depth of the material. To preserve this depth, we place the \vbox in a \vtop, thus \vtop{\vbox{\hrule\hbox{#1}}\hrule}. The depth of the \vtop is the depth of the \vbox plus the height of the bottom rule. The whole thing is now placed in an \hbox to supply the two vertical rules. The result is

```
\def\depthit#1{\hbox{\vrule\vtop{\vbox{%
 \hrule\hbox{#1}}\hrule}\vrule}}
```

Note that the macro creates the final result in an \hbox. If a \vbox is needed, the definition can be changed to

```
\def\depthit#1{\vbox{\hbox{\vrule\vtop{\vbox{%
 \hrule\hbox{#1}}\hrule}\vrule}}}
```

It's better to place a \leavevmode at the start of each of these definitions, in case they happen to be the first thing in a paragraph (end of answer). ◀

3.44: First, here is a block diagram showing how the different boxes are nested:

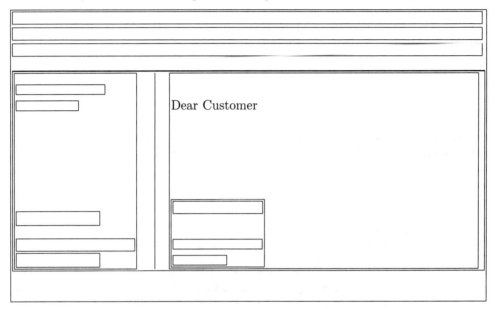

And here are the actual commands:

```
\vbox to 3in{% hsize is 5in
\centerline{\bf Lonely Publishers}
\centerline{\bf 1234 Box and Glue Blvd., Suite 56}
\centerline{\bf Stretch Village, Ill 56567}
\vskip0.15in\hrule
\line{%
\vbox to2in{% the hsize is that of the widest hbox
\vskip0.1in
\hbox{Caleb Summers}\hbox{\bf President}
\vfil
\hbox{(123) 555-1212}
\vskip0.1in
\hbox{\bf Technical Support}\hbox{(800) 555-1313}
            }
\hfil\vrule\hfil
\vbox to2in{\hsize=3.25in \parskip12pt \parindent0pt
% hsize has to be specified since this vbox contains text!
\vskip.1in
Dear Customer

Thank you for buying {\sl Lonely}$^{\scriptscriptstyle\rm TM}$, and
```

```
welcome ... customers!

\vfil
\vbox{
\hbox{Sincerely yours,}
\vskip.2in
\hbox{Caleb Summers}
\hbox{President}
    }
      } % end of vbox to2in
     } %end of line
\vfil} % end of vbox to3in
```

    (end of answer.) ◄

3.45: My solution uses toks registers, but they are not strictly necessary.

```
\newtoks\name \newtoks\address \newtoks\city
\name={\bf Preferred Publishers}
\address={\bf 1234 Moriah Woods Blvd., Suite 56}
\city={\bf Elk Grove Village, Ill 56567}
\setbox0=\vbox{
\hbox{\the\name}\hbox{\the\address}\hbox{\the\city}
             }
\vbox{
\centerline{\the\name}\centerline{\the\address}\centerline{\the\city}
\nointerlineskip\centerline{\vrule width\wd0 depth.4pt}
    }
```

    ◄

3.46:

```
\boxit{\vbox{\hsize=3.5in\parskip=6pt
\vbox{\hbox{\bf Your Comments Count}\kern2pt
\hrule height2pt}
\noindent1. What is your age group?

1--20\enspace\boxit{}{3}\enspace21--31\enspace%
\boxit{}{3}\enspace30--\enspace\boxit{}{3}

\noindent2. Which of the following have provided you with the most
helpful information about our company on your visit?
\vskip\parskip
\halign{#\ \hfil\boxit{}{3}\tabskip1em&#\hfil\boxit{}{3}\cr
Employee&Advertising\cr
Brochure&Agent\cr
Display&Other \vrule width.5in height.4pt depth0pt\ \cr}}}5
```

    But other solutions that create essentially the same thing are possible. ◄

3.47: This is straightforward.

```
\vbox{% No \offinterlineskip here (why?)
\hbox{%
% the left part
\vbox{\hbox{\sevenrm ref.~point $\rightarrow$}\kern26pt}
% the center part
\vbox{\offinterlineskip
\hrule width65pt\kern-.4pt
\hbox to65pt{\vrule height48pt\hfil\vrule}
\vbox to0pt{\vss % this \vss & the one below, guarantee that the
 \hbox to0pt{\kern-2.2pt\lower2.2pt\rlap{$\bullet$}\kern2.2pt%
 \vrule width 65pt height.2pt depth.2pt\hss}%                       \vrule
         \vss} % above will be vertically centered in the \vbox
% the bullet is moved to the left and down, to center it on the
\hbox to65pt{\vrule height30pt\hfil\vrule}%                   ref. point
\kern-.4pt\hrule width65pt
}
% the right part
\vbox{
 \vtop to48pt{\boxmaxdepth=0pt % the depth of the downarrow should
% become part of the total height of 48pt.
  \hbox{$\uparrow$}\vfil\hbox{\sevenrm height}\vfil%
   \hbox{$\downarrow$}}
 \vbox to30pt{% the \null below guarantees zero depth
  \hbox{$\uparrow$}\vss\hbox{\sevenrm depth}\vss%
   \hbox{$\downarrow$}\null}
     }
     }
% the bottom part
\hbox{\kern48pt%
\hbox to65pt{$\leftarrow$\hfil\sevenrm width\hfil$\rightarrow$}}
     }
```

The only subtle point is the `\vtop to 48pt{\boxmaxdepth=0pt`.... Without the `\boxmaxdepth` this `\vtop` would have `\ht=6.94pt` and `\dp=43pt`, a total of `49.94pt`. It is possible to limit the height to zero by `\vtop to48pt{\null`..., but that would still have generated the `\vtop` with a depth of `49.94pt`. The `\vbox to30pt` has the depth of the bottom component (the `\downarrow` ($\downarrow$)). The `\null` has thus been artificially added to become the bottom component, which guarantees a depth of zero. ◄

3.48: The solution uses the two vertical arrows available in math mode '$\uparrow$', '$\downarrow$'. A `\vbox` is created with three components. The first and third ones are the arrows, placed in boxes of height zero. The middle one is a `\vrule` to the required height. `\boxmaxdepth=0pt` is used, so that the depth of the $\downarrow$ is included in the height of the `\vbox`.

```
\def\varrowfill#1{%
\vbox{\offinterlineskip\boxmaxdepth=0pt
\setbox0=\hbox{$\uparrow$}%
\vbox to0pt{\hbox{$\uparrow$}\vss}%
\moveright.45\wd0\hbox{\vrule height#1}%
\vbox to0pt{\vss\hbox{$\downarrow$}}}}}
```

This solution is not ideal since the vertical arrows have a size of about 8.9pt each. The size of the whole thing cannot, therefore, be less than that size. For better results, it would be nice to have a font with arrowheads. It turns out that the LaTeX font line has precisely the right symbols, characters '066 and '077. LaTeX users can, of course, create many arrows with the \vector command. ◄

3.49: First, macro \Bline is defined, to build each of the four boxes.

```
\newdimen\Bwidth \newdimen\Bheight \newdimen\Bdepth
\Bwidth=60pt \Bheight=10pt \Bdepth=4pt
\newdimen\BWidth \BWidth=\Bwidth \advance\BWidth by50pt
\newdimen\Base \Base=2\Bdepth
\newdimen\Baseline \Baseline=22pt % 22=10+4+2*4. The baseline separa-
tion.
```

```
\def\Bline{\vbox{\offinterlineskip
\hrule\kern-0.4pt
\hbox to\Bwidth{\vrule height\Bheight\hfil\vrule}
\hbox to0pt{\kern-4pt\vbox to0pt{\hrule width\BWidth\vss}\hss}
\vtop{
\hbox to\Bwidth{\vrule depth\Bdepth\hfil\vrule}
\kern-0.4pt\hrule
     }  }   }
```

Note the use of \vtop to give the large box a depth. The diagram is now a large \hbox with three components. The first is the four boxes done with \Bline; the second is the arrows (using macro \varrowfill); and the third, the four words of text.

```
\hbox{
\vbox{\offinterlineskip
\kern\Base\Bline\kern\Base\Bline
\kern1.5\Base
\hbox to0pt{\kern-4pt\vbox to0pt{\hrule width\BWidth\vss}\hss}
\kern1.5\Base\Bline\kern\Base\Bline}
\kern10pt
\vbox{\offinterlineskip
\varrowfill{\Baseline}\varrowfill{\Baseline}
\varrowfill{2\Base}
\varrowfill{\Baseline}\varrowfill{\Baseline}}
\kern10pt
\vbox{\offinterlineskip
```

```
\kern\Baseline\smash{\hbox{\tt baselineskip}}
\kern\Baseline\smash{\hbox{\tt parskip}}
\kern2\Base\smash{\hbox{\tt baselineskip}}
\kern\Baseline\smash{\hbox{\tt baselineskip}}
\kern.5\Baseline}
}
```

◀

3.50: This is straightforward and uses the principles of \boxit above.

```
\def\leadbox#1#2{\vbox to#2{\hrule height\thick\kern-\thick
\hbox to#2{\vrule width\thick height#2 \hfil%
\vbox to #2{\vfil#1\vfil}\hfil\vrule width\thick}
\kern-\thick\hrule height\thick}}
```

◀

3.51: Register \dimen0 is temporarily used to save the quantity #1-#2. The \moveleft, \moveright, \raise, and \lower commands are used to move the rules away from their normal positions. Note that the \moveright moves by a different amount. The rest is a simple application of boxes.

```
\def\stagbox#1#2{\dimen0=#1 \advance\dimen0 by-#2
\vbox to#1{\offinterlineskip
\moveleft\dimen0\vbox{\hrule width#2 height\thick}\kern-\thick
\vbox to#1{\vss
\hbox to#1{\lower\dimen0\hbox{\vrule height#2 width\thick}%
\hfil \raise\dimen0\hbox{\vrule height#2 width\thick}}
\vss}
\advance\dimen0\thick
\kern-\thick\moveright\dimen0\vbox{\hrule width#2 height\thick}
}}
```

◀

3.52: The first rule is placed in \box0. The height and width of the box are then explicitly set to zero, and it is typeset (the reference point is not moved). The same thing is done with the second rule. Only the third rule moves the reference point. This is how the three rules are superimposed on each other:

```
\def\crossrules#1{\vbox{\thick=#1 \offinterlineskip
\setbox0=\vbox to#1{\vfil
\hbox to#1{\hfil\vrule height.3\thick
depth.3\thick width.6\thick\hfil}\vfil}
\ht0=0pt \wd0=0pt \box0
\setbox0=\vbox to#1{\vfil
\hbox to#1{\hfil\vrule height.5\thick depth.5\thick
width.2\thick\hfil}\vfil}
```

```
\ht0=0pt \wd0=0pt \box0
\vbox to#1{\vfil\hrule width#1 height.1\thick depth.1\thick\vfil}
}}
```

◄

3.53: This uses no special tricks.

```
\def\di{\hbox{$\diamond$}}
\def\fourdiamonds{\vbox{\offinterlineskip
\hbox to10.33333pt{\hfil\di\hfil}\kern-1.5pt
\hbox to10.33333pt{\di\hfil\di}\kern-1.5pt
\hbox to10.33333pt{\hfil\di\hfil}}}
```

◄

3.54: It can be opened and then transferred back to itself by means of \setbox1=\vbox{\unvbox1}. This will set its height to its natural value. ◄

3.55: Because the badness is still less than \vbadness. (It should be $100 \times 1.5^3 = 337$, but it actually comes out as 336.) ◄

3.56: A \loop is necessary. This is discussed in detail in (Ref. 12, p. 212). ◄

3.57: This is one of the typical blunders of the ubiquitous B. L. User. The test terminating a loop cannot contain an \else, because of the way macro \loop is defined. ◄

4.1: \clubpenalty+\widowpenalty+\interlinepenalty+\brokenpenalty as can be verified by the experiment:

```
\clubpenalty=1000 \widowpenalty=100
\interlinepenalty=10 \brokenpenalty=1
\setbox0=\vbox{\hsize=.5in\noindent aa bb final dd}
\showbox0
```

It results in a small, 2-line box, with a hyphenated first line. A penalty of 1111 is shown between the lines. ◄

4.2: Each font has parameters called \fontdimen. Their number is not fixed (although it seems to be limited), and they are used to transfer any desired information from METAFONT to TeX. Parameters 2, 3, and 4 specify the natural size, stretch, and shrink, respectively, of the interword glue of the font. Parameter 7 specifies the extra space. Their values can be "shown" by, e.g., \showthe\fontdimen7\xyz, where \xyz is a font name. The values can also be changed. Several \fontdimen values for common CM fonts are listed on [433]. ◄

4.3: The first \showthe displays 1250, the \sfcode of a comma. The second one, however, displays 1000 instead of the expected 3000, because f has been changed to 1 just before the period. ◄

4.4: When defining `\raggedleft` it is not enough to copy `\raggedright` and make `\leftskip` flexible. The user has to consider two points:

1. The last line of the paragraph has `\parfillskip` glue on the right side. This makes sense for raggedright, but for raggedleft this glue has to be set to zero.

2. Consider a paragraph that consists of a single word. When typeset in raggedright, the word is flush left (following the paragraph indentation) but when typeset in raggedleft, the word should be flush right. The `\raggedleft` macro should, therefore, set `\leftskip` to `\0pt plus\hsize`, so it could expand and fill the entire line, if necessary. The definition of `\raggedleft` is therefore:

```
\def\raggedleft{\spaceskip=.3333em \xspaceskip=.5em
  \parfillskip=0pt \leftskip=0pt plus\hsize}
```

◄

4.5: A value for `\baselineskip` or `\lineskip` that's appropriate for the huge font. Without it, the lines of a long heading would be much too close. ◄

4.6: By temporarily assigning small sfcodes to 'a', 'b', and 'c'. With sfcodes of 100, the shrinkability of the interword glue becomes $1.11111 \times (1000/100) = 11.1111$pt. With two pieces of glue, the total shrinkability of the box becomes 22.2222pt, which is why

```
{\sfcode`a=100\sfcode`b=100\sfcode`c=100
\hbox spread-22.2222pt{a b c}}
```

will not be considered overfull. Note that the natural width of the box above is 21.66669pt. After shrinking by 22.2222pt, its width becomes -0.55551pt; negative! Such a box is probably not very practical, but the idea can be used in cases where a line of text is slightly overfull and we want to shrink it a bit more than normal.
◄

4.7: Saying `\narrower\raggedright` first advances `\rightskip` and then sets it to '0pt plus 6em'. The effect of the `\narrower` is preempted. On the other hand, saying `\raggedright\narrower` first sets `\rightskip` to '0pt plus 6em' and then advances it by `\parindent` to '20pt plus 6em'. The two combinations behave differently. ◄

4.8: Setting `\hangafter=0 \hangindent=-\parindent` indents *all* the lines of the paragraph on the right. An `\rlap{\kern\hsize...}` takes us temporarily to the right margin, where we typeset the label, overlapping the indentation, by an `\llap{#1\enspace}`. The result is

```
\def\rItem#1{\par\hangafter=0\hangindent=-\parindent%
\noindent\rlap{\kern\hsize\llap{\enspace#1}}\ignorespaces}
```

◄

4.9: This is easy:

```
\def\lNote#1#2{\vadjust{\llap{\vbox to0pt{\sevenrm
```

\hbox{#1}\hbox{#2}\vss}\thinspace}}}

◄

4.10: With \everypar. ◄

4.11: Leaving room on the page for a diagram. Hanging indentation is usually used for this purpose, but it is reset after each paragraph. If we want hanging indentation over several paragraphs, we have to count lines (see also [ex. 14.24]). ◄

5.1: The \catcode, \\, and \everypar are three control sequence tokens. The "'=12 {*}" are eight character tokens. ◄

5.2: The six tokens '\def', '\abc', '#1', '{', '#1', and '}'. ◄

5.3: It is '$A_1, A_2, \ldots, A$'. The second argument is empty. However, if the macro is defined with two parameters, each expansion must supply two arguments. An expansion such as '\vec a', with only one argument, is invalid. ◄

5.4: Just saying '\vec x' is not enough. TeX will read ahead and absorb the next token as the second parameter. Even moving the expansion to the end of the document and saying '\vec a \end' will not constitute an expansion with one argument. The '\end' would, in such a case, be considered the second argument, and TeX will then complain of a missing '\end'. The only ways of expanding '\vec' with just one argument are

  1. to say '\vec a' as the very last thing in the input file. This will cause the message 'file ended while scanning use of \vec'.

  2. to say {\vec a} in a group. The error message in this case will be '! Argument of \vec has an extra }.'. ◄

5.5: It's a match! ◄

5.6: An expansion of \step creates either \advance\temp by 1 or \advance\temp by 2, so a \relax is necessary. The macro should be defined as
\def\step{\advance\temp...\fi\relax}
Note that a space is not enough in this case. When \step is defined as
\def\step{\advance\temp...\fi }
expansions of \step will not include the space. ◄

5.7: The error message '! Extra }, or forgotten $.' ◄

5.8: Because a control word must consist of letters only. ◄

5.9: This is fully discussed in (Ref. 16). ◄

5.10: It is used as a parameter delimiter, to gobble up the ';', which would otherwise be typeset. ◄

5.11: It is 5 and 1. The space following the 'z' is not counted, since it is absorbed when TeX scans the \z. Note also that \z is not expanded and may, therefore, be undefined. ◄

5.12: Without this space, an expansion such as '\scan␣ab␣c*d;' would yield 4 and 1 (the space between the 'b' and the 'c' would be absorbed instead of being counted). ◄

5.13: The simple test 'A\scan␣ab␣c*;B' shows that there are none. Such tests are important because seemingly innocent spaces may end up being typeset, thereby affecting the rest of the document. ◄

5.14:

```
\newcount\others \newcount\spaces \newcount\letters
\def\scan{\afterassignment\nextoken\let\next= }
\def\nextoken{\ifx\next;\let\next\relax
 \else\expandafter\if\space\next\advance\spaces by1
 \else\ifcat\next a\advance\letters by1
             \else\advance\others by1
  \fi\fi\let\next\scan
\fi \next}
```

To identify a token as a space, an \if is used instead of an \ifx, and the **plain** macro \space is expanded before the \if is executed. To identify a token as a letter, its catcode is compared to that of an 'a'. ◄

5.15: The new value of \t would be the token string {\the\t *}. ◄

5.16: It would be the lowercase roman numeral value of \n. \uppercase expects an argument that is a list of tokens. It converts the character tokens to their uppercase equivalents and ignores all control sequence tokens. ◄

5.17: By the simple test

```
\def\y{\a}
\toks0=\expandafter{\expandafter\x\y}
\showthe\toks0
```

The \showthe displays \x\a in the log file. ◄

5.18: This defines a control sequence whose name is \=76 and whose definition is abc. ◄

5.19: The following two definitions are due to G. Lamaître (private communication):

```
\def\ifdefined#1 {%
  \ifodd\expandafter\ifx\csname#1\endcsname\relax 0 \else 1 \fi}
```

```
\def\ifdefined#1 {%
 \expandafter\ifx\csname#1\endcsname\relax
     \csname iffalse\expandafter\endcsname \else
     \csname iftrue\expandafter\endcsname \fi}
```

In the first one, the \expandafter and \csname result in \ifodd 0 or in \ifodd 1, which should be completed outside. The second approach results in an \iffalse or an \iftrue. ◄

5.20: Yes. It can be defined in the preamble, but then it cannot have any parameters, and it is local to the preamble. To expand it in the table, it should be defined with a \gdef. It can also be defined (with parameters) in any row or in a \noalign. Generally, it is better to define any necessary macros outside the \halign. ◄

5.21: Macro \Uppercase starts by redefining '\i' from a dotless 'ı' to uppercase 'I'; it than expands \uppercase. The definition is:
'\def\Uppercase#1{{\def\i{I}\uppercase{#1}}}'. The extra pair of braces is needed to keep the redefinition local. The expansion
'\'\i\Uppercase{\'\i{}a\'Ib}\i' results in 'íIAÍBı'. ◄

5.22: By using two pairs of ##. Inside the definitions, each pair ## counts for one #. Thus in the definition
\def\a#1{'#1'\def\b##1{[#1,##1]\def\x####1{(#1,##1,####1)}\x X}\b Y}
the notation ####1 stands for the parameter of macro \x. The expansion '\a Z' will result in 'Z'[Z,Y](Z,Y,X) ◄

5.23: \end and expansions of \outer macros. ◄

5.24: Yes. The expansion '\compndArg ;' will cause \pickup to be expanded with a null argument. ◄

5.25:

\def\separg#1{\singarg#1\end}
\def\singarg#1{\ifx#1\end \let\next=\relax
 \else '#1'\ \let\next=\singarg \fi\next}

This is similar to the last example on [219]. ◄

5.26: 'viiithe letter A\abc\relax' ◄

5.27: The ones that are valid responses to an error (see Chapter 15). ◄

5.28: If the file name is 'X', then \jobtitle receives the string X\\. The first parameter is 'X', the second one is '\\', and then TEX starts scanning the text, looking for an '\\' to delimit the third one. The macro will not work, and more sophisticated parameter delimiters should be used to make it more robust. ◄

5.29: This is straightforward. ◄

6.1: A plain macro, defined [351] as \def\empty{}. ◄

6.2: It is a success, proving that the macros compared aren't required to have the same status with respect to \def and \edef. ◄

6.3: (a) displays 'no' on the terminal since it compares the macro \\ to the letter 'm'; (b) typesets 'yes', since it compares two identical macros. ◄

6.4: Macro \ifundefined supplies the \ifx, and the matching \else and \fi are provided outside. We want \ifdefined also to suppply an if that can be completed outside. We start with the test for an undefined macro
'\expandafter\ifx\csname#1\endcsname\relax'

and create either an `\iffalse` (if the macro is undefined) or an `\iftrue` (in case it is defined) to be matched outside. The first version is

```
\def\ifdefined#1 {%
 \expandafter\ifx\csname#1\endcsname\relax
 \let\next=\iffalse\else\let\next=\iftrue\fi
 \next}
```

But it fails! The reason is explained at the bottom of [211]. The next, working, version is

```
\def\maca{\let\next=\iffalse}
\def\macb{\let\next=\iftrue}
\def\ifdefined#1 {%
 \expandafter\ifx\csname#1\endcsname\relax
\maca\else\macb\fi \next}
```

after which, we can say: `\ifdefined a \message{yes}\else\message{no}\fi`.

A better version has been communicated to the author by Bernd Raichle. It has the advantage that it is completely executed in the mouth.

```
\def\ifdefined#1 {%
 \csname if\expandafter\ifx\csname #1\endcsname\relax
  false\else true\fi\endcsname}
```

(End of answer.) ◄

6.5: The `\font` command takes an optional parameter (`scaled` or `at`), so TeX executes the `\ifx` in an attempt to find such a parameter. The `\relax` terminates the `\font`. ◄

6.6: Because `\let\a=~` defines `\a` as an active character, whereas `\def\a{~}` defines `\a` as a macro (whose value is the active character '`~`'). The `\ifx` does not look too deeply into the meaning of its comparands, so it decides that a macro is not equal to an active character. In contrast, the `\if` comparison (section 6.4) which looks deeper into the meaning of its comparands, returns a 'yes' for both tests. ◄

6.7: Just do it. It's worth it. Then do the similar test

```
\if\the\count90\the\count90
 \message{yes}\else\message{no}\fi
```

◄

6.8: A success, since it compares the catcodes of the two letters 't', 'r'. It also typesets 'ue'. ◄

6.9: Macro `\tmp` expands to the first character of the parameter.

```
\def\suppose#1{\def\tmp##1##2\\{##1}%
 \ifcat A\tmp#1\\...\else...\fi...}
```

◄

6.10:   The \ifx compares \b and \b, and they, of course, match. The '1' is thus the first token left for the outer \if to compare. The rest of the \ifx (\else\if\a\b2\fi\fi) is skipped. Next comes the '+', followed by the *else* part of the outer \if, with the '3'. The outer \if can now be written as \if1+\else3\fi which, of course, typesets the '3'. ◄

6.11:   Answer not provided. ◄

6.12:   We place an expansion of \isnextspace at the end of \foo. This sets \next to the token following the parameter of \foo. The \ifcat can then be used to compare the category of \next to that of a space. The following test

```
\def\spacecheck{%
 \ifcat\next\space
 \message{yes}\else\message{no}\fi}
\def\isnextspace{\futurelet\next\spacecheck}
\def\foo#1{#1\isnextspace}
\foo{A} \foo{B}.\foo{C} D
```

produces 'yes no yes' on the terminal. ◄

6.13:   The \ifnum evaluates to true, so the \iffalse starts. It is always false, so TEX executes what follows the \else. It is an \iftrue, which is always true. At this point the three conditions, \ifnum, \iffalse, and \iftrue are nested. Three \fis are necessary to terminate them. The first \fi found is the last token in the definition of \newtest, and it terminates the \iftrue. The second \fi is the one in '\newtest1ABC\fi', and it terminates the \iffalse (note that the text 'ABC' will be typeset). There is no \fi to terminate the \ifnum, so the final error message is (\end occurred when \ifnum on line 2 was incomplete). The reader should try to figure out the details of the expansion '\newtest2ABC\fi'. ◄

7.1:   Experiments seem to suggest a size of .75pt for a 300dpi printer, and .25pt for a 1200 dpi one. ◄

7.2:   This is straightforward:

```
\def\slantB#1#2{\hbox\bgroup\slantBaux{#1}{#2}}
\def\slantBaux#1#2{\raise\step\hbox{\Dot}\advance\step by#2pt
 \advance\loup by1
\ifnum\loup<#1 \slantBaux{#1}{#2}\else\egroup\step=0pt \loup=0 \fi}
```

  ◄

7.3:   The new #4 parameter should be added, and the vrule on line 5 should have height#4pt. That's all. ◄

7.4:   This is straightforward:

```
\newcount\kount \newdimen\vlength \newdimen\hlength
\def\gridb#1 #2 #3.{%
\vlength=#1pt \advance\vlength by -1pt \multiply\vlength by #3
\kount=1 \setbox1=\hbox{%
```

```
\loop
\vrule height\vlength
\ifnum\kount<#2\advance\kount by 1\kern-0.4pt\kern #3pt
\repeat}%
\hlength=\wd1 \wd1=0pt\ht1=0pt
%     box 1 is now loaded with vertical lines
\kount=1 \setbox2=\vbox{
\loop                   % over horizontal lines
\hrule width\hlength
\ifnum\kount<#1 \advance\kount by 1 \kern-0.4pt\kern #3pt
\repeat}%
\ifvmode\wd2=0pt\fi       % don't move the ref. point in v mode
\hbox{\box1\box2}\ifvmode\nointerlineskip\fi}
```

Note the '%' signs following \repeat}%. They are important. ◄

7.5: Because \count0 is used. Normally \count0 is the page number and should not be changed inside a macro. The extra pair makes everything inside the macro local, so the global \count0 is not disturbed. ◄

7.6:

```
\vbox{\hbox{\phantom{Wood}\verti{Ed}}
\nointerlineskip\hbox{Woodward}}
```

        ◄

7.7: This is easy, just change \vbox to \vtop in the definition of \vertcl.

```
\def\vertcl#1#2\endd{\ifx\end#1\box\ver
                \else\setbox0=\hbox{#1}%
\setbox\ver=\vtop{\unvbox\ver\kern.3ex\box0}%
                \vertcl#2\end\endd\fi}
```

        ◄

7.8: This is straightforward. Only \vertcl needs to be modified.

```
\def\vertcl#1#2\endd{\ifx\end#1\box\ver
                \else\setbox0=\hbox{#1}%
\setbox\ver=\vbox{\unvbox\ver\kern.3ex
                \hbox to1em{\hss\box0\hss}}%
                \vertcl#2\end\endd\fi}
```

        ◄

7.9: This is easy but may not be that useful since the final \box0 may include unwanted kernings.

```
\def\revers#1{\setbox0=\null\macb#1\\}
\def\macb#1#2\\{%
\setbox0=\hbox{#1\unhbox0}
```

```
\def\tmp{#2}
\ifx\tmp\empty\let\next=\macc\else\let\next=\macb\fi\next#2\\}
\def\macc\\{\showbox0}
```

◄

7.10: Because the control sequence \toks can be followed by more than one digit. Without the space TeX would have expanded \end, trying to look for more digits. ◄

7.11: If \Let is the last control sequence (or the only one) in a string, our macros can handle it. ◄

7.12: The string \endverbatim is only assigned its special meaning when it appears on a line by itself, with no preceding spaces, so in our case there was no problem. It is, however, possible to list an \endverbatim anywhere by saying |\endverbatim|. ◄

7.13: If we place it inside a macro, then the space following \let would get catcode 10 when the macro is defined. When the macro is expanded later, the \let command would fail, because it is followed by a catcode 10 token instead of by an active character. ◄

7.14: To terminate the glue specification. Without the \relax, if #1 happens to be one of the words plus or minus, TeX would consider it part of the glue assigned to \xspaceskip and would expect it to be followed by a number. ◄

7.15: Sanitize the space instead of using \obeyspaces (also the special settings of \spaceskip and \xspaceskip are no longer necessary). ◄

7.16: \readOptions is expanded in H mode, where spaces are sometimes significant, and \readoptions, in V mode, where spaces are ignored. The rule is that a space that's necessary as a separator is not typeset (does not become spurious). There is no strict need for a space after the '#1' since the # can be followed by one digit only (there can be at most nine parameters). ◄

7.17: Easy, just turn it temporarily into a letter (catcode 11). The following {\makeletter\[|[...|} works nicely because the \ifx[\nextc compares \nextc to a left bracket with catcode 12. ◄

7.18: No effect, since \numbered only changes \everypar, which is not used during vertical bar verbatim listing anyway. Since the change is done in a group, it is local. ◄

7.19: Normally, a space following a number is considered a terminator and is not printed. However, at this point, because of the \obeyspaces, the space is active (has catcode 13 instead of the normal 10) and is defined as a control space. It would therefore be typeset as a spurious space. This is especially annoying if the verbatim macros are part of a format file that is eventually dumped. We don't want such a file to create any typeset material. ◄

7.20:

```
\def\verbfile{\medskip\begingroup\tt\sanitize%
 \makeother\|\futurelet\nextc\oPtions}
\def\oPtions{\ifx[\nextc\let\next=\getoptions
 \else\let\next=\preverbfile\fi\next}
\def\getoptions[#1]{#1 \preverbfile}
\def\preverbfile#1 {\def\par{\leavevmode\endgraf}%
 \lastasks\obeylines\ifvispace\makeother\ \else\obeyspaces\fi%
 \input#1\endgroup\medskip}
```

A typical expansion is '\verbfile[\numbered\vispacetrue]test', where test is
the name of the file (no space between the] and the file name).   ◄

7.21:   We start by measuring the size of the argument. If it has less than two
characters, we treat this case separately, by invoking \macc immediately:

```
\def\ignoreLast#1{\length{#1}%
\ifnum\count20<2 \macc#1\\
\else\toks0={}\macb#1\\ \fi}
\def\macb#1#2\\{\toks0=\expandafter{\the\toks0#1}\length{#2}
\ifnum\number\count20<2
 \let\next=\macc\else\let\next=\macb\fi\next#2\\}
\def\length#1{\count20=0 \glength#1\end }
\def\glength#1{\ifx#1\end \let\nxt=\relax \else
 \advance\count20by1 \let\nxt=\glength\fi \nxt}
\def\macc#1\\{\showthe\toks0}
```

◄

7.22:   The necessary changes are shown below. Note that the second argument is
absorbed by \Aux.

```
\def\Aux#1#2{\dimen0=#2 \setbox0=\null \aux#1\end}
\def\aux#1{%
\ifx\end#1%
 \hbox to\dimen0{\unhbox0 \unskip}%
...
...
```

◄

7.23:   Parameters #1, #2 are the left and right brackets, and #3 is the text.

```
\def\bracket#1#2#3\endbracket{\hbadness=10000%
 \hfuzz=\maxdimen\brack={#3}
\setbox\brkt=\vbox{\hsize=\nwidth \the\brack}
\ifdim\ht\brkt<\baselineskip \setbox\brkt=\hbox{\the\brack} \fi
$$\left#1\matrix{\box\brkt\cr}\right#2$$}
```

Typical expansions are \bracket\vert. text...\endbracket, \bracket()
text..\endbracket. We only allocate 0.105in for each bracket, but certain large
operators are wider than this, which may cause overfull boxes. Hence the special
settings of \hbadness & \hfuzz. ◄

7.24: To suppress the height of the lowered \vrule. Eliminating the \smash shows
its effect very clearly. ◄

7.25: One way of doing it is

```
\def\leftdropshadow#1#2{\setbox0=\boxit{#1}3%
\vbox{\offinterlineskip
 \hbox{{\lower#2\hbox{\vrule width#2height\ht0}}\copy0}
 \kern-#2\hbox{\vrule width\wd0height#2}}}
```

◄

7.26: Place a \smallskip, instead of a blank line or a \par, after each paragraph
in the third argument. ◄

7.27: Such a macro was developed by Raymond Chen (*TeXmag*, vol. 5, no. 2),
with bug fixes by Donald Arseneau (*TeXmag*, vol. 5, no. 3). ◄

8.1: Yes, with a category code of 10. ◄

8.2: It will be redefined as the name of \toks1, and its old meaning will become
inaccessible. ◄

8.3: Five. The space and the control sequence \xyz (which may be undefined)
become one token each. ◄

8.4: Three. The third one is the '2'. ◄

8.5: With a \def, the replacement text of \tmp becomes the string '\the\toks2
\the\toks3'. With the \edef, it becomes 'abc123'. Generally, when \tmp is expanded, both cases produce the same result. However, when \tmp is expanded
inside a token register, control sequences are not expanded further. ◄

8.6: The \write16 is not immediate, whereas \message is [227]. Also \write (or
\immediate\write) starts a new line, while several messages can be displayed on
the same line [228]. ◄

8.7: Because macro \aste requires a numeric quantity, and \mtst is not numeric
(it is essentially a macro). ◄

8.8: Yes. If \ctst is zero, no asterisks are typeset. ◄

8.9: Because \number is expandable [213] and \ctst is unexpandable (it's just
the name of a count register). ◄

9.1: The macro is fairly simple. Note the \global. It is necessary because
the macro is expanded inside an equation. The test for a null macro parameter
(\ifx\tmp\empty) is discussed on page 147.

\newcount\eqnum

```
\def\EqNum#1{\global\advance\eqnum by 1 \eqno(\the\eqnum)
 \def\tmp{#1} \ifx\tmp\empty\else \label{\the\eqnum}{#1}\fi}
```

◄

9.2: Because '\def\lblitem{\chnum.\the\secnum}' would define \lblitem as '\chnum.\the\secnum', whereas \edef would define it as 12.3 or something similar.
◄

9.3: Yes, unless the text has been changed in the first two passes. ◄

9.4: Just place the command '\write\toc{\string\nobreak}' before macro \chapter is expanded for the last chapter. However, if another chapter is ever written, to become the new last chapter, this command will have to be moved to the new chapter. We thus say that the \nobreak command is inserted, by this method, only semi-automatically, rather than automatically. ◄

10.1: It gives the \vbox its width. If the box is the only thing on its line, its width is irrelevant. However, if the box is combined with other structures on its left and right, its width is very important. ◄

10.2: Answer not provided. ◄

10.3: The assignment \rnda=\rndval\unit converts the \count register \rndval (in the range $0\ldots99$) into a dimension in the range $0\ldots346\,$pt. The way \unit is used is explained on [61]. Since $346\,$pt $\approx 4.78\,$in, the resulting dimensions can be used as coordinates, without danger of overflowing the page. ◄

11.1: To eliminate its height (which is slightly more than 1pt in cmr10). Without the \smash, the leaders may be too high for certain applications. ◄

11.2: Because \stagleaders is mostly used in a box, and we want \stagger to be advanced globally. ◄

11.3:

```
\def\vdotlead{\leaders\vrule\vfil}
\vbox{
 \hrule
 \hbox to.3in{\vbox to.3in{\vdotlead}\hfil%
  \vbox to.3in{\vdotlead}}
 \hrule}
```

◄

11.4:

```
\def\vdotlead{\leaders\hbox to0pt{\hss.\hss}\vfil}
\setbox0=\vbox to.3in{\vfil\hbox to.3in{\hss lead\hss}\vfil}
\vbox{
 \hrule
 \hbox to.3in{\vbox to.3in{\vdotlead}\box0%
```

```
  \vbox to.3in{\vdotlead}}
 \hrule}
```

◄

11.5: This is a simple application of horizontal and vertical leaders.

```
\def\hdotlead#1{\leaders\hbox to12pt{#1\hfil}\hfil}
\def\vdotlead#1{\leaders\vbox to12pt{\vfil#1\vfil}\vfil}
\line{
\vbox to48pt{\hbox to48pt{\hdotlead{\stagbox{10pt}{9pt}}}
\vdotlead{\stagbox{10pt}{9pt}}}
\hfil
\vbox to48pt{\hbox to48pt{\hdotlead{\crossrules{10pt}}}
\vdotlead{\crossrules{10pt}}}
\hfil
\vbox to48pt{\hbox to48pt{\hdotlead{\fourdiamonds}}
\vdotlead{\fourdiamonds}}
\hfil
\vbox to48pt{\hbox to48pt{\hdotlead{\leadbox{%
 \hbox{$\bullet$}}{10pt}}}
\vdotlead{\leadbox{\hbox{$\bullet$}}{10pt}}}
\hfil
\vbox to48pt{\hbox to48pt{\hdotlead{\leadbox{%
 \hrule height4pt width4pt}{10pt}}}
\vdotlead{\leadbox{\hrule height4pt width4pt}{10pt}}}
}
```

◄

12.1: In the latter case the columns are typeset left justified. In the former case each entry is spread, if possible, to fill up its column. To understand this, the reader should try example 2 with 'birdsmen' instead of 'birds', and 'nest x' instead of 'nest'. ◄

12.2: Either place a \bf in every preamble spec, or say {\bf \halign...}. ◄

12.3: Yes. Note that the current \tabskip will apply inside the inner table, unless modified by it. The following example shows how this works.

```
{\tt\halign{#\tabskip=6pt&#\cr1&2\cr
\noalign{\sevenrm\tabskip=4pt\halign{#&#\cr3&4\cr5&6\cr}}
7&8\cr}}
```

 (Run it to see the results!) ◄

12.4: The heading will be centered on the last three columns instead of on the second and third ones. ◄

12.5: Because otherwise the \hrules would stick out on the left and on the right of the table. The width of the \hrule is set to that of the table, but the width of a table includes the two extreme tabskips. ◄

12.6: The strut separates consecutive baselines by 16pt instead of by the normal 12pt. Its effect can thus be eliminated by a `\noalign{\kern-4pt}` placed between the two lines. ◀

12.7:

```
\vbox{\offinterlineskip \tabskip=0in \halign{#&#&#&#\cr
&.&.&.\cr.& &.&.\cr.&.& &.\cr.&.&.& \cr}}
```

◀

12.8: By using special struts. The following:

```
\def\topStrut{\vrule height18pt depth4pt width0pt}
\def\botStrut{\vrule height12pt depth10pt width0pt}
\vbox{\offinterlineskip\hrule\tabskip=0pt
\halign{\vrule\quad#\tabskip0.1in&#&\vrule#&#&%
 \hfil#\hfil\quad\vrule\tabskip=0pt\cr
*&*&&*&*\topStrut\cr*&*&&*&*\cr*&*&&*&*\botStrut\cr
}\hrule}
```

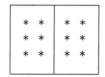

produces   ◀

12.9: This is the result of the `\noalign{\kern-2pt}` command that was inserted between the two rows. ◀

12.10: Because `\halign` is V mode material and thus cannot appear in RH mode, it is possible (but probably useless) to say `$$\hbox{\vbox{\halign{...}}}$$`. (Incidentally, when seen in H mode, the `\halign` terminates the paragraph.) ◀

12.11: By placing the entire table in a box, and showing it. Thus:

```
\setbox0=\vbox{\halign{...}}
\showbox0
```

◀

12.12: The second tabskip overrides the first one. ◀

12.13: A `\phantom)` should be used on the top row to occupy the space taken by the ')' on the bottom row.

```
\halign{\hfil$#$&&\tabskip=1pt#\cr
x\phantom)=&1&2&3&4&5\cr
f(x)=&1&1&2&3&5\cr}
```

◀

12.14: There is a difference between the two cases if the data item has no decimal point. An item such as '67' would be typeset by the first preamble as '67', and by the second one, as '67.'. ◄

12.15: The definition of \offinterlineskip [352] is

```
\def\offinterlineskip{\baselineskip=-1000pt \lineskip=0pt%
\lineskiplimit=\maxdimen}
```

The three quantities are advanced by 1 point, which, practically speaking, affects only \lineskip. Thus a \lineskip glue of 1 point will be placed between boxes stacked vertically. ◄

13.1: This would be treated as a display since the expansion of \tlp provides the necessary two consecutive, catcode-3 tokens. ◄

13.2: Only a horizontal list can be placed in an \hbox. A single '$' can be an element of a horizontal list [95], but a double '$$' is treated as an empty inline formula [287]. ◄

13.3: In an expansion such as 'A\indentDisplay{1in}{x^2}B', we want the 'A' to terminate the current line and the 'B' to be typeset below the display, on a new line. Without the \hfil\break, the 'B' would normally appear on the same line as the 'A'. ◄

13.4: It is not used, but it is necessary for proper parameter delimiting. ◄

13.5: By $\displaystyle A^2$. ◄

13.6: The simple expression 'A^B\sqrt{A^B}', typeset in both inline and display modes. ◄

13.7:

```
$\leaders\hbox{$\Xi\mkern-2mu\Xi\mkern-2mu$} \hskip.5in$
$\leaders\hbox{$\top\mkern-6mu\bot\mkern-6mu$} \hskip.5in$
```

◄

13.8: \mathchardef\prime="0230. Character in position "30 of family 2, declared an Ord (class 0). ◄

13.9: None. The \mathord command is superfluous. ◄

13.10: The \font command is ignored in math mode, so it is necessary to temporarily get out of math mode, which is done by the \hbox. ◄

13.11: Because it is defined as a box, not a single character. To prove this, try:
\def\bigF{\mathop{f}} $$A+\bigF\nolimits_0^2x$$ ◄

13.12: It is ⟂ instead of the expected ⟂. The top symbol is treated as a superscript, so it is typeset in script style and is raised. ◄

13.13: $\buildrel\circ\lower4pt\hbox{$\angle$}\bullet\over\smile$. ◄

13.14: '`$a+\left(b\right)\showlists$`'. ◄

13.15: Because the character used for the weird sign is narrower than its box. ⫟
◄

13.16: This information must be placed in the `.tfm` file of the font by the font designer. The `charlist` command (Ref. 3, p. 317) is used for this purpose (see also [433]). ◄

13.17:

`$\overbrace{x_a+x_b+x_c+\cdots}^{\overbrace{a+b+\cdots=n}^{n>0}}$`

◄

13.18: We start with the display. The first try is to typeset the '1' and follow with an empty formula where the '2' and the '8' are the super- and subscript, respectively. Following that, the O should be declared a large operator, so the '3' and '7' can be its limits. The rest is obvious. We therefore write `$$1{}_8^2\mathop{O}^3_7{}^4_65$$`, which produces $1{}_8^2\overset{3}{\underset{6}{O}}{}^4 5$. Two things are easy to correct. The '1' and the '5' should be typeset in scriptstyle and should be separated from the rest by thin skips. The result
`$${\scriptstyle1}\,{}_8^2\mathop{O}^3_7{}^4_6\,\scriptstyle5$$`
produces $1{}_8^2\overset{3}{\underset{6}{O}}{}^4 5$, which is much better. The only remaining problem is to move the '2' and the '8' apart a bit (and the same for the '4' and the '6'). This is done by placing them in a box and using `\raise`. A simple macro is defined for this purpose `\def\posit#1 #2{{\raise#1pt\hbox{$\scriptstyle#2$}}}`, and the final try is

`$${\scriptstyle1}\,{}_\posit-2 8^\posit1 2`
`\mathop{O}^3_7{}^\posit1 4_\posit-2 6\,\scriptstyle5$$`

which produces the desired result. For inline mode, the only necessary change is to say `\mathop{O}\limits`, to place the limits above and below the O, rather than to the right. ◄

13.19:

`\setbox0=\hbox{$\dashv\joinrel\mathrel{\phantom{\top}}%`
` \joinrel\vdash$}`
`\vbox{\offinterlineskip`
`\hbox to\wd0{\hfil$\bot$\hfil}`
`\copy0`
`\hbox to\wd0{\hfil$\top$\hfil}}`

◄

13.20: It is created by '`$ab\surd$`'. ◄

13.21: `$a^{2^{2^2}}{}^{^2}{}^2a$`. ◄

14.1: They are pure numbers. The badness is defined as the quotient of two dimensions. Penalty values are always numbers. ◄

14.2: It corresponds to a glue set ratio of 50%, since $100 \times .5^3 = 12.5$. ◄

14.3: Every line except the last one. The last line of a paragraph has \parfillskip glue on the right. This glue can stretch to infinity, so the badness of this line is (artificially) zero (see [ex. 6.5]). ◄

14.4: The paragraph will be set with the minimum number of tolerable lines. ◄

14.5: \parshape and \hangindent are set to zero. \hangafter is set to 1. ◄

14.6: Experiment 5 is similar to 3, resulting in a loose line with badness 22. Experiment 6 is similar to 4, resulting in a tight line with badness 31. ◄

14.7: The only difference would be the result of experiment 2, since it is the only one with a large badness. It will result in:

x be it as it

may

The first line is now tolerable, since its badness (232) no longer exceeds the tolerance. ◄

14.8: An \hbox is indivisible, so there are no line breaks in it and, consequently, no hyphenations. ◄

15.1: First a '2', to skip over the 'pl', then 'i\hskip1pt'. ◄

15.2: It is 'Missing $ inserted'. The best response is to type 2 and, after the '$^' have been removed, to type i\^. ◄

15.3: (Tongue in cheek) B. L. User, of course. The original error, however, was made by J. H. Quick [Ex. 27.4]. ◄

15.4: Yes, when the \write is done during the \shipout. Our example says \immediate\write, but any error message in a nonimmediate write contains both indicators. ◄

16.1: Answer not provided. ◄

18.1: Because the command \vrule is supposed to be followed by a height⟨dimen⟩ [281–282]. If we use glue, such as \skip0, only the natural size is used, and the stretch and shrink components are ignored. ◄

20.1: Macro \subhead starts \obeylines, in which each line is a paragraph. ◄

20.2: It is invalid to write '^{{}' or '^|}|'. A good idea is to write '^[+@$\{$ (left brace)]'. The part on the left of the '@' will be the sort key (a '+' is close to braces in the ASCII collating sequence), and the part on the right will be typeset (the print key). ◄

20.3: To properly close the last index item (each item opens up a group that's closed by its successor). ◄

Exercise—Something performed or practiced in order to develop, improve or display a specific power or skill.

— Webster's Dictionary

Index

Page numbers in boldface indicate the most definitive source of information about an item.

ASCII, 20, 21, 23, 24, 137, 197, 283, 469
assignments, 26
\ast (∗), 262
asterisks, 145, 199, 230, 463
  counting them, 120
\asymp (≍), 262
at (font spec.), 8
atoms in math mode, 256
Auden, W. H., 383

\backslash (\), 259, 262
bad box, 61
badness, 7, **61–64**, 271, 273, 276, 278, 281,
  284, 299, 469
  definition of, 62
  infinite, 409
Ballou, H., x
Barnhart, E., 436
Barrie, J. M., 85, 224
baseline separation, 349
\baselineskip, 23, 27, 44, 64, 65, 83, 88,
  95, 100, 201, 246, 284, 302, 303, 320,
  321, 333, 336, 349, 391, 409, 454
\batchmode, 287
Batzinger, R., 296, 336, 338, 351
Bechtolsheim, S., 436
\begingroup, 10, **116**, 394
\beginsection, 58
\belowdisplayshortskip, 250
\belowdisplayskip, 244, 250
best fit, 274
\beta (β), 259
\bf, 9, 127
\bgroup, **115**, 121, 154, 371, 394
\Big, 266
\big, 260, 266
big point, 11
\bigcap (⋂), 260
\bigcirc (◯), 262
\bigcup (⋃), 260
\Bigg, 266
\bigg, 266
\bigl, *see* big, Big, etc.
\bigodot (⨀), 260
\bigoplus (⨁), 260
\bigotimes (⨂), 260
\bigr, *see* big, Big, etc.
\bigskip, 395
\bigsqcup (⨆), 260
\bigtriangledown (▽), 262

\bigtriangleup (△), 262
\biguplus (⨄), 260
\bigvee (⋁), 260
\bigwedge (⋀), 260
Billawala, N., 10, 435
bin (a math atom), 257
binary operations, 262
binary search, 189
bitmap file, 2
blank line, 18, **89**, 103, 171, 463
block structure, 2
\bloop, 191
boolean variables, 145
bordered page, 312
bordered text (leaders example), 234, 431
\bordermatrix, 17
\bot (⊥), 259, 270
\botmark, 345, 349
\bowtie (⋈), 262
box, 4, 16, 26
  advanced examples, 74–77
  bad, 61
  classification of, 63–64, 272
  contents sticking out, 35, 59
  definition of, 4, 33, 296
  dimensions, 33
  emptying of, 38
  examples, 52–53
  exercises, 77–79
  good, 63
  its contents, 33
  loose, 61, 63
  negative dimensions, 67
  opening, 51–52
  overfull, 35, 47, 50, 61, 62, **63**, 79, 81
  tight, 61, 63
  underfull, 35, 37, 47, 58, 61, **63**, 81
  unsplittable, 81
  uses of, 33
  writing on file, 364, 375
\box, 195
box registers, 16–17, 38–41
  dimensions of, 40
\box0 and plain, 17, 40
\box255, 16, 18, 87, 295–307, 313–315, 317,
  318, 320, 321, 328, 329, 331–334, 336–
  338, 341, 343, 345, 346, 351, 353, 356,
  358, 372, 388, 392, 398, 399, 401–403,
  406, 409, 410
  as a logical page, 313
  breaking up of, 327, 331, 351

Indexing requires decision making of a far
higher order than computers are yet capable
of.

> —*The Chicago Manual of Style (1982)*

Ay me, what act, that roars so loud, and
thunders in the index?

> —*William Shakespeare, Hamlet*

Colophon

This book was typeset with TeX (of course), using Textures on the Macintosh computer. The diagrams were done with Superpaint, a simple drawing program, also on the Macintosh. Here are a few features that illustrate the amount of work that went into it:

- The main body of the text is in cmr10, but about 30 fonts were used. See Chapter 20 for the complete list
- . The book uses about 80 formatting macros, which are discussed in Chapter 20.
- . The auxiliary file contains about 150 cross references.
- . The format file is about 500 lines long.
- The raw index file contains more than 4100 items.
- Since high-quality typesetting is what TeX is all about, typesetting this book involved a lot of fine-tuning. The author had to manually check for bad page- and line-breaks and correct them by moving, adjusting, and rewriting parts of the text.

Of particular interest to aspiring authors are the quotations at the end of every chapter. They are discussed in the Preface.